Environmental Biotechnology

Environment Biotechnology

Edited by **Emma Layer**

R Callisto Reference

New York

Published by Callisto Reference,
106 Park Avenue, Suite 200,
New York, NY 10016, USA
www.callistoreference.com

Environmental Biotechnology
Edited by Emma Layer

International Standard Book Number: 978-1-63239-310-4 (Hardback)

Printed in the United States of America.

Contents

Preface

The main aim of this book is to educate learners and enhance their research focus by presenting diverse topics covering this vast field. This is an advanced book which compiles significant studies by distinguished experts in the area of analysis. This book addresses successive solutions to the challenges arising in the area of application, along with it; the book provides scope for future developments.

Descriptive information regarding the field of environmental biotechnology has been provided in this book. Considering the importance of studying and applying the biological methods to redeem or mitigate the harmful effects of global pollution on the natural environment, as direct consequences of quantitative expansion and qualitative diversification of persistent and hazardous pollutants, the present book provides useful information regarding latest approaches and prospective applications in environmental biotechnology. This book covers important aspects of environmental biotechnology. It provides comprehensive information deeming scientific experiments that were carried out in different parts of the world to test different procedures and methods designed to remove or mitigate the impact of hazardous pollutants on environment. The book targets researchers and students with specialties in biotechnology, bioengineering, ecotoxicology, environmental engineering and all those readers who are interested in improving their knowledge in order to keep the Earth healthy.

It was a great honour to edit this book, though there were challenges, as it involved a lot of communication and networking between me and the editorial team. However, the end result was this all-inclusive book covering diverse themes in the field.

Finally, it is important to acknowledge the efforts of the contributors for their excellent chapters, through which a wide variety of issues have been addressed. I would also like to thank my colleagues for their valuable feedback during the making of this book.

Editor

Biotechnology for Conversion of Organic Wastes

Environmental Biotechnology for Bioconversion of Agricultural and Forestry Wastes into Nutritive Biomass

Marian Petre and Violeta Petre

Additional information is available at the end of the chapter

1. Introduction

The cellulose is the most widely distributed skeletal polysaccharide and represents about 50% of the cell wall material of plants. Beside hemicellulose and lignin, cellulose is a major component of agricultural wastes and municipal residues. The cellulose and hemicellulose comprise the major part of all green plants and this is the main reason of using such terms as "cellulosic wastes" or simply "cellulosics" for those materials which are produced especially as agricultural crop residues, fruit and vegetable wastes from industrial processing, and other solid wastes from canned food and drinks industries.

The cellulose biodegradation using fungal cells is essentially based on the complex interaction between biotic factors, such as the morphogenesis and physiology of fungi, as the cellulose composition and its complexness with hemicellulose and lignin (Andrews & Fonta, 1988; Carlile & Watkinson, 1996).

An efficient method to convert cellulose materials, in order to produce unconventional high-calorie foods or feeds, is the direct conversion by cellulolytic microorganisms. Theoretically, any microorganism that can grow as pure culture on cellulose substrata, used as carbon and energy sources, should be considered a potential organism for "single-cell protein" (SCP) or "protein rich feed" (PRF) producing.

2. Biotechnology of mycelia biomass producing through submerged bioconversion of agricultural crop wastes

The submerged cultivation of mushroom mycelia is a promising method which can be used in novel biotechnological processes for obtaining pharmaceutical substances of anticancer, antiviral, immuno-modulating, and anti-sclerotic action from fungal biomass and cultural liquids and also for the production of liquid spawn (Breene, 1990).

The researches that were carried out to get nutritive supplements from the biomass of *Ganoderma lucidum* species (Reishi) have shown that the nutritive value of its mycelia is owned to the huge protein content, carbohydrates and mineral salts. *Lentinula edodes* species (Shiitake) is a good source of proteins, carbohydrates (especially polysaccharides) and mineral elements with beneficial effects on human nutrition (Wasser & Weis, 1994; Mizuno et al., 1995).

It is well known the anti-tumor activity of polysaccharide fractions extracted from mycelia of *Pleurotus ostreatus*, known on its popular name as Oyster Mushroom (Mizuno et al., 1995; Hobbs, 1996).

The main purpose of this research work consists in the application of biotechnology for continuous cultivation of edible and medicinal mushrooms by submerged fermentation in agro-food industry which has a couple of effects by solving the ecological problems generated by the accumulation of plant wastes in agro-food industry through biological means to valorise them without pollutant effects as well as getting fungal biomass with high nutritive value which can be used to prepare functional food (Carlile & Watkinson, 1996; Moser, 1994).

The continuous cultivation of medicinal mushrooms was applied using the submerged fermentation of natural wastes of agro-food industry, such as different sorts of grain by-products as well as winery wastes that provided a fast growth as well as high biomass productivity of the investigated strains (Petre & Teodorescu, 2012; Petre & Teodorescu, 2011).

2.1. Materials and methods

Ganoderma lucidum (Curt. Fr.) P. Karst, *Lentinula edodes* (Berkeley) Pegler and *Pleurotus ostreatus* (Jacquin ex Fries) Kummer were used as pure strains. The stock cultures were maintained on malt-extract agar (MEA) slants, incubated at 25°C for 5-7 d and then stored at 4°C. The seed cultures were grown in 250-ml flasks containing 100 ml of MEA medium (20% malt extract, 2% yeast extract, 20% agar-agar) at 23°C on rotary shaker incubator at 100 rev.min^{-1} for 7 d (Petre & Petre, 2008; Petre et al., 2007).

The fungal cultures were grown by inoculating 100 ml of culture medium using 3-5% (v/v) of the seed culture and then cultivated at 23-25°C in rotary shake flasks of 250 ml. The experiments were conducted under the following conditions: temperature, 25°C; agitation speed, 120 rev. min^{-1}; initial pH, 4.5–5.5.

After 10–12 d of incubation the fungal cultures were ready to be inoculated aseptically into the glass vessel of a laboratory-scale bioreactor (Fig. 1).

For fungal growing inside the culture vessel of this bioreactor, certain special culture media were prepared by using liquid nutritive broth, having the following composition: 15% cellulose powder, 5% wheat bran, 3% malt extract, 0.5% yeast extract, 0.5% peptone, 0.3% powder of natural argillaceous materials. After the steam sterilization at 121°C, 1.1 atm., for 15 min. this nutritive broth was transferred aseptically inside the culture vessel of the laboratory scale bioreactor shown in figure 1.

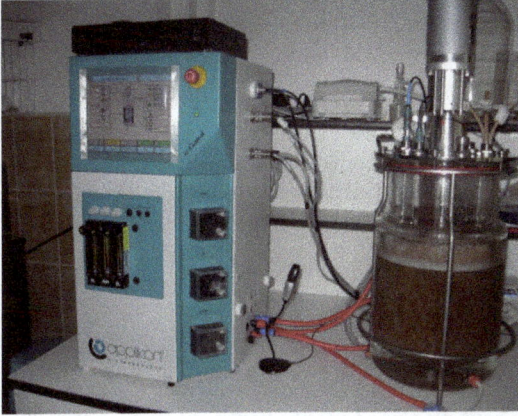

Figure 1. Laboratoy-scale bioreactor for submerged cultivation of edible and medicinal mushrooms

The culture medium was aseptically inoculated with activated spores belonging to *G. lucidum*, *L. edodes* and *P. ostreatus* species. After inoculation into the bioreactor vessel, a slow constant flow of nutritive liquid broth was maintained inside the nutritive culture medium by recycling it and adding from time to time a fresh new one.

The submerged fermentation was set up at the following parameters: constant temperature, 23°C; agitation speed, 80-100 rev. min^{-1}; pH level, 5.7–6.0 units; dissolved oxygen tension within the range of 30-70%. After a period of submerged fermentation lasting up to 120 h, small fungal pellets were developed inside the broth (Petre & Teodorescu, 2010; Petre & Teodorescu, 2009).

The experimental model of biotechnological installation, represented by the laboratory scale bioreactor shown in figure 1, was designed to be used in submerged cultivation of the mentioned mushroom species that were grown on substrata made of wastes resulted from the industrial processing of cereals and grapes (Table 1).

Variants of culture substrata	Composition
S1	Mixture of winery wastes and wheat bran 2.5%
S2	Mixture of winery wastes and barley bran 2.5%
S3	Mixture of winery wastes and rye bran 2.5%
Control	Pure cellulose

Table 1. The composition of compost variants used in mushroom cultures

2.2. Results and discussion

The whole process of mushroom mycelia growing lasts for a single cycle between 5-7 days in case of *L. edodes* and between 3 to 5 days for *G. lucidum* and *P. ostreatus*. All experiments regarding the fermentation process were carried out by inoculating the growing medium volume (15 L) with secondary mycelium inside the culture vessel of the laboratory-scale bioreactor (see Fig. 1).

The strains of these fungal species were characterized by morphological stability, manifested by its ability to maintain the phenotypic and taxonomic identity. Observations on morphological and physiological characters of these two tested species of fungi were made after each culture cycle, highlighting the following aspects:

• sphere-shaped structure of fungal pellets, sometimes elongated, irregular, with various sizes (from 7 to 12 mm in diameter), reddish-brown colour of *G. lucidum* specific culture (Fig. 2a);

• globular structures of fungal pellets, irregular with diameters of 5 up to 10 mm or mycelia congestion, which have developed specific hyphae of *L. edodes* (Fig. 2b);

• round-shaped pellets with diameter measuring between 5 and 15 mm, having a white-cream colour and showing compact structures of *P. ostreatus* mycelia (Fig. 2c).

The experiments were carried out in three repetitions. Samples for analysis were collected at the end of the fermentation process, when pellets formed specific shapes and characteristic sizes. For this purpose, fungal biomass was washed repeatedly with double distilled water in a sieve with 2 mm diameter eye, to remove the remained bran in each culture medium (Petre at al., 2005a).

Biochemical analyses of fungal biomass samples obtained by submerged cultivation of edible and medicinal mushrooms were carried out separately for the solid fraction and extract fluid remaining after the separation of fungal biomass by pressing and filtering. Also, the most obvious sensory characteristics (color, odor, consistency) were evaluated and presented at this stage of biosynthesis taking into consideration that they are very important in the prospective view of fungal biomass using as raw matarials for nutraceuticals producing. In each experimental variant the amount of fresh biomass mycelia was analyzed.

Percentage amount of dry biomass was determined by dehydration at 70° C, until constant weight. The total protein content was investigated by using the biuret method, whose principle is similar to the Lowry method, being recommended for the protein content ranging from 0.5 to 20 mg/100 mg sample (Bae et al., 2000; Lamar et al., 1992).

The principle method is based on the reaction that takes place between copper salts and compounds with two or more peptides in the composition in alkali, which results in a red-purple complex, whose absorbance is read in a spectrophotometer in the visible domain (λ 550 nm). In addition, this method requires only one sample incubation period (20 min) eliminating the interference with various chemical agents (ammonium salts, for example).

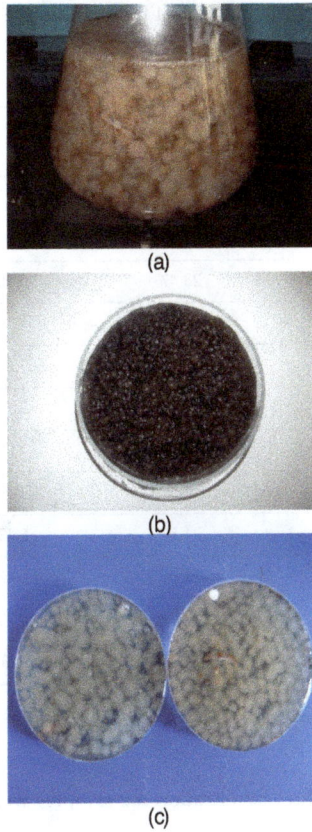

Figure 2. Fungal pellets of *G. lucidum,* b. Fungal pellets of *L. Edodes,*. c. Fungal pellets of *P. ostreatus*

In table 2 are presented the amounts of fresh and dry biomass as well as the protein contents for each fungal species and variants of culture media.

According to registered data, using a mixture of wheat bran 2.5% and winery wastes the growth of *G. lucidum* biomass was stimulated, while the barley bran led to increased growth of *L. edodes* mycelium and *G. lucidum* as well.

In contrast, the dry matter content was significantly higher when using barley bran 2.5% mixed with winery wastes for both species used. Protein accumulation was more intense when using barley bran compared with those of wheat bran and rye bran, at both mushroom species.

The sugar content of dried mushroom pellets collected after the biotechnological experiments was determined by using Dubois method. The mushroom extracts were prepared by immersion of dried pellets inside a solution of NaOH pH 9, in the ratio 1:5. All dispersed solutions containing the dried pellets were maintained 24 h at the precise temperature of 25 °C, in full darkness, with continuous homogenization to avoid the oxidation reactions.

Mushroom species	Culture variants	Fresh biomass (g)	Dry biomass (%)	Total proteins (g % d.w.)
G. lucidum	I	25.94	9.03	0.67
G. lucidum	II	22.45	10.70	0.55
G. lucidum	III	23.47	9.95	0.73
	Control	5.9	0.7	0.3
L. edodes	I	20.30	5.23	0.55
L. edodes	II	23.55	6.10	0.53
L. edodes	III	22.27	4.53	0.73
	Control	4.5	0.5	0.2
P. ostreatus	I	21.50	5.73	0.63
P. ostreatus	II	23.95	7.45	0.55
P. ostreatus	III	23.25	4.79	0.75
	Control	4.7	0.5	0.3

Table 2. Fresh and dry biomass and protein content of G. lucidum, L. edodes and P. ostreatus mycelia grown by submerged fermentation

After the removal of solid residues by filtration the samples were analyzed by the previous mention method (Wasser & Weis, 1994).

The nitrogen content of mushroom pellets was analyzed by Kjeldahl method. All the registered results are related to the dry weight of mushroom pellets that were collected at the end of each biotechnological culture cycle (Table 3).

Comparing all the registered data, it could be noticed that the correlation between the dry weight of mushroom pellets and their sugar and nitrogen contents is kept at a balanced ratio for each tested mushroom species.

From these mushroom species that were tested in biotechnological experiments G. lucidum (variant III) showed the best values concerning the sugar and total nitrogen content. On the very next places, L. edodes (variant I) and G. lucidum (variant II) could be mentioned from these points of view.

The registered results concerning the sugar and total nitrogen contents have higher values than those obtained by other researchers (Bae et al., 2000; Jones, 1995; Moo-Young, 1993). The nitrogen content in fungal biomass is a key factor for assessing its nutraceutical potential, but the assessing of differential protein nitrogen compounds requires additional investigations.

Mushroom species	Culture variants	Mushroom pellets d. w. (%)	Sugar content of dried pellets (mg/ml)	Kjeldahl nitrogen of dried pellets (%)
G. lucidum	I	17.64	4.93	5.15
G. lucidum	II	14.51	3.70	5.35
G. lucidum	III	20.16	5.23	6.28
	Control	0.7	0.45	0.30
L. edodes	I	19.67	4.35	6.34
L. edodes	II	17,43	3.40	5.03
L. edodes	III	15.55	4.75	6.05
	Control	0.5	0.45	0.35
P. ostreatus	I	19.70	5.15	6.43
P. ostreatus	II	14.93	4.93	6.25
P. ostreatus	III	15.63	5.10	5.83
	Control	0.55	0.50	0.35

Table 3. The sugar and total nitrogen contents of dried mushroom pellets

3. Laboratory-scale biotechnology of edible mushroom producing on growing composts of apple and winery wastes

The agricultural works as well as the industrial activities related to apple and grape processing have generally been matched by a huge formation of wide range of cellulosic wastes that cause environmental pollution effects if they are allowed to accumulate in the environment or much worse they are burned on the soil (Petre, 2009; Verstrate & Top, 1992).

The solid substrate fermentation of plant wastes from agro-food industry is one of the challenging and technically demanding biotechnology that is known so far (Petre & Petre, 2008; Carlile & Watkinson, 1996).

The major group of fungi which are able to degrade lignocellulose is represented by the edible mushrooms of Basidiomycetes Class. Taking into consideration that most of the edible mushrooms species requires a specific micro-environment including complex nutrients, the influence of physical and chemical factors upon fungal biomass production and mushroom fruit bodies formation were studied by testing new biotechnological procedures (Petre & Petre, 2008; Moser, 1994; Beguin & Aubert, 1994; Chahal & Hachey, 1990).

The main aim of research was to find out the best biotechnology of recycling the apple and winery wastes by using them as a growing source for edible mushrooms and, last but not least, to protect the environment (Petre et al., 2008; Smith, 1998; Raaska, 1990).

3.1. Materials and methods

Two fungal species of Basidiomycetes group, namely *Lentinula edodes* (Berkeley) Pegler (folk name: Shiitake) as well as *Pleurotus ostreatus* (Jacquin ex Fries) Kummer (folk name: Oyster Mushroom) were used as pure mushroom cultures isolated from the natural environment and now being preserved in the local collection of the University of Pitesti.

The stock cultures were maintained on malt-extract agar (MEA) slants (20% malt extract, 2% yeast extract, 20% agar-agar). Slants were incubated at 25°C for 120-168 h and stored at 4°C. The pure mushroom cultures were expanded by growing in 250-ml flasks containing 100 ml of liquid malt-extract medium at 23°C on rotary shaker incubators at 110 rev. min^{-1} for 72-120 h. To prepare the inoculum for the spawn cultures of *L. edodes* and *P. ostreatus* the pure mushroom cultures were inoculated into 100 ml of liquid malt-yeast extract culture medium with 3-5% (v/v) and then maintained at 23-25°C in 250 ml rotary shake flasks.

After 10–12 d of incubation the fungal cultures were inoculated aseptically into glass vessels containing sterilized liquid culture media in order to produce the spawn necessary for the inoculation of 10 kg plastic bags filled with compost made of winery and apple wastes.

These compost variants were mixed with other needed natural ingredients in order to improve the enzymatic activity of mushroom mycelia and convert the cellulose content of winery and apple wastes into protein biomass. The best compositions of five compost variants are presented in Table 4.

Compost variants	Compost composition
S1	Winery and apple wastes (1:1)
S2	Winery wastes + wheat bran (9:1)
S3	Winery wastes and rye bran (9:1)
S4	Apple wastes and wheat bran (9:1)
S5	Apple wastes + rye bran (9:1)
Control	Poplar, beech and birch sawdust (1:1:1)

Table 4. The composition of five compost variants used in mushroom culture cycles

In this way, the whole bags filled with compost were steam sterilized at 121°C, 1.1 atm., for 30 min. In the next stage, all the sterilized bags were inoculated with liquid mycelia, and then, all inoculated bags were transferred into the growing chambers for incubation. After 10-15 d, on the surface of sterilized plastic bags filled with compost, the first buttons of mushroom fruit bodies emerged. For a period of 20-30 d there were harvested between 1.5–3.5 kg of mushroom fruit bodies per 10 kg compost of one bag (Petre et al., 2012; Oei, 2003; Stamets, 1993; Wainwright, 1992; Ropars et al., 1992).

3.2. Results and discussion

To increase the specific processes of winery and apple wastes bioconversion into protein of fungal biomass, there were performed experiments to grow the mushroom species of *P. ostreatus* and *L. edodes* on the previous mentioned variants of culture substrata (see Table 1).

During the mushroom growing cycles the specific rates of cellulose biodegradation were determined using the direct method of biomass weighing the results being expressed as percentage of dry weight (d.w.) before and after their cultivation (Stamets, 1993; Wainwright, 1992).

In order to determine the evolution of the total nitrogen content in the fungal biomass there were collected samples at precise time intervals of 50 h and they were analyzed by using Kjeldahl method. The registered results concerning the evolution of total nitrogen content in *P. ostreatus* biomass are presented in figure 3 and the data regarding *L. edodes* biomass could be seen in figure 4.

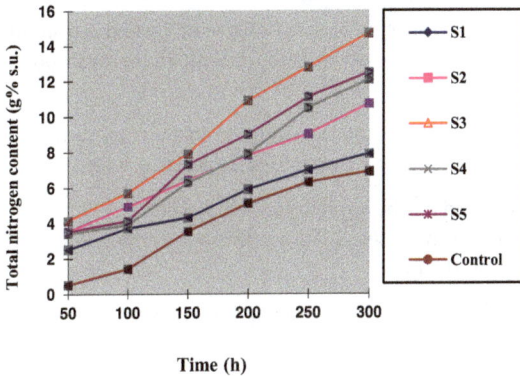

Figure 3. The evolution of total nitrogen content in *P. ostreatus* biomass

During the whole period of fruit body formation, the culture parameters were set up and maintained at the following levels, depending on each mushroom species:

- air temperature, 15–17°C;
- the air flow volume, 5–6m³/h;
- air flow speed, 0.2–0.3 m/s;
- the relative moisture content, 80–85%;
- light intensity, 500–1,000 luces for 8–10 h/d.

According to the registered results of the performed experiments the optimal laboratory-scale biotechnology for edible mushroom cultivation on composts made of marc of grapes and apples was established (Fig. 5).

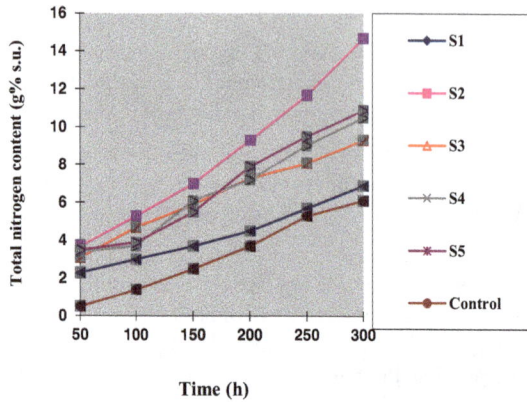

Figure 4. The evolution of total nitrogen content in *L. edodes* biomass

As it is shown in figure 5, two technological flows were carried out simultaneously until the first common stages of the inoculation of composts with liquid mushroom spawn followed by the mushroom fruit body formation.

The whole period of mushroom growing from the inoculation to the fruit body formation lasted between 30–60 d, depending on each fungal species used in experiments.

The registered data revealed that by applying such biotechnology, the winery and apple wastes can be recycled as useful raw materials for mushroom compost preparation in order to get significant mushroom production.

In this respect, the final fruit body production of these two mushroom species was registered as being between 20–28 kg relative to 100 kg of composts made of apple and winery wastes.

4. Biotechnology of forestry wastes recycling as growing composts for edible and medicinal mushroom cultures

The most part of wastes produced all over the world arise from industrial, agricultural and domestic activities. These wastes represent the final stage of the technical and economical life of products (Verstraete & Top 1992).

As a matter of fact, the forestry works as well as the industrial activities related to forest management and wood processing have generally been matched by a huge formation of wide range of waste products (Beguin & Aubert 1994, Wainwright 1992).

Many of these lignocellulosic wastes cause serious environmental pollution effects, if they are allowed to accumulate in the forests or much worse to be burned for uncontrolled domestic purposes. So far, the basis of most studies on lignocellulose-degrading fungi has been eco-

```
┌─────────────────────────┐      ┌─────────────────────────┐
│   Pure mushroom cultures │      │ Mechanical pre-treatment of │
│   (L. edodes, P. ostreatus) │      │   winery and apple wastes │
└─────────────────────────┘      └─────────────────────────┘
            │                                │
            ▼                                ▼
┌─────────────────────────┐      ┌─────────────────────────┐
│  Expanding of pure mushroom │    │ Adding carbon, nitrogen and mineral │
│ cultures by growing in liquid media │ │ sources to the compost variants │
└─────────────────────────┘      └─────────────────────────┘
            │                                │
            ▼                                ▼
┌─────────────────────────┐      ┌─────────────────────────┐
│  Inoculum preparation and │      │ Transfer of each compost variant │
│   growing on culture media │      │      to 1000 ml jars │
└─────────────────────────┘      └─────────────────────────┘
            │                                │
            ▼                                ▼
┌─────────────────────────┐      ┌─────────────────────────┐
│ Growing of submerged mushroom │  │  Steam sterilization of the │
│  spawn in nutritive media │      │       filled jars │
└─────────────────────────┘      └─────────────────────────┘
            │                                │
            └────────────┬───────────────────┘
                         ▼
┌──────────────────────────────────────────────────────────┐
│  Inoculation of the filled jars with liquid mushroom spawn │
└──────────────────────────────────────────────────────────┘
                         │
                         ▼
┌──────────────────────────────────────────────────────────┐
│ Spawn growing on the composts made of winery and apple wastes │
└──────────────────────────────────────────────────────────┘
                         │
                         ▼
┌──────────────────────────────────────────────────────────┐
│       Mushroom fruit body formation and growing │
└──────────────────────────────────────────────────────────┘
                         │
                         ▼
┌──────────────────────────────────────────────────────────┐
│            Mushroom fruit bodies cropping │
└──────────────────────────────────────────────────────────┘
```

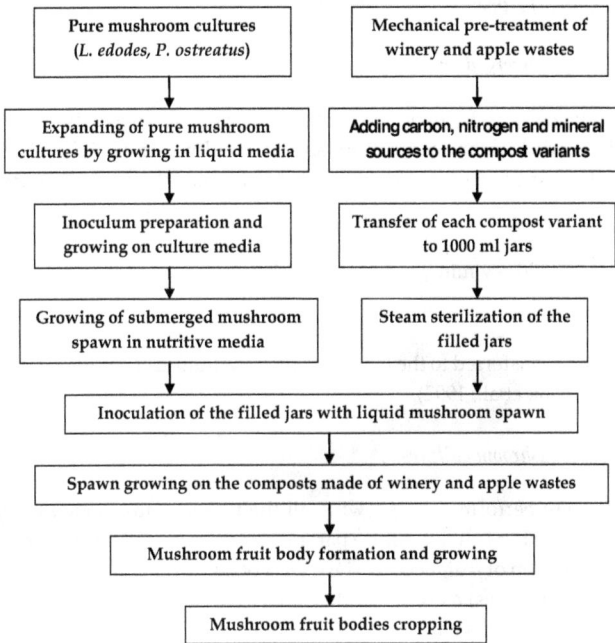

Figure 5. Scheme of laboratory-scale biotechnology for edible mushroom producing on winery and apple wastes

nomic rather than ecological, with emphasize on the applied aspects of lignin and cellulose decomposition, including biodegradation and bioconversion (Carlile & Watkinson 1996).

In this respect, the main aim of this work was focused on finding out the best way to convert the wood wastes into useful food supplements, such as mushroom fruit bodies, by using them as growing sources for the edible and medicinal mushrooms (Smith, 1998).

4.1. Materials and methods

4.1.1. Fungal species and culture media

According to the main purpose of this work, three fungal species from Basidiomycetes, namely *Ganoderma lucidum* (Curt.:Fr.) P. Karst, *Lentinus edodes* (Berkeley) Pegler and *Pleurotus ostreatus* (Jacquin ex Fries) Kummer were used as pure mushroom cultures during all experiments. The stock mushroom cultures were maintained by cultivating on malt-extract agar (MEA) slants. After that, they were incubated at 25° C for 5-7 d and then stored at 4° C. These pure mushroom cultures were grown in 250-ml flasks containing 100 ml of MEA medium (20% malt extract, 2% yeast extract) at 23°C on rotary shaker incubators at 110 rev min $^{-1}$ for 5-7 d.

4.1.2. Methods used in experiments

4.1.2.1. Preparation of submerged mycelia inoculum

The pure mushroom cultures for experiments were prepared by inoculating 100 ml of culture medium with 3-5% (v/v) of the seed culture and then cultivated at 23-25°C in rotary shake flasks of 250 ml. The experiments were conducted under the following conditions:

- temperature, 25°C;

- agitation speed, 90-120 rev min^{-1};

- initial pH, 4.5–5.5.

The seed culture was transferred to the fungal culture medium and cultivated for 7–12 d (Petre et al., 2005a; Glazebrook et al., 1992).

4.1.2.2. Incubation of mushroom cultures

The experiments were performed by growing all the previous mentioned fungal species in special culture rooms, where all the culture parameters were kept at optimal levels in order to get the highest production of fruit bodies. The effects of culture compost composition (carbon, nitrogen and mineral sources) as well as other physical and chemical factors (such as: temperature, inoculum size and volume and incubation time) on mycelial net formation and especially, on fruit body induction were investigated (Petre & Petre, 2008).

All the culture composts for mushroom growing were inoculated using liquid inoculum with the age of 5–7 days and the volume size ranging between 3-7% (v/w). During the period of time of 18–20 d after this inoculation, all the fungal cultures had developed a significant biomass on the culture substrata made of wood wastes, such as: white poplar and beech wood sawdusts. These woody wastes were used as main ingredients to prepare natural composts for mushroom growing. The optimal temperatures for incubation and mycelia growth were maintained between 23–25°C. The whole period of mushroom growing from the inoculation to the fruit body formation lasted between 30–60 days, depending on each fungal species used in experiments (Petre & Teodorescu, 2010).

4.1.2.3. Preparation of mushroom culture composts

The lignocellulosic materials were mechanical pre-treated to breakdown the lignin and cellulose structures in order to induce their susceptibility to the enzyme actions during the mushroom growing. All these pre-treated lignocellulosic wastes were disinfected by steam sterilization at 120° C for 60 min (Petre et al., 2005b; Leahy & Colwell 1990).

The final composition of culture composts was improved by adding the following ingredients: 15-20% grain seeds (wheat, rye, rice) in the ratio 2:1:1, 0.7–0.9% $CaCO_3$, 0.3–0.5% $NH_4H_2PO_4$, each kind of culture medium composition depending on the fungal species used to be grown. As control samples for each variant of culture composts used for the experimental growing of all these fungal species were used wood logs of white poplar and beech that were kept in water three days before the experiments and after that they were steam sterilized to be disinfected.

4.1.2.4. Preparation of mushroom spawn

3000 g of white poplar sawdust and 1500 g of beech sawdust were mixed with cleaned and ground rye grain, 640 g of $CaCO_3$, 50 g of $NH_4H_2PO_4$ and 3550 ml of water, in order to obtain the growth substratum for mushroom spawn. The ingredients of such smal compost were mixed and then they were sterilized at 121° C, for 20 min. and allowed to cool until the mixture temperature decreased below 35° C. The spawn mixture was inoculated with 100-200 ml of liquid fungal inoculums and mixed for 10 min. to ensure complete homogeneity. Sterile polyethylene bags, containing microporus filtration strips, were filled with the smal composts and incubated at 25° C, until the spawn fully colonized the whole composts. At this point the spawn may be used to inoculate the mushroom growing substrate or alternatively it may be stored for up to 6 months at 4° C before use (Chahal & Hachey, 1990).

All the culture composts were inoculated using inoculum with the age of 5–7 d and the volume size ranging between 3-7% (v/w). The optimal temperatures for incubation and mycelia growth were maintained between 23–25°C. The whole period of mushroom growing from the inoculation to the fruit body formation lasted between 30–50 days.

4.1.2.5. Mushroom cultivation

The experiments were carried out inside such *in vitro* growing rooms, where the main culture parameters (temperature, humidity, aeration) were kept at optimal levels to get the highest production of mushroom fruit bodies (Moser, 1994).

In order to find a suitable carbon source for the mycelia growth and consequently for fungal biomass synthesis, the pure cultures of *P. ostreatus* (Oyster Mushroom), as well as *L. edodes* (Shiitake) and *G. lucidum* (Reishi) were cultivated in different nutritive culture media containing various carbon sources, and each carbon source was added to the basal medium at a concentration level of 1.5% (w/v) for 7-12 d (Raaska, 1990).

To investigate the effect of nitrogen sources on mycelia growth and fungal biomass production, the pure cultures of these two fungal species were cultivated in media containing various nitrogen sources, where each nitrogen source was added to the basal medium at a concentration level of 10 g/l. At the same time, malt extract was one of the better nitrogen sources for a high mycelia growth. Peptone, tryptone and yeast extract are also known as efficient nitrogen sources for fungal biomass production by using the pure cultures of such fungal species (Chang & Hayes, 1978). In comparison with organic nitrogen sources, inorganic nitrogen sources gave rise to relatively lower mycelia growth and fungal biomass production (Bae et al., 2000).

The influence of mineral sources on fungal biomass production was examined at a standard concentration level of 5 mg. In order to study the effects of initial pH correlated with the incubation temperature upon fruit body formation, *G. lucidum*, *P. ostreatus* and *L. edodes* were cultivated on substrates made of wood wastes of white poplar and beech at different initial pH values (4.5–6.0). The experiments were carried out for 6 days at 25°C with the initial pH 5.5. Similar observations were made by Stamets (1993), during the experiments. K_2HPO_4 could improve the productivity through its buffering action, being favourable for mycelia growth. The experiments were carried out between 30-60 days at 25°C.

4.2. Results and discussion

The effects of carbon, nitrogen and mineral sources as well as other physical and chemical factors on mycelial net formation and especially, on fruit body induction were investigated by adding them to the main composts made of white poplar and beech sawdusts in the ratio 2:1. For the experimental growing of all these fungal species white poplar and beech logs were used as control samples.

4.2.1. The effect of carbon sources upon mushroom mycelia growth

When the cells were grown in the maltose medium, the fungal biomass production was the highest among the tested variants. Data presented in the following table are the means ± S.D. of triple determinations (Table 5).

Carbon source (g/l)	Fresh Fungal Biomass Weight (g/l)			Final pH		
	G. lucidum	L. edodes	P. ostreatus	G. l	L. e	P. o
Glucose	27±0.10	41±0.05	43±0.03	5.5	5.3	5.1
Maltose	27±0.14	45±0.12	49±0.05	5.8	5.4	5.3
Sucrose	25±0.23	35±0.03	37±0.09	5.1	5.1	5.7
Xylose	26±0.07	38±0.07	35±0.07	5.3	5.5	5.9

Table 5. The effect of carbon sources upon the mycelia growth of pure mushroom cultures on white poplar and beech composts

What is very important to be noticed is that the maltose has a significant effect upon the increasing of mycelia growth and fungal biomass synthesis. The experiments were carried out for 12 days at 25 °C with the initial pH 5.5 (Petre, 2002).

4.2.2. The effect of nitrogen sources upon mushroom mycelia growth

Among five nitrogen sources examined, rice bran was the most efficient for mycelia growth and fungal biomass production. The experiments were carried out for 12 days at 25 °C with the initial pH 5.5 (Table 6).

Nitrogen sources (1%, w/v)	Fresh Fungal Biomass Weight (g/l)			Final pH		
	G. lucidum	L. edodes	P. ostreatus	G. l	L. e	P. o
Rice bran	37±0.21	57±0.05	73±0.23	5.5	5.5	5.1
Malt extract	36±0.12	55±0.03	69±0.20	5.3	5.2	5.7
Peptone	35±0.03	41±0.12	57±0.15	4.6	4.9	5.3
Tryptone	36±0.15	38±0.07	55±0.17	5.1	5.3	5.9
Yeast extract	37±0.20	30±0.01	61±0.14	4.3.	5.1	5.1

Data presented in table 6 are the means ± S.D. of triple determinations.

Table 6. The effect of nitrogen sources upon the mycelia growth of pure mushroom cultures on white poplar and beech composts

4.2.3. The effect of mineral sources upon mushroom mycelia growth

Among the various mineral sources examined, K_2HPO_4 yielded good mycelia growth as well as fungal biomass production and for this reason it was recognized as a favourable mineral source (Table 7). Data presented in table 7 are the means ± S.D. of triple determinations

Mineral Sources (5 mg)	Fresh Fungal Biomass Weight (g/l)			Final pH		
	G. lucidum	L. edodes	P. ostreatus	G. l	L. e	P. o
KH_2PO_4	37±0.15	45±0.07	53±0.12	5.5	5.3	5.9
K_2HPO_4	45±0.07	57±0.05	59±0.07	5.1	5.1	5.7
$MgSO_4 \cdot 5H_2O$	35±0.25	55±0.09	63±0.28	5.6	5.4	6.1

Table 7. The effect of mineral source upon mycelia growth of pure mushroom cultures on white poplar and beech composts

4.2.4 The influence of initial pH and temperature upon mushroom fruit body formation

The optimal pH and temperature levels for fungal fruit body production were 5.0–5.5 and 21–23°C (Table 8).

To find the optimal incubation temperature for mycelia growth, these fungal species were cultivated at different temperatures ranging from 20-25°C, and, finally, the optimum level of temperature was found at 23°C, being correlated with the appropriate pH level 5.5, at it is shown in Table 8. All data presented in the previous table are the means ± S.D. of triple determinations

Initial pH (pH units)	Initial temperature (t°)	Final Weight of Fresh Mushroom Fruit Bodies (g / kg substratum)		
		G. lucidum	L. edode	P. ostreatus
4.5	18	175±0.23	191±0.10	180±0.02
5.0	21	193±0.15	203±0.05	297±0.14
5.5	23	198±0.10	195±0.15	351±0.23
6.0	26	181±0.12	179±0.12	280±0.03
6.5	29	173±0.09	105±0.23	257±0.15

Table 8. The effects of initial pH and temperature upon mushroom fruit body formation on white poplar and beech composts

4.2.5. The influence of inoculum age and inoculum volume upon mushroom fruit body formation

Amongst several fungal physiological properties, the age and volume of mycelia inoculum may play an important role in fungal hyphae development as well as in fruit body formation (Petre & Teodorescu, 2012).

To examine the effect of inoculum age and inoculum volume, mushroom species *G. lucidum*, *P. ostreatus* and *L. edodes* were grown on substrates made of vineyard wastes during different time periods between 30 and 60 days, varying the inoculum volume (5 - 7 v/w).

All the experiments were carried out at 25°C and initial pH 5.5. As it is shown in Tables 9 and 10, the inoculum age of 120 h as well as an inoculum volume of 6.0 (v/w) have beneficial effects on the fungal biomass production.

Inoculum age (h)	Final Weight of Fresh Mushroom Fruit Bodies (g /kg substratum)		
	G. lucidum	L. edodes	P. ostreatus
264	123±0.14	128±0.05	135±0.23
240	141±0.10	150±0.28	157±0.17
216	154±0.12	195±0.90	193±0.15
192	155±0.23	221±0.25	215±0.05
168	169±0.37	235±0.78	241±0.07
144	210±0.20	248±0.03	259±0.12
120	230±0.15	253±0.05	264±0.21
96	215±0.09	230±0.15	253±0.10
72	183±0.05	205±0.23	210±0.05

Table 9. The effect of inoculum age upon mushroom fruit body formation on white poplar and beech composts

Inoculum Volume (v/w)	Final Weight of Fresh Mushroom Fruit Bodies (g /kg substratum)		
	G. lucidum	L. edodes	P. ostreatus
7.0	234±0.12	215±0.20	220±0.05
6.5	245±0.15	248±0.23	251±0.20
6.0	253±0.1	257±0.07	280±0.15
5.5	243±0.12	235±0.03	247±0.07
5.0	255±0.23	215±0.15	235±0.03

Table 10. The effect of inoculum volume upon mushroom fruit body formation on white poplar and beech composts

From all these fungal species tested, *P. ostreatus* was registered as the fastest mushroom (25–30 days), then *L. edodes* (35–45 days) and eventually, *G. lucidum* as the longest mushroom culture (40–50 days).

The registered data revealed that the white poplar and beech wood wastes have to be used as substrates for mushroom growing only after some mechanical pre-treatments (such as grinding) that could breakdown the whole lignocellulose structure in order to be more susceptible to the fungal enzyme action (Chahal, 1994).

Due to their high content of carbohydrates and nitrogen, the variants of culture composts supplemented with wheat grains at the ratio 1:10 and rice grains at the ratio 1:5 as well as a water content of 60% were optimal for the fruit body production of *P. ostreatus* and, respectively, *L. edodes*. The mushroom culture of *G. lucidum* does not need such supplements (Ropars et al., 1992; Lamar et al., 1992).

So far, lignocellulose biodegradation made by mushroom species of *Ganoderma* genus had been little studied, mostly because of their slow growth, difficulty in culturing as well as little apparent biotechnological potential. Only, Stamets (1993) reported a few experimental data concerning the cultivation of such fungal species in natural sites and he noticed its slowly growing.

In spite of these facts, some strains of *G. lucidum* were grown in our experiments on culture substrates made of wood wastes of white poplar and beech mixed with rye grains at the ratio 1:7 and a water content of 50%.

Higher ratio of rye grains might lead to an increase of total dry weight of fruit body, but also could induce the formation of antler branches and smaller fruit bodies than those of the control samples.

The final fruit body mushroom production ranged between 15 and 20 kg relative to 100 kg of compost made of wood, depending on the specific strains of those tested mushroom species.

5. Conclusions

1. The cereal by-products and winery wastes used as substrata for growing the fungal species *G. lucidum*, *L. edodes* and *P. ostreatus* by controlled submerged fermentation showed optimal effects on the mycelia development in order to get high nutritive biomass.

2. The dry matter content of fungal biomass produced by submerged fermentation of barley bran was higher for both tested species.

3. The protein accumulation is more intense when using barley bran compared with those of wheat and rye, at both fungal species.

4. *G. lucidum* (variant III) registered the best values of sugar and total nitrogen contents, being followed by *L. edodes* (variant I)

5. The winery and apple wastes can be recycled as useful raw materials for mushroom compost preparation in order to get significant mushroom fruit body production and protect the natural environment surrounding apple juice factories as well as wine making industrial plants.

6. By applying the biotechnology of recycling the grape and apple wastes can be produced between 20–28 kg of mushroom fruit bodies relative to 100 kg of composts made of winery and apple wastes.

7. From all these fungal species tested in experiments, *P. ostreatus* was registered as the fastest mushroom culture (25–30 days), then *L. edodes* (35–45 days) and finally, *G. lucidum* as the longest mushroom culture (40–50 days).

8. The registered data revealed that when the cells were grown in the maltose medium, the fungal biomass production was the highest among the tested variants.

9. From five nitrogen sources examined, rice bran was the most efficient for mycelia growth and fungal biomass production

10. Among the various mineral sources examined, K_2HPO_4 yielded good mycelia growth as well as fungal biomass production and for this reason it was as a favourable mineral source.

11. The inoculum age of 120 h as well as an inoculum volume of 6.0 (v/w) have beneficial effects on the fungal biomass production and the optimal pH and temperature levels for fungal fruit body production were 5.0–5.5 and 21–23° C.

12. The final fruit body mushroom production ranged between 15 and 20 kg relative to 100 kg compost made of wood, depending on the specific strains of those tested mushroom species.

Acknowledgements

The authors express their highest respect and deepest gratefulness for the professional competence and outstanding scientific contribution which were proven by Dr. Paul Adrian during so many research works.

Author details

Marian Petre[1] and Violeta Petre[2]

1 Department of Natural Sciences, Faculty of Sciences, University of Pitesti, Romania

2 Department of Fruit Growing, Faculty of Horticulture, University of Agronomic Sciences and Veterinary Medicine-Bucharest, Romania

References

[1] Bae, J. T, Sinha, J, Park, J. P, Song, C. H, & Yun, J. W. (2000). Optimization of sub-merged culture conditions for exo-biopolymer production by *Paecilomyces japonica*. *Journal of Microbiology and Biotechnology*, 1017-7825, 10, 482-487.

[2] Beguin, P, & Aubert, J. P. (1994). The biological degradation of cellulose. *FEMS Microbiol. Rev.*, , 13, 25-58.

[3] Breene, W. M. (1990). Nutritional and medicinal values of mushrooms. *J. Food Prot.*, , 53, 833-894.

[4] Carlile, M. J, & Watkinson, S. C. (1996). Fungi and biotechnology. In: *The Fungi*, M.J. Carlile, S.C. Watkinson (Eds.), Academic Press, 0-12159-960-4England, 253-264.

[5] Chahal, D. S. (1994). Biological disposal of lignocellulosic wastes and alleviation of their toxic effluents. In: *Biological Degradation and Bioremediation of Toxic Chemicals*, G.R. Chaudry (Ed.), Chapman & Hall, 978-0-41262-290-8London, England, 347-356.

[6] Chahal, D. S, & Hachey, J. M. (1990). Use of hemicellulose and cellulose system and degradation of lignin by *Pleurotus sajor-caju* grown on corn stalks. *Am. Chem. Soc. Symp.*, , 433, 304-310.

[7] Chang, S. T, & Hayes, W. A. Eds.), ((1978). The Biology and Cultivation of Edible Mushrooms. Academic Press, New York, , 147-156.

[8] Jones, K. (1995). *Shiitake- The Healing Mushroom*. Healing Arts Press, Rochester, 0-89281-499-3USA

[9] Glazebrook, M. A, Vining, L. C, & White, R. L. (1992). Growth morphology of *Strepto-myces akiyoshiensis* in submerged culture: influence of pH, inoculum, and nutrients. *Canadian Journal of Microbiology,* , 38, 98-103.

[10] Hobbs, C. (1996). Medicinal mushrooms. Santa Cruz, Botanika Press, , 251-270.

[11] Lamar, R. T, Glaser, J. A, & Kirk, T. K. (1992). White rot fungi in the treatment of haz-ardous chemicals and wastes. In: Leatham, G.F. (ed.), *Frontiers in industrial mycology,* Chapman & Hall, New York, , 223-245.

[12] Leahy, J. G, & Colwell, R. R. (1990). Microbial Degradation of Hydrocarbons in the Environment, *Microbial Rev.,* , 54, 305-315.

[13] Moo-young, M. (1993). Fermentation of cellulose materials to mycoprotein foods, *Bio-technology Advances,* 0734-9750, 11(3), 469-482.

[14] Moser, A. (1994). Sustainable biotechnology development: from high-tech to eco-tech. *Acta Biotechnologica,* 0138-4988, 12(2), 10-15.

[15] Oei, P. (2003). Mushroom Cultivation. 3rd Edition, Backhuys Publishers, 9-05782-137-0The Netherlands

[16] Petre, M, & Teodorescu, A. (2012). Biotechnology of Agricultural Wastes Recycling Through Controlled Cultivation of Mushrooms. In: Advances in Applied Biotechnol-ogy (M. Petre Editor), InTech Open Access Publisher, 978-9-53307-820-5, 3-23.

[17] Petre, M, Teodorescu, A, & Giosanu, D. (2012). Advanced Biotechnological Proce-dures of Mushroom Cultivation. In: Mushroom Biotechnology and Bioengineering, M. Petre and M. Berovic Editors, CD Press, Bucharest, 978-6-06528-146-2, 1-21.

[18] Petre, M, & Teodorescu, A. (2011). Recycling of Vineyard and Winery Wastes as Nu-tritive Composts for Edible Mushroom Cultivation. Proc. of the International Confer-ence on Advances in Materials and Processing Technologies AMPT, American Institute of Physics (978-0-73540-871-5

[19] Petre, M, & Teodorescu, A. (2010). Handbook of submerged cultivation of eatable and medicinal mushrooms. CD Press, 978-6-06528-087-8Bucharest, Romania

[20] Petre, M. & Petre, V. (2008). Environmental Biotechnology to Produce Edible Mush-rooms by Recycling the Winery and Vineyard Wastes. *Journal of Environmental Protec-tion and Ecology,* Vol. 9, No.1, pp. 88-95, ISSN: 1311-5065

[21] Petre, M, & Petre, V. (2008). Environmental Biotechnology to Produce Edible Mush-rooms by Recycling the Winery and Vineyard Wastes. *Journal of Environmental Protec-tion and Ecology,* 1311-5065, 9(1), 88-95.

[22] Petre, M, Peng, M-X, & Mao, L-X. (2005a). The influence of culture conditions on fun-gal pellets formation by submerged fermentation of *Cordyceps sinensis* (*Paecilomyces hepiali*)- Cs 4. *Acta Edulis Fungi,* Suppl.), , 12, 345-353.

[23] Petre, M, Teodorescu, A, & Dicu, G. (2005b). The Growing Effect of Vineyard and Winery Wastes on the Production of Mycelia and Fruit Bodies of Edible and Medicinal Fungi. *International Journal of Medicinal Mushrooms*, 1521-9437, 7(3), 444-445.

[24] Petre, M, Zarnea, G, Adrian, P, & Gheorghiu, E. (2001). Biocontrol of cellulose waste pollution by using immobilized filamentous fungi. In: Environmental Monitoring and Biodiagnostics of Hazardous Contaminants (Healy, M., Wise, D.L. Moo-Young, M, eds), Kluwer Academic Publishers, The Netherlands, 079236869, 227-241.

[25] Raaska, L. (1990). Production of *Lentinus edodes* mycelia in liquid media: Improvement of mycelial growth by medium modification. *Mushroom Journal of the Tropics*, , 8, 93-98.

[26] Ropars, M, Marchal, R, Pourquie, J, & Vandercasteele, J. P. (1992). Large scale enzymatic hydrolysis of agricultural lignocellulosic biomass. *Biores. Technol.*, , 42, 197-203.

[27] Smith, J. (1998). Biotechnology. 3rd Edition. Cambridge University Press, 0-52144-911-1England

[28] Stamets, P. (2000). Growing Gourmet and Medicinal Mushrooms. Ten Speed Press, 1-58008-175-4Toronto, Canada

[29] Verstraete, W, & Top, E. (1992). Holistic Environmental Biotechnology. Cambridge University Press, 0-52142-078-4England

[30] Wainwright, M. (1992). An Introduction to Fungal Biotechnology. Wiley, Chichester, , 103-115.

[31] Wasser, S. P, & Weis, A. L. (1994). Therapeutic effects of substances occurring in higher Basidiomycetes mushrooms: a modern perspective. *Critical Reviews in Immunology*, 1040-8401, 19, 65-96.

Chapter 2

Biochemical Processes for Generating Fuels and Commodity Chemicals from Lignocellulosic Biomass

Amy Philbrook, Apostolos Alissandratos and
Christopher J. Easton

Additional information is available at the end of the chapter

1. Introduction

Fuels and chemicals derived from biomass are regarded as an environmentally friendly alternative to petroleum based products. The concept of using plant material as a source for fuels and commodity chemicals has been embraced by governments to alleviate dependence on the volatile petroleum market. This trend is driven not only by economics but also by social and political factors. Global warming has been associated with CO_2 emissions largely originating from the combustion of fossil fuels.[1] This, together with depleting and finite carbon fossil fuel resources, and insecurity of petroleum supplies has prompted a shift towards biofuels and biomaterials.[1] The use of biomass as an economically competitive source of transport fuel was initiated by the fuel crisis in 1970 and its commercialization was led by the USA and Brazil.[2] In 2010, the USA and Brazil processing corn and sugarcane, respectively, produced 90% of the world's bioethanol. In 2008, the "food for fuel" debate emerged sparked by concerns that the use of arable land for bioethanol and biodiesel crops was placing pressure on food demand for a growing world population.[3] In June 2011, the World Bank and nine other international agencies produced a report advising governments to cease biofuel subsidies as the use of food stock for fuel production was linked to increasing food prices.[4] Subsidies were thus ended in the USA when their Senate voted overwhelmingly to end billions of dollars in bioethanol subsidies.[5] This reform resulted in USA bioethanol plants recording losses in the first quarter of 2012[6] and is foreseen as the end of bioethanol production from corn at least in the USA.

Emerging from the "food for fuel" debate, the concept of commercializing second generation biofuels was embraced by governments as a route to produce biofuels without diminishing global food supplies.[7]. Second generation biofuels address concerns over designating arable land to grow food crops for fuel production as lignocellulosic biomass may consist of waste materials such as plant residues.[8] In many proposed biorefinery setups, the food portion of the crop is to be used for human consumption and the waste residues, for example, the leaves and stalks, are to be processed for biofuels and chemicals.[8] An illustration of the processes for 1st, 2nd and 3rd generation biofuel production is shown in Figure 1. Third generation biofuel production, the generation of biodiesel from algae, is included in the diagram for completeness.

All three of the processes outlined in Figure 1 rely on biotechnology for the conversion of biomass to fuels. First generation bioethanol production traditionally incorporates two biological transformations. The first stage uses commercialized saccharification biotechnology which depolymerises starch into fermentable glucose units. The second stage is the fermentation of sugar units to ethanol and again uses commercialized biotechnology generally with yeast extracts.[1] Although the use of lignocellulosic biomass is socially widely supported, the processes for its conversion are more complex and therefore more costly. The major cost-adding component of 2nd generation bioethanol production compared to the 1st is the pretreatment step as the removal of the lignin is required for cellulose accessibility.[9] Whilst 1st generation bioethanol production converts substrates high in starch (mainly corn, sugarbeet and sugarcane), the effective utilization of lignocellulosic biomass requires at least separation, if not complete conversion, of all plant components. The composition of plant material includes lignin, cellulose and hemicelluloses and a diagram illustrating how these components relate is shown in Figure 2. The percentage of these three plant components varies with species (Figure 2) further complicating the processing of such biomass.

It has been reported that the separation and use of all plant components is required for environmentally and economically viable biorefineries.[1, 8, 9] The application of biotechnology for all aspects of biomass conversion avoids toxic by-products and high energy inputs encountered with chemical, thermal and mechanical processes often used. It is due to these energy and environmental concerns that biochemical methods are feverously being investigated, as enzymatic processes are largely environmentally benign and low in energy demand. Processes for the transformation of biomass need to be carbon efficient, otherwise the environmental objectives of biomass utilization are negated. It is with this in mind that the current chapter is focused on advances using environmentally benign biocatalysts.

The use of biochemical techniques for processing of lignocellulosic biomass is covered herein. This includes the bioprocessing of the plant components, lignin, cellulose and hemicellulose and is focused on progress made in their biochemical conversion not only to ethanol but also to value-added chemicals according to biomass fraction. The review of the literature is concentrated on biocatalytic advances in the past decade and is delineated by the plant-derived substrate. Strategies for the commercialization of 2nd generation biofuels and commodity chemicals are discussed.

2. Biochemical pretreatment

Pretreatment of lignocellulosic biomass is required to increase holocellulose (cellulose and hemicellulose) accessibility for its hydrolysis into fermentable sugars and only 20% of the

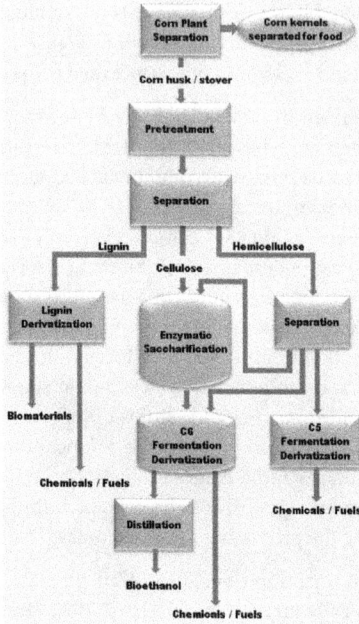

Figure 1. Examples of processes to produce 1st generation bioethanol from corn, 2nd generation bioethanol from corn waste residues and 3rd generation biodiesel from algae.

theoretical sugar yield can be obtained from lignocellulose without pretreatment.[10, 11] Currently, the biomass pretreatment step for producing 2nd generation bioethanol is the most expensive component of the process after the raw material.[12] Thermal and mechanical methods are energy intensive and therefore carbon costly as they indirectly produce CO_2. Chemical techniques result in contamination of the biomass producing biochemical inhibitors as by-products[13, 14] and require costly neutralization processes.[15] Removal of the lignin fraction using microorganisms has several advantages compared to other pretreatment methods. Firstly, microorganisms function under ambient conditions thus eliminating thermal and electrical energy inputs. When compared to chemical pretreatment methods, biochemical pretreatment does not result in chemical by-products that often inhibit cellulose hydrolysis.

In nature, fungi are responsible for the biodegradation of lignin, thus the majority of research into biochemical pretreatments has focused on fungi for the delignification of biomass. Early research in the area was led by the pulp and paper industry and focussed on fungal treatment as a method for removing the lignin fraction from wood to facilitate cellulose accessibility and to lower pulping energy costs. In 1982, Eriksson and Vallander were able to achieve a 23% reduction in refining energy by incubating wood chips with the white-rot fungus

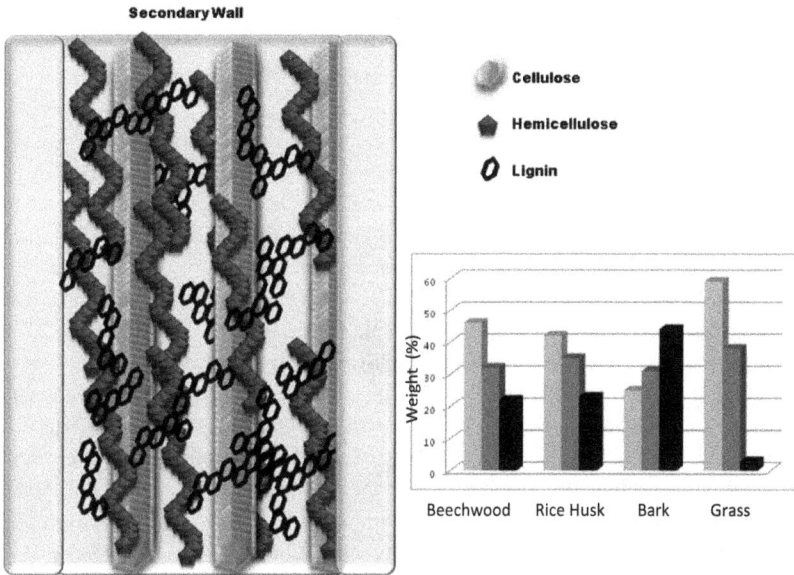

Figure 2. Diagram of plant components cellulose, hemicelluloses and lignin and a graphical representation of their weight percentage according to biomass source[10].

Phanerochaete chrysosporium for 2 weeks.[16] Messner and Srebotnik[17] studied the same species of fungus and reported similar results. Later studies by Akhtar et al.[18] also found substantial energy savings in pulping when the wood was treated with *Ceriporiopsis subvermispora*. Others[19] studied soft and hard wood processes and reported a reduction in refining energy from incubation with strains of white-rot fungi of 33% for soft-wood pulp and more than 50% for hardwood. More recently, Liew et al.[20] reported a lignin loss of 26.9% in biopulping studies with *Acacia mangium* wood chips when incubated with the white-rot fungi *Trametes versicolour*.

The amount of lignin present in the biomass directly affects enzymatic digestion of the holocellulose fraction. For example, a decrease from 22% to 17% lignin in biomass samples doubles the sugar yield and samples with 26% lignin result in virtually no sugar.[21] However, the effectiveness of pretreatment is not only measured by the decrease in lignin content but also by holocellulose recovery and ultimately the saccharification percentage. Table 1 summarizes recent studies conducted on the pretreatment of different biomass substrates. As stated earlier and depicted in Figure 2, the amount of lignin, cellulose and hemicellulose varies greatly with biomass source and it is therefore logical to assess the effectiveness of fungal pretreatment according to substrate. It is important to note that direct comparisons are not always possible as the experimental techniques and measurements vary within many of the cited studies. For example, the fungal incubation times listed in Table 1 vary from 2 to 120 days.

Different species of white-rot fungi, *Echinodontium taxodii*,[22, 23], *Coriolus versicolor*[24] and *Trametes versicolor*[23], have been studied for their ability to degrade lignin to promote cellulose digestibility in bamboo residues. Zeng, Yang et al.[22] recently reported a 29% decrease in lignin content in bamboo treated with *Echinodontium taxodii* however the aim of the work was to improve the thermal decomposition of the bamboo and not to recovery and utilize the holocellulose. Zhang, Xu et al.[24] reported an increase in saccharification rate of 37% when bamboo residues were incubated with *Coriolus versicolor*. Zhang, Yu et al.[23] compared two species of white-rot fungus for their effectiveness in increasing sugar yields and found that incubation with both *Trametes versicolor* and *Echinodontium taxodii* improved sugar yields 5.15 and 8.75 times respectively. Cornstalk and corn stover have been pretreated with different fungal strains for delignification. Impressive results were reported by Wan and Li[25] who pretreated corn stover with *Ceriporiopsis subvermispora* and measured a 31.59% reduction in lignin with only a 6% loss in cellulose. In 2010, Dias et al.[26] reported a nearly 4-fold increase in saccharification of wheat straw treated with basidiomycetous fungi Euc-1 and *Irpex lacteus*. *Dichomitus squalens*,[27] *Pleurotus ostreatus*[28] and *Phaerochaete chrysosporium*[28] were applied to rice straw with varying effects (Table 1), with the most notable reported by Bak et al.,[27] being a 58.1% theoretical glucose yield of rice straw treated with *Dichomitus squalens*. The biochemical pretreatment of cotton stalks was studied by Shi et al. who reported 33.9% lignin reduction[29] using submerged fungus cultivation and 27.6% lignin reduction[30] using solid state cultivation of the same fungus, *Phanerochaete chrysosporium*. Hideno et al.[31] applied *Grifola frondosa* for the pretreatment of sawdust matrix and reported a 21% reduction in lignin with 90% cellulose recovery.

Substrate	Species	Findings	Duration	Ref
Bamboo	*Echinodontium taxodi*	29% reduction in lignin	30 days	[22]
Bamboo residues	*Coriolus versicolor*	Enhanced saccharification rate of 37%	2 days	[24]
Bamboo culms	*Echinodontium taxodii*	5.15-fold increase in sugar yields	120 days	[23]
Bamboo culms	*Trametes versicolor*	8.75-fold increase in sugar yields	120 days	[23]
pCornstalk	*Phanerochaete chrysosporium*	34.3% reduction in lignin with a maximum enzyme saccharification of 47.3%	15 days	[32]
Corn Stover	*Ceriporiopsis subvermispora*	Lignin degradation reached 45%	30 days	[33]
Corn Stover	*Irpex lacteus CD2*	66.4% saccharification ratio	25 days	[34]
Corn Stover	*Ceriporiopsis subvermispora*	31.59% lignin degradation with less than 6% cellulose loss	18 days	[25]
Corn Stover	*Cyanthus stercoreus*	3- to 5-fold improvement in enzymatic digestibility	29 days	[12]
Wheat straw	Basidiomycetous fungi Euc-1	4-fold increase in saccharification	46 days	[26]
Wheatstraw	*Irpex lacteus*	3-fold increase in saccharification	46 days	[26]
Rice straw	*Dichomitus squalens*	58% theoretical glucose yield for remaining glucan	15 days	[27]
Rice straw	*Pleurotus ostreatus*	39% degradation of lignin with 79% cellulose retention	48 days	[35]
Rice straw	*Phaerochaete chrysosporium*	64.9% of maximum glucose yield from recovered glucan	15 days	[28]
Cotton stalks	*Phaerochaete chrysosporium*	33.9% lignin degradation with 18.4% carbohydrate availability	14 days	[29]
Cotton stalks	*Phaerochaete chrysosporium*	27.6% lignin degradation	14 days	[30]
Sawdust matrix	*Grifola frondosa*	21% reduction in lignin and 90% recovery of cellulose	2 days	[31]

Table 1. Fungal strains studied for pretreatment of lignocellulosic biomass.

3. Bioconversion of lignin to chemicals and fuels

During biochemical pretreatment, the lignin fraction is metabolized by the microorganism. In chemical and thermal pretreament processes the lignin fraction often remains intact and is thus able to be separated and utilized. After separation, microorganisms could in principle transform the lignin into materials, chemicals and fuels. Despite efforts over a long period of time, research into the bioconversion of lignin into economically viable products is still in its infancy, primarily because of the complex and irregular structure of lignin (Figure 3). However, advancements for the valorization of lignin are actively being pursued, as lignin is the second most abundant carbon source in nature and contains valuable phenolic building blocks within its structure.[36]

Although lignin has traditionally been burned as an inefficient energy source by-product from bioethanol or pulping production, lignin derived value-added products are necessary to improve biomass conversion economics.[37] Lignin has been used in the manufacture of wood adhesives as a component of phenol-formaldehyde resins (LPF resins).[38, 39] Lignin-derived commodity chemicals have been targeted through chemical and biological routes (Figure 3). Vanillin and cinnamic acid are subunits of the complex lignin structure and are used commercially as food sweeteners, as additives for fragrances and as precursors for pharmaceuticals. Phenol is the most widely used starting material in the plastic and resin industry and phenolic monomers have also been targeted from lignin. After the depolymerisation of lignin into monomeric units, the substituted monomers are precursors of a range of products including fuels such as cyclohexane.

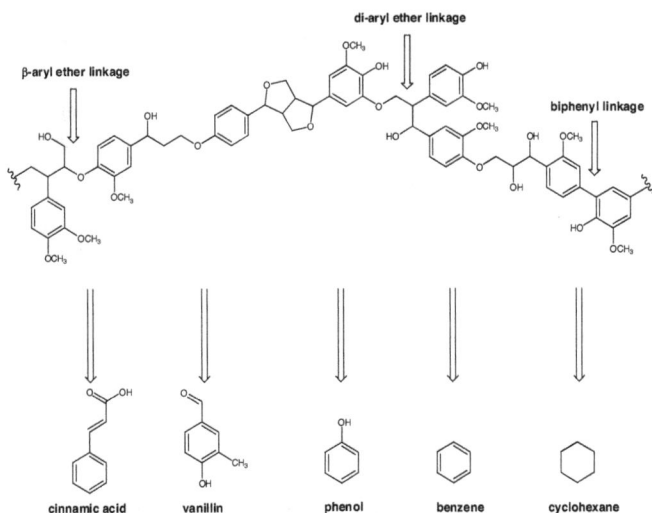

Figure 3. Examples of chemicals targeted from lignin.

Although biochemical pretreatment methods generally use fungi as the lignin degrading microorganism, it is unlikely that usable lignin-derived materials and chemicals will result from fungal processes as white-rot fungi are known to mineralize lignin.[40] Thus bacterial conversion of lignin into chemicals and fuels constitutes an attractive method for the valorization of lignin. Classes of bacteria capable of degrading lignin have been identified as Actinomycetes, α-Proteobacteria and γ-Proteobacteria.[41-44] Recently, a range of metabolites (Figure 4) have been isolated from the bacterial degradation of lignocelluloses. [40] Metabolites A and B have been observed from lignocelluloses processed by the bacteria *Pseudomonas putida* mt-2[43], *Rhodococcus jostii* RHA1[43] and *Sphinogobium* sp. SYK-6.[42] Furthermore *Sphinogobium* sp. metabolizes β-aryl ether linked aromatics to vanillin.[42] Compounds C, D, E and G were identified using GC-MS as bacterial degradation products of Kraft lignin.[45] Ferulic acid as well as compounds F, H, I and J were identified by GC-MS as products of waste paper effluent treated with *Aeromonas formicans*.[46] There are established chemical and biochemical methods for converting lignin derived monomers, like those observed in the bacterial degradation of lignin (Figure 4), into simpler aromatics like phenol (Figure 3) as well as hydrocarbons like cyclohexane (Figure 3).

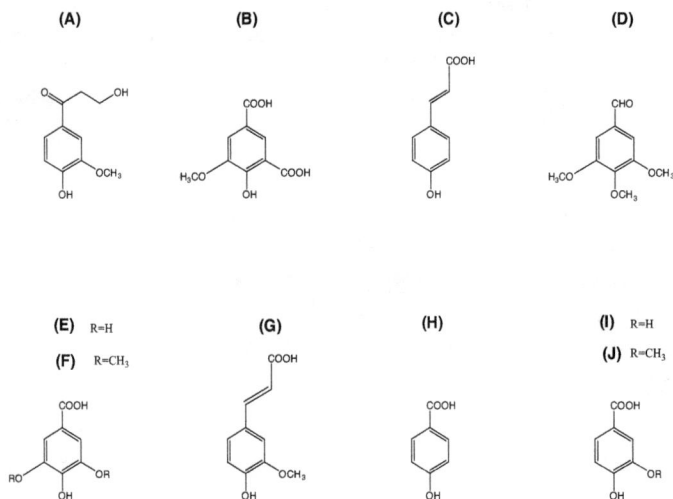

Figure 4. Compounds isolated after bacterial conversion of lignin.[40]

4. Biochemical conversion of cellulose

The use of starchy feedstock, such as corn and sugar cane, is problematic in relation to food sustainability and biodiversity. Therefore, as mentioned, the focus in second generation biofuel production processes has been on biomass consisting mainly of cellulose. A high percentage of cellulose (usually 35-50% dry weight) is consistently found in all plants despite the vast genetic diversity that is observed within the plant kingdom.[47] For the production of ethanol, cellulose is exposed during pretreatment, hydrolysed by either chemical or enzymatic hydrolysis and then fermented into ethanol. The material, once stripped from other biopolymers surrounding it within the plant structure, also appears to have characteristics independent of plant taxa. Cellulose is a linear polymer, composed of glucose monomers held together by β-1,4-glucosidic bonds (Figure 5), in contrast to α-1,4-bonds found in other common glucans such as starch and glycogen. Through interchain and intrachain hydrogen bonding as well as Van der Waals forces, cellulose chains self-assemble on biosynthesis into protofibrils, then microfibrils, which are in turn packed into fibres with high crystallinity, imparting the material with high tensile strength and water insolubility.[48] These very properties that make it a suitable structural polysaccharide are the cause of the main difficulties associated with the use of biomass rich in cellulose, for the generation of products through fermentation. The additional energy required to break down the rigid structure of cellulose is one of the main obstacles towards commercialization of lignocellulosic biomass processing.

Cellulases, the enzymes responsible for cellulose hydrolysis, differ from other glucoside hydrolases in that they are able to catalyse hydrolysis of β-1,4-glucosidic bonds. Cellulases vary significantly and belong to several glycoside hydrolase families.[49] The main differences between them relate to their mode of action. While endoglucanases are thought to randomly hydrolyse the amorphous fraction of cellulose, exoglucanases process the polysaccharide preferentially from a reducing or non-reducing end, releasing cellobiose (cellobiohydrolases) or glucose (glucanohydrolases).[47] An important feature of the exoglucanase structure is a distinct domain termed the carbohydrate binding module (CBM), which allows the enzyme to remain attached to the cellulose chain during catalytic action. This aids enzymatic action upon crystalline material by bringing the catalytic domain closer to the substrate and has been suggested to also help catalysis by peeling fragments of cellulose from the cellulosic surface. [50] β-Glucosidases are the third general category of cellulolytic enzymes; they act upon bonds in soluble cellobiose or cellodextrins formed by the action of the other two types of cellulases. The different types of cellulases act in coordination to efficiently hydrolyse cellulose, displaying synergy and, depending on the host, may or may not form stable complexes of high-molecular weight.[51] These complexes, although beneficial to penetration of cellulosic material *in vivo*, when used in bioprocessing are generally considered problematic.[52]

The microorganism to receive by far the most attention in relation to sourcing of cellulolytic enzymes has been *Trichoderma reesei*.[53] This fungus was identified by E.T. Reese as the culprit for the rapid destruction of allied forces' cotton tents during WWII. Since its isolation, it has been extensively studied in relation to its cellulolytic capability and various cellulase hyper-producing strains have been developed, with RUT C30 currently the benchmark strain for production of cellulases in high yields.[54] One of the problems associated with this fungus has been the low expression of β-glucosidases, the enzymes responsible for liberation of glucose from short oligosaccharides. This, however, has been overcome with genetic engineering and supplementation of commercial preparations with foreign β-glucosidases.[55] Other promising fungal sources for cellulases exist, such as *Acremonium*, *Penicillium* and

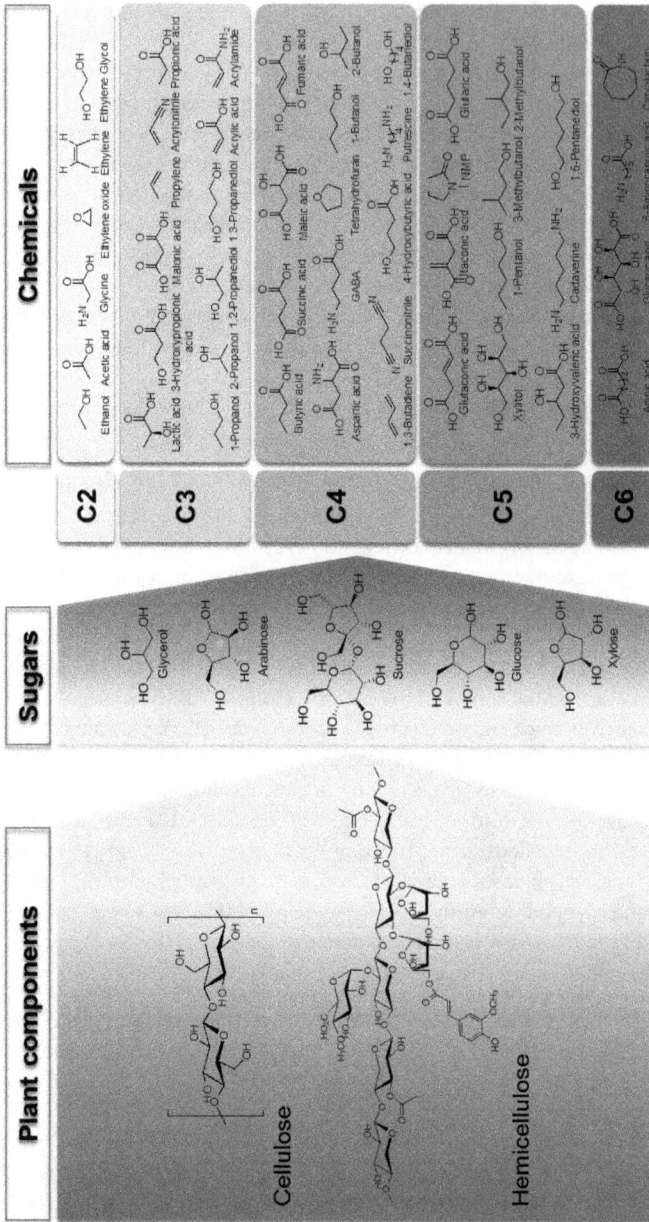

Figure 5. Cellulose and hemicellulose, their sugar units and some potential chemical targets organized by carbon chain number.

Chrysosporium strains.[52] Their cellulase properties are comparable to those of *T. reesei*, however, they are unlikely to replace it as the standard enzyme source due to the amount of improvement already achieved with the latter. Bacterial cellulases have been the focus of some attention due to the higher robustness observed with some hyperthermophilic enzymes, making them more adaptable to the harsh conditions of industrial processes.[56] However, the production of cellulases as part of complexed systems (cellulosomes) in anaerobes, as well as the much lower protein yields in bacteria, means that interest in these enzymes is mainly restricted to their heterologous expression in fungi and use in consolidated bioprocessing (CBP, see *Consolidated fermentation*).[47, 56]

5. Biochemical conversion of hemicellulose

Hemicellulose (Figure 5) is a mixture of several different polysaccharides, the composition of which varies from plant to plant as well as within the same plant.[57] While cellulose is built from a single building block, a number of different monomers compose hemicellulosic heteropolymers including pentoses, hexoses and sugar acids. Commonly xylans, glucomannans, arabinogalactans and different types of glucans are found in hemicellulose. Xylans are comprised of β-1,4-linked xyloses interspersed with arabinose and glucuronic acid, while glucomannans are a mixture of β-1,4-linked glucose with α-1,6-substituted mannose side chains (Figure 5]. The presence of acetyl substitutions on hydroxyl groups of carbohydrates in hemicellulose is not completely understood, but may pose difficulties in hydrolysis due to the generation of acetate which acts as an enzyme and microorganism inhibitor.[58] The network of hemicellulosic chains is highly branched, cross-linking with cellulose microfibrils and lignin, creating a very compact material from which plant cell walls are composed. It is generally agreed that economically viable bioprocessing of lignocellulosic biomass requires efficient extraction and conversion of the hemicellulosic sugars.

As is the case with cellulases, hemicellulases constitute a useful tool for the generation of fermentable sugars from hemicellulose and are sometimes classed as cellulases themselves. Due to the diversity of components and complexity of structure found in this polymer it is only natural that a myriad of enzymes with different catalytic functions have been produced by microorganisms to effectively attack this matrix.[53, 59] Therefore, for example, endoxylanases, exoxylanases and β-xylosidases have been identified to break the linkages between xylose moieties, while esterases releasing acetyl and ferulic acid groups are also found amongst this category of enzymes. These enzymes sometimes display relative promiscuity towards the type of bond they hydrolyse, making it extremely difficult to measure a specific enzymatic activity. They also display significant synergy between themselves as well as with other lignocellulose hydrolysing enzymes.[52] As expected, microorganisms that express cellulose degrading enzymes also possess the ability to degrade hemicellulosic polymers. Accordingly it has been highlighted that *Trichoderma* and *Penicillium* fungi contain efficient hemicellulolytic catalysts.[59] Another group of fungi identified for their important xylan degrading capabilities have been *Aspergillus* spp.[60]

6. Fermentation to fuels and chemicals

The vast amount of available know-how, due to the fact that this process is one of man's earliest biotechnological applications, continues to set the use of *Saccharomyces cerevisiae* for the production of ethanol as the benchmark fermentation system employed for second generation biofuel processes. This yeast's properties, particularly in relation to robustness, toxicity, ethanol productivities approaching the theoretical maximum and ease of genetic manipulation make it an extremely suitable microorganism for the fermentation step of lignocellulose conversion.[61, 62] As a result, much has already been accomplished in production of efficient yeast strains for the conversion of hexoses from starchy feedstock in first generation biofuel production. The issues that require addressing for carrying over these microorganisms to second generation biofuel production processes relate to tolerance to by-products of lignocellulose pretreatment and digestion, and the ability to ferment pentoses generated by the hemicellulosic fraction of the biomass. Furthermore, the possibility of combining efficient pentose and hexose utilisation as well as production of lignocellulose hydrolysing enzymes within a single host would allow the combination of hydrolysis and fermentation steps, greatly reducing the overall cost of the production process.

Other types of microorganisms have also been investigated as alternatives, mainly for the coproduction of other compounds. A recent review by Jang et al.[63] lists organisms according to their corresponding C2-C6 platform chemical products (Figure 3 and Table 2). Anaerobic clostridial strains have been of particular interest due to their ability to efficiently generate butanol as well as their tolerance to other common metabolites (acetate, lactate) which, they are able to use as nutrients for the further production of alcohols.[64, 65] As a result the use of microorganisms such as *Clostridium acetobutylicum* has been proposed for acetone-butanol-ethanol (ABE) bioprocesses, since butanol is an attractive alternative to ethanol as a biofuel due to its lower vapour pressure and higher energy density.[66] *Clostridia* are also interesting because of the broad spectrum of chemicals that they are able to produce, as well as recent advances in their genetic manipulation.[67]

Related types of yeast have also been investigated as alternatives in order to produce microorganisms with superior properties. Thermophilic yeasts show increased ability to work at elevated temperatures which may present great advantages, particularly in relation to *in situ* evaporative removal of the product in batch processes, a procedure that is generally considered essential for the reduction of down-stream costs as well as minimising product toxicity issues.[123]

The pretreatment and hydrolysis of the complex mixture of lignocellulose, unlike the simple hydrolysis of starch, yields a number of additional by-products. These may pose problems to the growth of the microorganism fermenting the simple sugars as feedstock for fuel or chemical production. Toxic compounds encountered in lignocellulosic hydrolysates normally consist of phenolic compounds, weak organic acids and furan aldehydes.[61] Complex strategies have been employed to combat the effects of the presence of these compounds. Genetic engineering approaches have aimed at overexpression of pathways which, metabolise the inhibitors.[62]

Platform chemical	Leading host	Substrates and/or conditions	Titer (g/L)	Yield (g/g)	Productivity (g/L/h)	Ref
C2 Ethanol	S. cerevisiae	Ammonia fiber expansion (AFEX)-corn stover (CS)-hydrolysates, batch fermentation	40	0.46	0.8	[68]
	S. cerevisiae	Cellobiose, xylose, and glucose, batch fermentation	48	0.37	0.8	[69]
	Z. mobilis	Glucose and xylose, batch fermention	62	0.46	1.29	[70]
	E. coli	Xylose, batch fermentation	23.5	0.48	n/a	[71]
Acetic acid	A. aceti	Ethanol, batch fermentation	111.7	n/a	0.6	[72]
C3 Propionic acid	P. acidipropionici	Glycerol, fed-batch fermentation in fibrous bed bioreactor	106	0.56	0.035	[73]
Lactic acid	Sporolactobacillus	Glucose, fed-batch supplemented with 40g/L peanut meal	207	0.93	3.8	[74]
	E.coli	Glucose, fed-batch fermentation	138	0.86	3.5	[75]
3-Hydroxypropionic acid	K. pneumonia	Glycerol, fed-batch fermentation	16	n/a	0.01	[76]
	E. coli	Glycerol, fed-batch fermentation	38.7	0.34	0.53	[77]
Propanol	E. coli	Glucose, flask culture	3.9	n/a	0.04	[78]
Iso-propanol	C. acetobutylicum	Glucose, anaerobic flask culture	5.1	n/a	n/a	[79]
	E. coli	Glucose, batch-fed fermentation	13.6	0.15	0.28	[80]
1,2 propanediol	C. thermosaccharolyticum	Glucose, anaerobic batch fermentation	9.1	0.20	0.35	[81]
	E. coli	Glycerol, batch fermentation	5.6	0.21	0.077	[82]
1,3 propanediol	C. acetobutylicum	Glycerol, anaerobic fed-batch fermentation	83.6	0.54	1.70	[83]
	E. coli	Glucose, fed-batch 10L fermentation	135	0.51	3.5	[84]
C4 Butyric acid	C. tyrobutyricum	Glucose fed-batch fermentation	32.5-41.7	0.38-0.42	0.24-0.68	[85, 86]
	C. tyrobutyricum	Glucose, repeated fed-batch fermentation by immobilized cells in a fibrous bed bioreactor	86.9	0.46	1.1	[87]
Succinic acid	Engineered rumen bacteria	Glucose, anaerobic fed-batch fermentation	52-106	0.76-0.88	1.8-2.8	[88, 89]
	E. coli	Glucose, fed-batch fermentation	73-87	0.8-1.0	0.7-0.9	[90-92]
	C. glutamicum	Glucose, fed-batch fermentation	140-146	0.92-1.1	1.9-2.5	[93, 94]
Malic acid	Aspergillus flavus	Glucose, batch fermentation	113	0.95	0.59	[95]
	S. cerevisia	Glucose, fed-batch fermentation	59	0.31	0.19	[96]
	E. coli	Glucose, two-stage fermentation	33.9	0.47	1.06	[97]
Fumaric acid	R. arrhizus NRRL 2582	Glucose, batch fermentation	97.7	0.81	1.02	[98]
GABA	L. brevis NCL912	Glucose and glutamate, fed-batch fermentation	103.7	n/a	n/a	[99]
	C. glutamicum	Glucose, batch fermentation	2.2	n/a	0.01	[100]
1-butanol	C. acetobutylicum	Glucose, anaerobic batch fermentation	16.7	n/a	0.31	[101]
	E. coli	Glucose, batch cultivation	14-15	0.33-0.36	0.20-0.29	[102, 103]
Isobutanol	E. coli	Glucose, batch cultivation	20	n/a	n/a	[104]
	C. glutamicum	Glucose, fed-batch fermentation	13.0	0.20	0.33	[105]
1,4-butanediol	E. coli	Glucose, microaerobic fed-batch fermentation	18	n/a	0.15	[106]
2,3-butanediol	K. pneumonia SDM	Glucose, fed-batch fermentation	150	0.48	4.21	[107]
	S. marcescens	Glucose, fed-batch fermentation	152	0.46	2.67	[108]
Putrescine	E. coli	Glucose, fed-batch culture	24.2	n/a	0.75	[109]
C5 Itaconic acid	Aspergillus terreus IFO-6365	Glucose and corn steep, flask and 100 L batch fermentation	82-85	0.54	0.57	[110]
	E. coli	Glucose, flask batch culture	6	0.61	0.06	[111]
3-hydroxyvalerate	P. putida	Glucose and levulinic acid, flask batch cultivation	5.3	n/a	n/a	[112]
	E. coli	Glucose and threonine, flask batch cultivate	1.3	n/a	n/a	[112]
	E. coli	Glucose, flask batch cultivation	0.81	n/a	n/a	[112]
1-pentanol	E. coli	Glucose	0.5	n/a	n/a	[113]
2-methyl-1-butanol	E. coli	Glucose	1.25	n/a	0.17	[114]
3-methyl-1-butanol	E. coli	Glucose	1.28	n/a	0.11	[115]
Xylitol	C. tropicalis	Xylose, oxygen-limited condition with cell recycling	1.82	0.85	12.0	[116]
	E. coli	Glucose and xylose, fed-batch fermentation	38	n/a	n/a	[117]
Cadaverine	E. coli	Glucose, fed-batch fermentation	9.61	n/a	0.12	[118]
C6 Glucaric acid	E. coli	Glucose, flask culture	2.5	n/a	n/a	[119]
Anthranilic acid	E. coli	Glucose, fed-batch cultivation	14	0.20	0.41	[120]
Phenol	P. putida S12	Glucose, flask batch culture	0.14	3.5	0.006	[121]
Catechol	P. putida ML2	3-Dehydroshikimate	4.2	n/a	0.12	[122]

Table 2. Current status of the production of platform chemicals using microorganisms. Duplicated with permission.[63]

Another approach is adaptation of the microorganism to an inhibitor rich environment through evolutionary processes. It has been observed that the stress imposed stimulates changes in the resulting strains, usually in relation to glycolytic enzyme activity, levels of intracellular materials and expression of inhibitor metabolising enzymes, which impart increased tolerance. The new strains are generally able to grow at higher hydrolysate, and consequently inhibitor, concentrations thus reducing processing time and cost.[124] Cell viability is also threatened by the target products of fermentation, as these may cause damage to cell membranes and interfere with physiological processes. Tolerance to these compounds without decreasing the process yield may be achieved by regulation of membrane transporters such as efflux pumps, modification of the membrane composition or regulation of heat shock proteins that have been found to be linked to stress response in cells.[125] An added benefit to the increase of tolerance in some cases may be an increase in product yield.[126, 127]

Strain	Inhibitor	Approach	Reference
S. cerevisiae	acetate	Deletion of HRK1 gene regulating membrane transporter activity	[128]
S. cerevisiae PK113-7D	formate, acetate	Expression of formate dehydrogenase structural gene FAHD2	[129]
S. cerevisiae	vanillin	Overexpression of laccase gene lacA from Trametes sp. AH28-2	[130]
E. coli	biodiesel, biogasoline	RND efflux pumps heterologously expressed	[131]
C. acetobutylicum	butanol	Overexpression of GroESL heat shock protein	[127]
E. coli	isobutanol	Simultaneous disruption of five unrelated genes	[132]
S. cerevisiae	ethanol, glucose	Global transcriptional machinery engineering, also improved ethanol yield by 15%	[126]

Table 3. Examples of engineering microorganisms for improved tolerance to inhibitors in lignocellulosic biomass processing.

Microorganisms naturally capable of fermenting pentoses such as *Pichia stipitis, Kluyveromyces marxianus, Clostridium saccharolyticum* and *Thermoanaerobacter ethanolicus* exist and may well be employed in processes for the production of ethanol as well as other chemicals.[62] However, considerable effort has been put into engineering pentose fermentation capability into strains traditionally used for ethanol production, such as *S. cerevisiae*, with great success. This yeast is able to take up pentose with hexose transporters, however the ability to metabolise these sugars had to be introduced with expression of bacterial and fungal gene insertion. This has also led towards engineering hexose/pentose efficient cofermentation, something that has not been identified in native microorganisms. The ability to coferment xylose, arabinose and glucose has been successfully introduced to *S. cerevisiae*, however modern approaches to metabolic engineering need to be employed in order to improve on this, concentrating more on non-traditional aspects of cell engineering, such as catabolism repression mechanisms and stress response.[133]

7. Consolidated fermentation

One of the great advantages of biochemical methods of biomass conversion is that they all require mild conditions, which makes them relatively compatible, allowing for potential consolidation of processing steps. This has been identified as an area of great potential in relation to process optimization, cost reduction and ultimately biorefinery commercialization. Instead of applying four distinct biochemical processing steps (cellulose production, cellulose hydrolysis, hexose fermentation, pentose hydrolysate fermentation), a setup termed Separate Hydrolysis and Fermentation (SHF), two or more steps may be consolidated leading to alternate process configurations for biomass conversion.[47] This requires generation of biocatalysts with properties suited to the optimum processing conditions, or engineering of microorganisms with more than one processing capability. Simultaneous Saccharification and Fermentation (SSF) involves performing cellulase-catalysed cellulose hydrolysis in the cellulose hydrolysate fermenter, after the enzymes are produced in a separate fermentation. Further consolidation may include cofermentation of the hemicellulose hydrolysate, either by a separate pentose utilising microorganism or by an engineered strain capable of efficient cofermentation of hexoses and pentoses. This configuration is termed Simultaneous Saccharification and Cofermentation (SSCF). The most desirable setup that minimises utility costs is direct fermentation of biomass to the product of choice with the aid of a cellulase expressing, hexose/pentose cofermenting microorganism. This approach was first introduced in 1996 as consolidated bioprocessing (CBP).[53, 134]

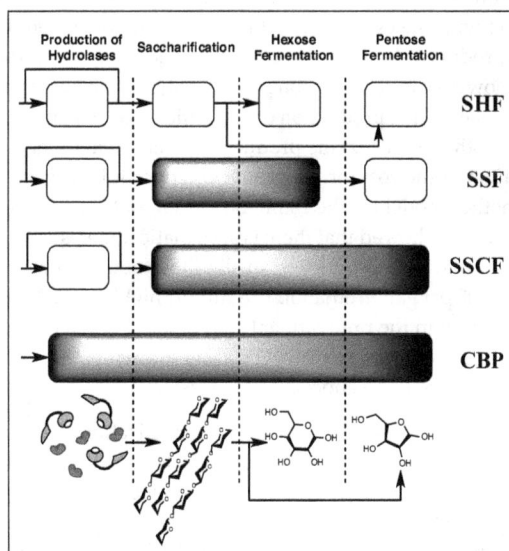

Figure 6. Consolidated fermentation processes.

8. Future outlook for commercialization

Inedible crops are a renewable and sustainable source of fuels and chemicals and it has been estimated that replacing corn with cellulosic stock would result in an 82% increase in bioethanol production. Despite this, 2nd generation biofuel and chemical production is yet to be commercialized.

Finding economical pretreatment methods has been recognized as one of the hurdles to commercializing 2nd generation biofuels and chemicals. The results listed in Table 1 show that fungal treatment can reduce lignin content of biomass and in most cases improve sugar yields. However, chemical methods are deemed more economical at present, mainly due to the long incubation times as well as the loss of holocellulose during biological pretreatment. With further screening studies, it is to be expected that fungal strains with selectivity for lignin and faster metabolic processing will be discovered, thus reducing the overall process cost. Bacteria present advantages for biotechnological applications in terms of growing times and being more prone to metabolic manipulation. As discussed above, bacterial strains have been identified that convert lignin into valuable phenolic monomers. The conversion of all plant components, and in particular the lignin fraction, is the basis for 2nd generation biorefineries, where a vast array of products may be prepared in conjunction with a central fermentation for biofuel production. Therefore integration of bacterial utilization of lignin will greatly contribute to the economic viability of these processes.

The cost of hydrolytic enzyme production greatly influences the overall cost of cellulosic biomass conversion thus hindering commercialization.[135] There have been great strides forward in this respect with the estimated cost being driven down from US$5.40 to US$0.20 per gallon of ethanol produced, according to claims of major enzyme producers.[136] The use of such information however in techno-economical analysis of biomass conversion is problematic. These values relate to the production of a specific target, usually ethanol, and depend on a range of variables other than enzyme production. Klein-Marcuschamer et al. prepared a model for the calculation of the cost for the production of cellulases from *T. reesei* that could then be applied to another model for the estimation of its contribution to the cost of ethanol production.[137] The results showed that there is systematic underestimation of the contribution of enzyme costs to biofuels production in the literature, as conservative calculations pointed to around US$1.00 per gallon ethanol. The authors highlighted that approaches aiming to decrease enzyme loading in the pretreatment steps should become a focus point. It seems that lowering the enzyme production cost will be a significant obstacle towards the commercialization of any process based on lignocellulosic feedstock.

Despite the hurdles that need to be overcome for commercialization, there is much anticipation from federal governements that biofuels and chemicals derived from lignocellulosic biomass will play a central role in overcoming fossil fuel dependence. In October 2012 the EU commission issued a proposal stating that advanced biofuel development has to be encouraged due to their high greenhouse gas savings and lower risk of land use change, and this should be mirrored in post-2020 renewable energy policies.[138] In accordance with this a directive was proposed to limit the allowed contribution of food crop derived biofuels, towards the 10%

2020 renewable transportation fuel objective, to only 5%. This means that a greater contribution from lignocellulosic biomass and especially agricultural waste derived biofuels will be required. Experts including those from Shell Corporation recognized that substantial research and development from industry and academia is still required in order to achieve this target. However, it is generally agreed upon that the rapid advances in enzyme, microbial and plant engineering as well as biocatalyst optimization suggest that biochemical processes are much more likely to provide the necessary breakthroughs that will propel second generation biofuels and chemicals into the marketplace.

Author details

Amy Philbrook, Apostolos Alissandratos and Christopher J. Easton*

*Address all correspondence to: easton@rsc.anu.edu.au

CSIRO Biofuels Research Cluster, Research School of Chemistry, Australian National University, Canberra ACT, Australia

References

[1] Kheshgi HS, Prince RC, Marland G. The potential of biomass fuels in the context of global climate change: Focus on transportation fuels. Annu Rev Energ Env. 2000;25:199-244.

[2] MacDonald T. Energy Fuel Use in Brazil. 2007 Jul 27 [cited 2012 Oct 25]. Available from: http://biomass.ucdavis.edu/newsletters

[3] Parker K. Investing in Agriculture: Far-Reaching Challenge, Significant Opportunity. 2009 Jun 25 [cited 2012 Oct 25]. Available from: http://www.dbcca.com/dbcca/EN/investment-research/investment_research_1735.jsp

[4] Price Volatility in Food and Agricultural Markets: Policy Responses. FAO, IFAD, IMF,OECD, UNCTAD, WFP, the World Bank, the WTO, IFPRI and the UN HLTF 2007.

[5] Doggett T. Senate vote marks start of end for ethanol subsidies. Reuters [Internet]. 2011 Jun 16 [cited 2012 Oct 25]. Available from: http://www.reuters.com/article/2011/06/16/us-usa-senate-ethanol-idUSTRE75F5IN20110616

[6] Fletcher O. Ethanol Makers' Long Hot Summer. The Wall Street Journal [Internet]. 2012 Jul 26 [cited 2012 Oct 25]. Available from: http://online.wsj.com/article/SB10001424052702303644004577522962336134368.html

[7]   Sierra R, Smith A, Granda C, Holtzapple MT. Producing fuels and chemicals from lignocellulosic biomass. Chem Eng Prog. 2008 Aug;104(8):S10-S8.

[8]   Sims REH, Mabee W, Saddler JN, Taylor M. An overview of second generation biofuel technologies. Bioresource Technol. 2010 Mar;101(6):1570-80.

[9]   Yang B, Wyman CE. Pretreatment: the key to unlocking low-cost cellulosic ethanol. Biofuel Bioprod Bior. 2008 Jan-Feb;2(1):26-40.

[10] Koullas DP, Christakopoulos P, Kekos D, Macris BJ, Koukios EG. Correlating the Effect of Pretreatment on the Enzymatic-Hydrolysis of Straw. Biotechnol Bioeng. 1992 Jan 5;39(1):113-6.

[11] Kim TH, Lee YY, Sunwoo C, Kim JS. Pretreatment of corn stover by low-liquid ammonia recycle percolation process. Appl Biochem Biotech. 2006 Apr;133(1):41-57.

[12] Keller FA, Hamilton JE, Nguyen QA. Microbial pretreatment of biomass - Potential for reducing severity of thermochemical biomass pretreatment. Appl Biochem Biotech. 2003 Spr;105:27-41.

[13] Roberto IC, Mussatto SI, Rodrigues RCLB. Dilute-acid hydrolysis for optimization of xylose recovery from rice straw in a semi-pilot reactor. Ind Crop Prod. 2003 May; 17(3):171-6.

[14] Klinke HB, Thomsen AB, Ahring BK. Inhibition of ethanol-producing yeast and bacteria by degradation products produced during pre-treatment of biomass. Appl Microbiol Biot. 2004 Nov;66(1):10-26.

[15] Sandra T. Merino JC, editor. Progress and Challenges in Enzyme Development for Biomass Utilization: Springer Berlin Heidelberg; 2007.

[16] Eriksson KE, Vallander L. Properties of Pulps from Thermomechanical Pulping of Chips Pretreated with Fungi. Sven Papperstidn. 1982;85(6):R33-R8.

[17] Messner K, Srebotnik E. Biopulping - an Overview of Developments in an Environmentally Safe Paper-Making Technology. Fems Microbiol Rev. 1994 Mar;13(2-3): 351-64.

[18] Akhtar M, Attridge MC, Myers GC, Blanchette RA. Biomechanical Pulping of Loblolly-Pine Chips with Selected White-Rot Fungi. Holzforschung. 1993;47(1):36-40.

[19] Kashino Y, Nishida T, Takahara Y, Fujita K, Kondo R, Sakai K. Biomechanical Pulping Using White-Rot Fungus Izu-154. Tappi J. 1993 Dec;76(12):167-71.

[20] Liew CY, Husaini A, Hussain H, Muid S, Liew KC, Roslan HA. Lignin biodegradation and ligninolytic enzyme studies during biopulping of Acacia mangium wood chips by tropical white rot fungi. World J Microb Biot. 2011 Jun;27(6):1457-68.

[21] Carroll A, Somerville C. Cellulosic Biofuels. Annu Rev Plant Biol. 2009;60:165-82.

[22] Zeng YL, Yang XW, Yu HB, Zhang XY, Ma FY. The delignification effects of white-rot fungal pretreatment on thermal characteristics of moso bamboo. Bioresource Technol. 2012 Jun;114:437-42.

[23] Zhang XY, Yu HB, Huang HY, Liu YX. Evaluation of biological pretreatment with white rot fungi for the, enzymatic hydrolysis of bamboo culms. Int Biodeter Biodegr. 2007 Oct;60(3):159-64.

[24] Zhang XY, Xu CY, Wang HX. Pretreatment of bamboo residues with Coriolus versicolor for enzymatic hydrolysis. J Biosci Bioeng. 2007 Aug;104(2):149-51.

[25] Wan CX, Li YL. Microbial pretreatment of corn stover with Ceriporiopsis subvermispora for enzymatic hydrolysis and ethanol production. Bioresource Technol. 2010 Aug;101(16):6398-403.

[26] Dias AA, Freitas GS, Marques GSM, Sampaio A, Fraga IS, Rodrigues MAM, et al. Enzymatic saccharification of biologically pre-treated wheat straw with white-rot fungi. Bioresource Technol. 2010 Aug;101(15):6045-50.

[27] Bak JS, Kim MD, Choi IG, Kim KH. Biological pretreatment of rice straw by fermenting with Dichomitus squalens. New Biotechnol. 2010 Sep 30;27(4):424-34.

[28] Bak JS, Ko JK, Choi IG, Park YC, Seo JH, Kim KH. Fungal Pretreatment of Lignocellulose by Phanerochaete chrysosporium to Produce Ethanol From Rice Straw. Biotechnol Bioeng. 2009 Oct 15;104(3):471-82.

[29] Shi J, Sharma-Shivappa RR, Chinn MS. Microbial pretreatment of cotton stalks by submerged cultivation of Phanerochaete chrysosporium. Bioresource Technol. 2009 Oct;100(19):4388-95.

[30] Shi J, Chinn MS, Sharma-Shivappa RR. Microbial pretreatment of cotton stalks by solid state cultivation of Phanerochaete chrysosporium. Bioresource Technol. 2008 Sep;99(14):6556-64.

[31] Hideno A, Aoyagi H, Isobe S, Tanaka H. Utilization of spent sawdust matrix after cultivation of Grifola frondosa as substrate for ethanol production by simultaneous saccharification and fermentation. Food Sci Technol Res. 2007 May;13(2):111-7.

[32] Zhao L, Cao GL, Wang AJ, Ren HY, Dong D, Liu ZN, et al. Fungal pretreatment of cornstalk with Phanerochaete chrysosporium for enhancing enzymatic saccharification and hydrogen production. Bioresource Technol. 2012 Jun;114:365-9.

[33] Cui ZF, Wan CX, Shi J, Sykes RW, Li YB. Enzymatic Digestibility of Corn Stover Fractions in Response to Fungal Pretreatment. Ind Eng Chem Res. 2012 May 30;51(21): 7153-9.

[34] Xu CY, Ma FY, Zhang XY, Chen SL. Biological Pretreatment of Corn Stover by Irpex lacteus for Enzymatic Hydrolysis. J Agr Food Chem. 2010 Oct 27;58(20):10893-8.

[35] Taniguchi M, Takahashi D, Watanabe D, Sakai K, Hoshino K, Kouya T, et al. Evaluation of Fungal Pretreatments for Enzymatic Saccharification of Rice Straw. J Chem Eng Jpn. 2010;43(4):401-5.

[36] Zakzeski J, Bruijnincx PCA, Jongerius AL, Weckhuysen BM. The Catalytic Valorization of Lignin for the Production of Renewable Chemicals. Chem Rev. 2010 Jun; 110(6):3552-99.

[37] Doherty WOS, Mousavioun P, Fellows CM. Value-adding to cellulosic ethanol: Lignin polymers. Ind Crop Prod. 2011 Mar;33(2):259-76.

[38] Tejado A, Pena C, Labidi J, Echeverria JM, Mondragon I. Physico-chemical characterization of lignins from different sources for use in phenol-formaldehyde resin synthesis. Bioresource Technol. 2007 May;98(8):1655-63.

[39] Turunen M, Alvila L, Pakkanen TT, Rainio J. Modification of phenol-formaldehyde resol resins by lignin, starch, and urea. J Appl Polym Sci. 2003 Apr 11;88(2):582-8.

[40] Bugg TDH, Ahmad M, Hardiman EM, Singh R. The emerging role for bacteria in lignin degradation and bio-product formation. Curr Opin Biotech. 2011 Jun;22(3): 394-400.

[41] Zimmermann W. Degradation of Lignin by Bacteria. J Biotechnol. 1990 Feb;13(2-3): 119-30.

[42] Ramachandra M, Crawford DL, Hertel G. Characterization of an Extracellular Lignin Peroxidase of the Lignocellulolytic Actinomycete Streptomyces-Viridosporus. Appl Environ Microb. 1988 Dec;54(12):3057-63.

[43] Ahmad M, Taylor CR, Pink D, Burton K, Eastwood D, Bending GD, et al. Development of novel assays for lignin degradation: comparative analysis of bacterial and fungal lignin degraders. Mol Biosyst. 2010 May;6(5):815-21.

[44] Masai E, Katayama Y, Fukuda M. Genetic and biochemical investigations on bacterial catabolic pathways for lignin-derived aromatic compounds. Biosci Biotech Bioch. 2007 Jan;71(1):1-15.

[45] Raj A, Reddy MMK, Chandra R. Identification of low molecular weight aromatic compounds by gas chromatography-mass spectrometry (GC-MS) from kraft lignin degradation by three Bacillus sp. Int Biodeter Biodegr. 2007 Jun;59(4):292-6.

[46] Gupta VK, Minocha AK, Jain N. Batch and continuous studies on treatment of pulp mill wastewater by Aeromonas formicans. J Chem Technol Biot. 2001 Jun;76(6): 547-52.

[47] Lynd LR, Weimer PJ, van Zyl WH, Pretorius IS. Microbial Cellulose Utilization: Fundamentals and Biotechnology. Microbiology and Molecular Biology Reviews. 2002 September 1, 2002;66(3):506-77.

[48] Brown RM, Saxena IM. Cellulose biosynthesis: A model for understanding the assembly of biopolymers. Plant Physiology and Biochemistry. 2000 Jan-Feb;38(1-2): 57-67.

[49] Himmel ME, Ruth MF, Wyman CE. Cellulase for commodity products from cellulosic biomass. Current Opinion in Biotechnology. 1999;10(4):358-64.

[50] Teeri TT, Koivula A, Linder M, Wohlfahrt G, Divne C, Jones TA. Trichoderma reesei cellobiohydrolases: why so efficient on crystalline cellulose? Biochemical Society Transactions. 1998 May;26(2):173-8.

[51] Din N, Damude HG, Gilkes NR, Miller RC, Warren RAJ, Kilburn DG. C-1-C-X Revisited - Intramolecular Synergism in a Cellulase. Proceedings of the National Academy of Sciences of the United States of America. 1994 Nov 22;91(24):11383-7.

[52] Gusakov AV. Alternatives to Trichoderma reesei in biofuel production. Trends in Biotechnology. 2011;29(9):419-25.

[53] Menon V, Rao M. Trends in bioconversion of lignocellulose: Biofuels, platform chemicals & biorefinery concept. Progress in Energy and Combustion Science. 2012;38(4):522-50.

[54] Le Crom S, Schackwitz W, Pennacchio L, Magnuson JK, Culley DE, Collett JR, et al. Tracking the roots of cellulase hyperproduction by the fungus Trichoderma reesei using massively parallel DNA sequencing. Proceedings of the National Academy of Sciences of the United States of America. 2009 Sep 22;106(38):16151-6.

[55] Nieves RA, Ehrman CI, Adney WS, Elander RT, Himmel ME. Survey and analysis of commercial cellulase preparations suitable for biomass conversion to ethanol. World Journal of Microbiology & Biotechnology. 1998 Mar;14(2):301-4.

[56] Wilson DB. Cellulases and biofuels. Current Opinion in Biotechnology. 2009 Jun; 20(3):295-9.

[57] Zhang Z, Donaldson AA, Ma X. Advancements and future directions in enzyme technology for biomass conversion. Biotechnology Advances. 2012 Jul-Aug;30(4): 913-9.

[58] Fujitomi K, Sanda T, Hasunuma T, Kondo A. Deletion of the PHO13 gene in Saccharomyces cerevisiae improves ethanol production from lignocellulosic hydrolysate in the presence of acetic and formic acids, and furfural. Bioresource Technology. 2012;111(0):161-6.

[59] van Gool MP, Toth K, Schols HA, Szakacs G, Gruppen H. Performance of hemicellulolytic enzymes in culture supernatants from a wide range of fungi on insoluble wheat straw and corn fiber fractions. Bioresource Technology. 2012;114(0):523-8.

[60] Sanchez C. Lignocellulosic residues: Biodegradation and bioconversion by fungi. Biotechnology Advances. 2009 Mar-Apr;27(2):185-94.

[61] Madhavan A, Srivastava A, Kondo A, Bisaria VS. Bioconversion of lignocellulose-derived sugars to ethanol by engineered Saccharomyces cerevisiae. Critical Reviews in Biotechnology. 2012 Mar;32(1):22-48.

[62] Laluce C, Schenberg ACG, Gallardo JCM, Coradello LFC, Pombeiro-Sponchiado SR. Advances and Developments in Strategies to Improve Strains of Saccharomyces cerevisiae and Processes to Obtain the Lignocellulosic Ethanol-A Review. Applied Biochemistry and Biotechnology. 2012 Apr;166(8):1908-26.

[63] Jang YS, Kim B, Shin JH, Choi YJ, Choi S, Song CW, et al. Bio-based production of C2-C6 platform chemicals. Biotechnol Bioeng. 2012 Oct;109(10):2437-59.

[64] Tracy BP, Jones SW, Fast AG, Indurthi DC, Papoutsakis ET. Clostridia: the importance of their exceptional substrate and metabolite diversity for biofuel and biorefinery applications. Current Opinion in Biotechnology. 2012;23(3):364-81.

[65] Jurgens G, Survase S, Berezina O, Sklavounos E, Linnekoski J, Kurkijarvi A, et al. Butanol production from lignocellulosics. Biotechnology Letters. 2012 Aug;34(8): 1415-34.

[66] Green EM. Fermentative production of butanol—the industrial perspective. Current Opinion in Biotechnology. 2011;22(3):337-43.

[67] Hartman AH, Liu H, Melville SB. Construction and Characterization of a Lactose-Inducible Promoter System for Controlled Gene Expression in Clostridium perfringens. Applied and Environmental Microbiology. 2011 January 15, 2011;77(2):471-8.

[68] Lau MW, Dale BE. Cellulosic ethanol production from AFEX-treated corn stover using Saccharomyces cerevisiae 424A(LNH-ST). P Natl Acad Sci USA. 2009 Feb 3;106(5):1368-73.

[69] Ha SJ, Galazka JM, Kim SR, Choi JH, Yang XM, Seo JH, et al. Engineered Saccharomyces cerevisiae capable of simultaneous cellobiose and xylose fermentation. P Natl Acad Sci USA. 2011 Jan 11;108(2):504-9.

[70] Joachimsthal E, Haggett KD, Rogers PL. Evaluation of recombinant strains of Zymomonas mobilis for ethanol production from glucose xylose media. Appl Biochem Biotech. 1999 Spr;77-9:147-57.

[71] Wang YZ, Manow R, Finan C, Wang JH, Garza E, Zhou SD. Adaptive evolution of nontransgenic Escherichia coli KC01 for improved ethanol tolerance and homoethanol fermentation from xylose. J Ind Microbiol Biot. 2011 Sep;38(9):1371-7.

[72] Nakano S, Fukaya M, Horinouchi S. Putative ABC transporter responsible for acetic acid resistance in Acetobacter aceti. Appl Environ Microb. 2006 Jan;72(1):497-505.

[73] Zhang A, Yang ST. Engineering Propionibacterium acidipropionici for Enhanced Propionic Acid Tolerance and Fermentation. Biotechnol Bioeng. 2009 Nov 1;104(4): 766-73.

[74] Wang LM, Zhao B, Li FS, Xu K, Ma CQ, Tao F, et al. Highly efficient production of D-lactate by Sporolactobacillus sp. CASD with simultaneous enzymatic hydrolysis of peanut meal. Appl Microbiol Biot. 2011 Feb;89(4):1009-17.

[75] Zhu Y, Eiteman MA, DeWitt K, Altman E. Homolactate fermentation by metabolically engineered Escherichia coli strains. Appl Environ Microb. 2007 Jan;73(2):456-64.

[76] Ashok S, Raj SM, Rathnasingh C, Park S. Development of recombinant Klebsiella pneumoniae Delta dhaT strain for the co-production of 3-hydroxypropionic acid and 1,3-propanediol from glycerol. Appl Microbiol Biot. 2011 May;90(4):1253-65.

[77] Rathnasingh C, Raj SM, Jo JE, Park S. Development and Evaluation of Efficient Recombinant Escherichia coli Strains for the Production of 3-Hydroxypropionic Acid From Glycerol. Biotechnol Bioeng. 2009 Nov 1;104(4):729-39.

[78] Atsumi S, Cann AF, Connor MR, Shen CR, Smith KM, Brynildsen MP, et al. Metabolic engineering of Escherichia coli for 1-butanol production. Metab Eng. 2008 Nov; 10(6):305-11.

[79] Lee J, Jang YS, Choi SJ, Im JA, Song H, Cho JH, et al. Metabolic Engineering of Clostridium acetobutylicum ATCC 824 for Isopropanol-Butanol-Ethanol Fermentation. Appl Environ Microb. 2012 Mar;78(5):1416-23.

[80] Jojima T, Inui M, Yukawa H. Production of isopropanol by metabolically engineered Escherichia coli. Appl Microbiol Biot. 2008 Jan;77(6):1219-24.

[81] Sanchezriera F, Cameron DC, Cooney CL. Influence of Environmental-Factors in the Production of R(-)-1,2-Propanediol by Clostridium-Thermosaccharolyticum. Biotechnol Lett. 1987 Jul;9(7):449-54.

[82] Clomburg JM, Gonzalez R. Metabolic Engineering of Escherichia coli for the Production of 1,2-Propanediol From Glycerol. Biotechnol Bioeng. 2011 Apr;108(4):867-79.

[83] Gonzalez-Pajuelo M, Meynial-Salles I, Mendes F, Andrade JC, Vasconcelos I, Soucaille P. Metabolic engineering of Clostridium acetobutylicum for the industrial production of 1,3-propanediol from glycerol. Metab Eng. 2005 Sep-Nov;7(5-6):329-36.

[84] Nakamura CE, Whited GM. Metabolic engineering for the microbial production of 1,3-propanediol. Curr Opin Biotech. 2003 Oct;14(5):454-9.

[85] Zhu Y, Yang ST. Adaptation of Clostridium tyrobutyricum for enhanced tolerance to butyric acid in a fibrous-bed bioreactor. Biotechnol Progr. 2003 Mar-Apr;19(2):365-72.

[86] Liu X, Zhu Y, Yang ST. Construction and characterization of ack deleted mutant of Clostridium tyrobutyricum for enhanced butyric acid and hydrogen production. Biotechnol Progr. 2006 Oct 6;22(5):1265-75.

[87] Jiang L, Wang JF, Liang SZ, Cai J, Xu ZN, Cen PL, et al. Enhanced Butyric Acid Tolerance and Bioproduction by Clostridium tyrobutyricum Immobilized in a Fibrous Bed Bioreactor. Biotechnol Bioeng. 2011 Jan;108(1):31-40.

[88] Guettler MV JM, Rumler D, inventor Methods for making succinic acid, bacterial variants for use in the process, and methods for obtaining variants. USA1996.

[89] Lee SJ, Song H, Lee SY. Genome-based metabolic engineering of Mannheimia succiniciproducens for succinic acid production. Appl Environ Microb. 2006 Mar;72(3): 1939-48.

[90] Jantama K, Haupt MJ, Svoronos SA, Zhang XL, Moore JC, Shanmugam KT, et al. Combining metabolic engineering and metabolic evolution to develop nonrecombinant strains of Escherichia coli C that produce succinate and malate. Biotechnol Bioeng. 2008 Apr 1;99(5):1140-53.

[91] Jantama K, Zhang X, Moore JC, Shanmugam KT, Svoronos SA, Ingram LO. Eliminating Side Products and Increasing Succinate Yields in Engineered Strains of Escherichia coli C. Biotechnol Bioeng. 2008 Dec 1;101(5):881-93.

[92] Thakker C, Martinez I, San KY, Bennett GN. Succinate production in Escherichia coli. Biotechnol J. 2012 Feb;7(2):213-24.

[93] Litsanov B, Brocker M, Bott M. Toward Homosuccinate Fermentation: Metabolic Engineering of Corynebacterium glutamicum for Anaerobic Production of Succinate from Glucose and Formate. Appl Environ Microb. 2012 May;78(9):3325-37.

[94] Litsanov B, Kabus A, Brocker M, Bott M. Efficient aerobic succinate production from glucose in minimal medium with Corynebacterium glutamicum. Microb Biotechnol. 2012 Jan;5(1):116-28.

[95] Battat E, Peleg Y, Bercovitz A, Rokem JS, Goldberg I. Optimization of L-Malic Acid Production by Aspergillus-Flavus in a Stirred Fermenter. Biotechnol Bioeng. 1991 May;37(11):1108-16.

[96] Zelle RM, de Hulster E, van Winden WA, de Waard P, Dijkema C, Winkler AA, et al. Malic acid production by Saccharomyces cerevisiae: Engineering of pyruvate carboxylation, oxaloacetate reduction, and malate export. Appl Environ Microb. 2008 May; 74(9):2766-77.

[97] Zhang X, Wang X, Shanmugam KT, Ingram LO. L-Malate Production by Metabolically Engineered Escherichia coli. Appl Environ Microb. 2011 Jan;77(2):427-34.

[98] Kenealy W, Zaady E, Dupreez JC, Stieglitz B, Goldberg I. Biochemical Aspects of Fumaric-Acid Accumulation by Rhizopus-Arrhizus. Appl Environ Microb. 1986 Jul; 52(1):128-33.

[99] Li HX, Qiu T, Huang GD, Cao YS. Production of gamma-aminobutyric acid by Lactobacillus brevis NCL912 using fed-batch fermentation. Microb Cell Fact. 2010 Nov 12;9.

[100] Shi F, Li YX. Synthesis of gamma-aminobutyric acid by expressing Lactobacillus brevis-derived glutamate decarboxylase in the Corynebacterium glutamicum strain ATCC 13032. Biotechnol Lett. 2011 Dec;33(12):2469-74.

[101] Harris LM, Desai RP, Welker NE, Papoutsakis ET. Characterization of recombinant strains of the Clostridium acetobutylicum butyrate kinase inactivation mutant: Need for new phenomenological models for solventogenesis and butanol inhibition? Biotechnol Bioeng. 2000 Jan 5;67(1):1-11.

[102] Dellomonaco C, Clomburg JM, Miller EN, Gonzalez R. Engineered reversal of the beta-oxidation cycle for the synthesis of fuels and chemicals. Nature. 2011 Aug 18;476(7360):355-U131.

[103] Shen CR, Lan EI, Dekishima Y, Baez A, Cho KM, Liao JC. Driving Forces Enable High-Titer Anaerobic 1-Butanol Synthesis in Escherichia coli. Appl Environ Microb. 2011 May;77(9):2905-15.

[104] Atsumi S, Hanai T, Liao JC. Non-fermentative pathways for synthesis of branched-chain higher alcohols as biofuels. Nature. 2008 Jan 3;451(7174):86-U13.

[105] Blombach B, Riester T, Wieschalka S, Ziert C, Youn JW, Wendisch VF, et al. Corynebacterium glutamicum Tailored for Efficient Isobutanol Production. Appl Environ Microb. 2011 May;77(10):3300-10.

[106] Yim H, Haselbeck R, Niu W, Pujol-Baxley C, Burgard A, Boldt J, et al. Metabolic engineering of Escherichia coli for direct production of 1,4-butanediol. Nat Chem Biol. 2011 Jul;7(7):445-52.

[107] Ma CQ, Wang AL, Qin JY, Li LX, Ai XL, Jiang TY, et al. Enhanced 2,3-butanediol production by Klebsiella pneumoniae SDM. Appl Microbiol Biot. 2009 Feb;82(1):49-57.

[108] Zhang LY, Sun JA, Hao YL, Zhu JW, Chu J, Wei DZ, et al. Microbial production of 2,3-butanediol by a surfactant (serrawettin)-deficient mutant of Serratia marcescens H30. J Ind Microbiol Biot. 2010 Aug;37(8):857-62.

[109] Qian ZG, Xia XX, Lee SY. Metabolic Engineering of Escherichia coli for the Production of Putrescine: A Four Carbon Diamine. Biotechnol Bioeng. 2009 Nov 1;104(4): 651-62.

[110] Yahiro K, Takahama T, Park YS, Okabe M. Breeding of Aspergillus-Terreus Mutant Tn-484 for Itaconic Acid Production with High-Yield. J Ferment Bioeng. 1995;79(5): 506-8.

[111] Liao JCLACA, Chang P-CHC, inventors; Genetically Modified Microorganisms for Producing Itaconic Acid with High Yields. US patent US 2010/0285546 A1. 2010.

[112] Tseng HC, Harwell CL, Martin CH, Prather KLJ. Biosynthesis of chiral 3-hydroxyvalerate from single propionate-unrelated carbon sources in metabolically engineered E. coli. Microb Cell Fact. 2010 Nov 27;9.

[113]   Zhang KC, Sawaya MR, Eisenberg DS, Liao JC. Expanding metabolism for biosynthe-
        sis of nonnatural alcohols. P Natl Acad Sci USA. 2008 Dec 30;105(52):20653-8.

[114]   Cann AF, Liao JC. Production of 2-methyl-1-butanol in engineered Escherichia coli.
        Appl Microbiol Biot. 2008 Nov;81(1):89-98.

[115]   Connor MR, Liao JC. Engineering of an Escherichia coli strain for the production of
        3-methyl-1-butanol. Appl Environ Microb. 2008 Sep;74(18):5769-75.

[116]   Granstrom TB, Izumori K, Leisola M. A rare sugar xylitol. Part II: biotechnological
        production and future applications of xylitol. Appl Microbiol Biot. 2007 Feb;74(2):
        273-6.

[117]   Cirino PC, Chin JW, Ingram LO. Engineering Escherichia coli for xylitol production
        from glucose-xylose mixtures. Biotechnol Bioeng. 2006 Dec 20;95(6):1167-76.

[118]   Qian ZG, Xia XX, Lee SY. Metabolic Engineering of Escherichia coli for the Produc-
        tion of Cadaverine: A Five Carbon Diamine. Biotechnol Bioeng. 2011 Jan;108(1):
        93-103.

[119]   Moon TS, Dueber JE, Shiue E, Prather KLJ. Use of modular, synthetic scaffolds for
        improved production of glucaric acid in engineered E. coli. Metab Eng. 2010 May;
        12(3):298-305.

[120]   Balderas-Hernandez VE, Sabido-Ramos A, Silva P, Cabrera-Valladares N, Hernan-
        dez-Chavez G, Baez-Viveros JL, et al. Metabolic engineering for improving anthrani-
        late synthesis from glucose in Escherichia coli. Microb Cell Fact. 2009 Apr 2;8.

[121]   Wierckx NJP, Ballerstedt H, de Bont JAM, Wery J. Engineering of solvent-tolerant
        Pseudomonas putida S12 for bioproduction of phenol from glucose. Appl Environ
        Microb. 2005 Dec;71(12):8221-7.

[122]   Wang CL, Takenaka S, Murakami S, Aoki K. Isolation of a benzoate-utilizing Pseudo-
        monas strain from soil and production of catechol from benzoate by transpositional
        mutants. Microbiol Res. 2001;156(2):151-8.

[123]   Hasunuma T, Kondo A. Consolidated bioprocessing and simultaneous saccharifica-
        tion and fermentation of lignocellulose to ethanol with thermotolerant yeast strains.
        Process Biochemistry. 2012;47(9):1287-94.

[124]   Heer D, Sauer U. Identification of furfural as a key toxin in lignocellulosic hydroly-
        sates and evolution of a tolerant yeast strain. Microbial Biotechnology. 2008;1(6):
        497-506.

[125]   Dunlop MJ. Engineering microbes for tolerance to next-generation biofuels. Biotech-
        nology for Biofuels. 2011 Sep 21;4.

[126]   Alper H, Moxley J, Nevoigt E, Fink GR, Stephanopoulos G. Engineering yeast tran-
        scription machinery for improved ethanol tolerance and production. Science. 2006
        Dec 8;314(5805):1565-8.

[127] Tomas CA, Welker NE, Papoutsakis ET. Overexpression of groESL in Clostridium acetobutylicum results in increased solvent production and tolerance, prolonged metabolism, and changes in the cell's transcriptional program. Applied and Environmental Microbiology. 2003 Aug;69(8):4951-65.

[128] Zhang J-G, Liu X-Y, He X-P, Guo X-N, Lu Y, Zhang B-r. Improvement of acetic acid tolerance and fermentation performance of Saccharomyces cerevisiae by disruption of the FPS1 aquaglyceroporin gene. Biotechnology Letters. 2011 Feb;33(2):277-84.

[129] Geertman J-MA, van Dijken JP, Pronk JT. Engineering NADH metabolism in Saccharomyces cerevisiae: formate as an electron donor for glycerol production by anaerobic, glucose-limited chemostat cultures. Fems Yeast Research. 2006 Dec;6(8):1193-203.

[130] Ji L, Shen Y, Xu L, Peng B, Xiao Y, Bao X. Enhanced resistance of Saccharomyces cerevisiae to vanillin by expression of lacA from Trametes sp AH28-2. Bioresource Technology. 2011 Sep;102(17):8105-9.

[131] Dunlop MJ, Dossani ZY, Szmidt HL, Chu HC, Lee TS, Keasling JD, et al. Engineering microbial biofuel tolerance and export using efflux pumps. Molecular Systems Biology. 2011 May;7.

[132] Atsumi S, Wu T-Y, Machado IMP, Huang W-C, Chen P-Y, Pellegrini M, et al. Evolution, genomic analysis, and reconstruction of isobutanol tolerance in Escherichia coli. Molecular Systems Biology. 2010 Dec;6.

[133] Young E, Lee S-M, Alper H. Optimizing pentose utilization in yeast: the need for novel tools and approaches. Biotechnology for Biofuels. 2010 Nov 16;3.

[134] Xu Q, Singh A, Himmel ME. Perspectives and new directions for the production of bioethanol using consolidated bioprocessing of lignocellulose. Current Opinion in Biotechnology. 2009;20(3):364-71.

[135] Blanch HW. Bioprocessing for biofuels. Current Opinion in Biotechnology. 2012;23(3):390-5.

[136] Zhang YHP, Himmel ME, Mielenz JR. Outlook for cellulase improvement: Screening and selection strategies. Biotechnology Advances. 2006 Sep-Oct;24(5):452-81.

[137] Klein-Marcuschamer D, Oleskowicz-Popiel P, Simmons BA, Blanch HW. The challenge of enzyme cost in the production of lignocellulosic biofuels. Biotechnology and Bioengineering. 2012 Apr;109(4):1083-7.

[138] COM(2012) 595. Proposal for a directive of the European Parliament and of the Council: amending Directive 98/70/EC relating to the quality of petrol and diesel fuels, and amending Directive 2009/28/EC on the promotion of the use of energy from renewable sources. Brussels: European Comission; 2012.

Chapter 3

Synergistic Effects of Pretreatment Process on Enzymatic Digestion of Rice Straw for Efficient Ethanol Fermentation

Prihardi Kahar

Additional information is available at the end of the chapter

1. Introduction

Lignocellulose, the most abundant renewable biomass produced by plants from photosynthesis, has a yearly supply of approximately 200 billion metric tons worldwide (Ragauskas et al., 2006; Zhang et al., 2006). Lignocellulosic biomass is widely expected to be a major resource for biorefineries, including bioethanol (Lin and Tanaka, 2006). The composition of lignocellulose varies depending on plant species, plant parts, growth conditions, etc. (Ding and Himmel, 2006; Zhang and Lynd, 2004), and their structures are rigid and low degradable against cellulase enzymes. In general, the lignocellulose structure is composed by three major components: crystalline cellulose, amorphous hemicellulose and non-sugar lignin. Cellulose microfibrils are coated with hemicellulose matrices building holocellulose structures and severely protected by lignin outside. The structures are rigidly packed to form a physical barrier for cellulase access to cellulose chains (Mansfield et al., 1999). To hydrolyze them efficiently into sugars, a high dosage of commercial available cellulase enzymes is required. At a current technical stage, 20 g-cellulase is needed to hydrolysis 1 kg cellulose at 70% for 5 days (Gusakov, 2011; Roche et al., 2009). However, the baseline production cost of cellulase is still expensive as reported to be $10.14/kg (Klein-Marcuschamer et al., 2012). If the recalcitrance problem remains unresolved, it is not feasible for high-solids enzymatic saccharification.

The ability of cellulase access to cellulose chains within microfibrils will be limited even if lignin is completely removed from the cellulose structure, because its ability is generally limited to accessing the outer layer of the microfibrils (Mansfield et al., 1999). Although cellulose can be slowly eroded by surface shaving or planning, cellulose chains in highly ordered and tightly packed regions of microfibrils must be disintegrated by delamination, disruption or loosening to increase the surface area and make individual molecule more accessible and available for interaction with cellulase (Ishizawa et al., 2009; Ragauskas et al., 2006). For this reason, the pretreatment must be implemented before enzymatic saccharification and also required to facilitate amorphogenesis as the initial stage in the enzymatic hydrolysis of cellulose (Coughlan, 1985; Din et al., 1991; Teeri et al., 1992).

Among various methods available for biomass pretreatment, chemical delignification and swelling were investigated in this study, because these processes could show the similar effect to amorphogenesis of cellulose fiber by cellulase. Most of the chemical processes presently used, however, might not be preferable to saccharification, because of incomplete lignin removal and degradation of polysaccharides and loss of hemicelluloses (Fang et al., 1999; Sun et al., 2004). Most processes require high chemical charges to attain the complete lignin removal, because the single pretreatment process is not effective when performed at low chemical charges. In some cases, rearrangement of the lignin structure occurs during the pretreatment process (Kumar and Wyman, 2009). Products released by the degradation of polysaccharides and hemicellulose extracts strongly inhibit the cellulose hydrolysis by cellulase (Jing et al., 2009).

The pretreatment process should be designed to remove lignin and to disintegrate the cellulose structure without loss (degradation) of cellulose and hemicellulose parts (Frey-Wyssling, 1954; Peterlin and Ingram, 1970; Morehead, 1950). Pretreatment with sodium chlorite acidified by acetic acid (acidified sodium chlorite) perhaps meets the requirement of delignification and effectively solubilizes lignin at moderate temperatures. It is noteworthy that the acidified sodium chlorite delignification causes only trace solubilization of glucan and xylan (Ahlgren and Goring, 1971). It is also reported that sodium bicarbonate is effective to disintegrate the cellulose structure and the swelling by carboxylation of produced fiber (Kwasniakova et al., 1996).

In this study, the advanced pretreatment process for enzymatic hydrolysis of rice straw has been demonstrated by combining the delignification by acidified sodium chlorite with the disintegration of cellulose structure and the alteration of crystalline structure by swelling with sodium bicarbonate. The efficiency of the pretreatment process on saccharification and fermentation (based on simultaneous saccharification and fermentation process) of rice straw was evaluated by using commercially available cellulase.

2. Materials and methods

2.1. Materials and microorganism

Sun-dried rice straw of Koshi-hikari (Niigata-ken, Japan) was used as a source of lignocellulosic biomass. Chemical composition of rice straw was generally ranged from 24% to 38% cellulose, 12% to 22% hemicellulose and 16% to 20% lignin, based on the dried weight. Microcrystalline cellulose (<20 μm particle size) was purchased from Merk (Darmstadt, Germany). Avicel PH-101 (<50 μm particle size) was purchased from Sigma-Aldrich (St. Louis,

MO). The digestive enzyme mixtures of Novozym 188 (372 β-glucosidase IU/g, source of β-glucosidase) and Celluclast 1.5L (64 FPU/g, 16 β-glucosidase IU/g, source of endo-/exo-type cellulase) obtained from Novozymes A/S (Bagsværd, Denmark) were used for enzymatic saccharification. Milli-Q grade water (18.2 MΩcm resistivity) was used throughout all the experiments. For fermentation, wild-type yeast strain *Saccharomyces cerevisiae* NBRC2114, obtained from National Bioresources Research Center (NBRC, Ibaraki, Japan), was used. This strain can produce ethanol anaerobically from glucose but not from pentose sugars such as xylose, arabinose and ribose.

2.2. Sample pretreatment

The pretreatment of rice straw with acidified sodium chlorite was performed in a water bath using sodium chlorite and acetic acid at 80 °C according to a modified literature method (Hubble and Ragauskas, 2010). Rice straw samples were ground using a laboratory cutting mill to a particle size on the order of 5 mm, impregnated by immersion in a flask containing deionized water (60 ml/g solid) at 25 °C for 3 days to form solids slurry. The delignification was started by addition of glacial acetic acid (0.04 ml/g solid) and sodium chlorite (0.4 g/g solid) to solids slurry. The mix was heated to 80 °C with gentle swirling at intervals. Fresh amounts of acetic acid and sodium chlorite were added until the samples were judged to be sufficiently delignified by the persistence of yellowish-green chlorine dioxide gas that was generated on mixing the reagents (normally after 1 h for one reaction).

For swelling of delignified rice straw, the samples were initially impregnated by immersion in a flask containing sodium bicarbonate solution at 0.5% (wt./vol.) at 25 °C for 24 h. After autoclaving at 122 °C for 20 min, the samples were washed until the solution was colorless and neutral in pH. All samples were sun-dried for at least 3 days and stored in desiccators at 25 °C until used.

2.3. Scanning electron microscopy

The electron microscopic study of pretreated rice straw was performed with FE-SEM (Hitachi S-4700 Type II; Hitachi, Tokyo) after the dried samples were placed on a conductive carbon tape and coated with Pt-Pd using a sputter coater (Hitachi E102 Ion Sputter; Hitachi) for 2 min at DC±20 mA as previously mentioned (Kahar et al., 2010).

2.4. XRD and FTIR analysis

X-ray diffraction (XRD) analysis on avicel, cellulose microcrystalline, untreated and pretreated rice straw were conducted according to a method described by Chang and Holtzapple (2000). Samples of particle size less than 125 mm were scanned on a RIGAKU-D/MAX instrument (Uitima III, Japan) at a speed of 1°/min, range from $2\theta = 0°-40°$, and with a step size of 0.04° at 25 °C. Crystallinity index (CrI) was calculated according to the method described by Segal et al. (1959).

For FTIR analysis, the ground samples were prepared by pressing 2 mg of cellulosic samples on 200 mg of spectroscopic grade potassium bromide (KBr). The spectra were recorded in the middle IR range 3500-750 cm^{-1} using a JASCO FT/IR4200 Spectrometer with detector at 4 cm^{-1} resolution and 40 scans per measurement. Essential FTIR (Operant LLC, Sydney, Australia) software was used as a tool for analysis of IR spectra.

2.5. Enzymatic saccharification

A batch enzymatic hydrolysis was conducted at 1% (wt./wt.) of dry solid loading in a 0.1 M acetate buffer (pH 4.8) containing 0.02% (wt./vol.) sodium azide. The total working volume was 100 in a 300 ml flask. Before the addition of cellulase enzymes, the mixture of substrate and buffer was preheated in an incubator shaker at 50 °C for 30 min to allow the substrate to disperse uniformly in the buffer. Celluclast 1.5L and Novozym 188 were added into tubes immediately to initiate enzymatic hydrolysis. The saccharification was occurred under the temperature of 50 °C for 24 h. To finish the reaction, the mixtures were immediately placed over a boiling water bath for 5 min to deactivate the enzymes as described by Helle et al. (1993) and Desai and Converse (1997). After enzyme inactivation, each sample was centrifuged for 5 min at 8,000 × g, and supernatants were collected. The supernatant samples were stored at 4 °C for subsequent sugar analysis.

2.6. Fermentation

Fermentation was performed anaerobically in 2-l jar fermentor, equipped with pH and dissolved oxygen concentration monitoring system (FermExpert, BEM, Ibaraki). Prior to fermentation, yeast culture was prepared by inoculating a single colony of NBRC2114 strain in YM medium, which containing bacto peptone (0.5%, wt./vol.), bacto yeast extract (0.3%, wt./vol.), bacto malt extract (0.3%, wt./vol.), glucose (1%, wt./vol.), xylose (1%, wt./vol.) and aerobically cultured at 30 °C overnight. For fermentation, minimal medium (MM) containing bacto yeast nitrogen (without amino acids and ammonium sulfate) (0.17%, wt./vol.) and ammonium sulfate (0.5%, wt./vol.) supplemented with pretreated rice straw was used. After transferring the yeast culture into MM, the fermentation started by the addition of cellulase enzyme mixture at a final loading of 10, 20, 100, 200 (g-biomass/g-enzymes). The solution of 5 N NaOH was used to keep the pH of culture at around 5. To maintain the anaerobic condition at the initial stage of fermentation, a continuous stream of sterile nitrogen gas (0.1 VVM [volume of air/volume of reactor × minutes]) was flowed through the sterilized membrane filter into the reactor. The gassing was stopped upon cell production of sufficient gases (positive headspace pressure), usually between 12 and 24 h

2.7. Analytical methods

The Klason lignin content of the samples was determined using the Laboratory Analytical Procedures (LAPs) provided by the National Renewable Energy Laboratory (NREL) (Sluiter et al., 2008). The amount of total sugars was determined as reducing sugars by 3,5-dinitrosalicylic acid (DNS) assay, as described by Miller (1959).

To determine the concentration of byproduct ions upon acidified sodium chlorite treatment, ion chromatography analyses were carried out with a Dionex ICS -1500 High

Performance Integrated Ion Chromatography System equipped with an Auto suppressor system. The columns of ION PAC AS23 and ION PAC AG23 were used as a main isolation column and a guard column, respectively, for the determination chloride, chlorite and chlorate ions were used at a flow rate of 1 mL/min, with the elution program consisting of an isocratic elution with 4 mM $NaHCO_3$/0.4 mM Na_2CO_3 buffer at 30 °C. The spectrophotometric analysis at 359 nm was used to determine the concentration of chlorine dioxide.

The protein concentration was measured by the Bradford protein assay using bovine serum albumin (BSA) as a standard (Bradford, 1976). All the experimental results were the average of triplicates, unless specified otherwise.

3. Results and discussion

3.1. Delignification of rice straw by pretreatments with acidified sodium chlorite and sodium bicarbonate

Since lignin is arranged in multiple lamellar sheets of lignocellulose matrices, a single batch reaction of chemical pretreatment is occasionally not sufficient to achieve a complete delignification (Wise et al., 1946; Klein and Snodgrass, 1993). In this study, the delignification of rice straw by acidified sodium chlorite was evaluated by repeating one hour batch reaction from one time to four times (1x to 4x). Figure 1A shows the residual lignin content of rice straw after pretreated and the control. The lignin content of rice straw decreased to about 38% (wt./wt.) of the control by one time treating (1x) and then decreased to 16% (wt./wt.) after four times repetition (4x). Interestingly, much higher rates of lignin removal were achieved when the sodium bicarbonate treatment was additionally applied after the chlorite treatment. For example, in case of three times repetition (3x) of the chlorite treatment, the additional processing with sodium bicarbonate (3x+swelling) resulted in more decreasing of lignin content at 8% (wt./wt.) against 20% (wt./wt.) in the original 3x.

A previous study (Hubbell and Ragauskas, 2010) reports that the acidified sodium chlorite treatment should be sufficient to remove lignin from cellulose samples with lignin content below 30% (wt./wt.) when the reaction was repeated two times, and it should be performed at least three times for higher lignin contents. However, the work was conducted on the pure cellulose matrix artificially coated with lignin at appropriate concentrations. This means that the reported process could be implemented for removal of surface lignin, but it was not clear whether the process was effective for removal of integrated lignin including internal lignin. According to our study, three times repeated delignification (3x) was not enough, resulting in only about 80% (wt./wt.) removal as shown in Fig. 1(A), unless the swelling by sodium bicarbonate was applied. Therefore, the swelling seems to play an important role, not only in removal of surface lignin, but also in removal of integrated lignin.

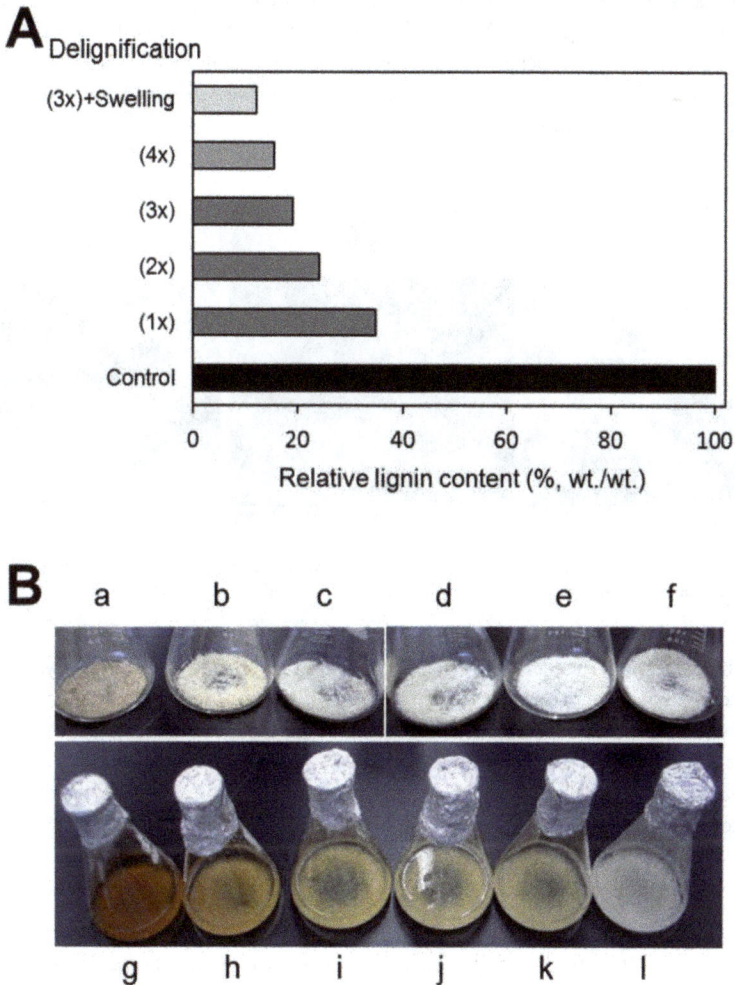

Figure 1. Delignification of rice straw. A) Effects of delignification by acidified sodium chlorite (treated 1x to 4x) and swelling by sodium bicarbonate on the lignin contents of rice straw. B) Photographs of dried and autoclaved rice straw fibers after treated with acidified sodium chlorite (1x)(b, h), (2x) (c, i), (3x) (d, j), (4x) (e, k), acidified sodium chlorite (3x) and sodium bicarbonate (f, l). As a control, untreated rice straw was used (a, g). Heat-treatment by autoclaving was performed at 122 °C for 20 min, just after impregnation of the samples in water.

To confirm the presence of internal lignin in the rice straw, the samples were immersed in water (2.5 g-solid/150 ml) and then autoclaved at 122 °C for 20 min. Lignocellulosic materials were tanned after processed with hot water due to the denaturation of lignin components to hot-water-extractable tannins (Allen et al., 1974). As shown in Fig. 1(B), the rice straw sample treated with sodium bicarbonate after chlorite were not tanned (l), in contrast to the control (g) and the sample treated with chlorite only for 2x (i) to 4x (k), even though they were bleached as colorless solids prior to hot-water processing (a, c, d, e, f). This result indicates that the internal lignin could be efficiently removed by swelling, not by acidified sodium chlorite treatment.

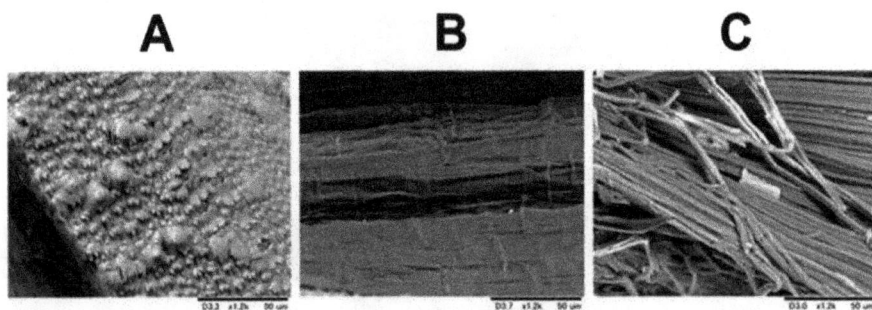

Figure 2. SEM images of untreated and pretreated rice straw. (A) Untreated rice straw. (B) Rice straw after treated with acidified sodium chlorite (3x). (C) Rice straw after treated with acidified sodium chlorite (3x) and sodium bicarbonate.

The delignification of rice straw was further visually evaluated by SEM analysis (Fig. 2). Observed by SEM were rice straws treated with acidified sodium chlorite three times (3x) (B), treated with acidified sodium chlorite plus sodium bicarbonate (3x+swelling) (C) and treated only by autoclaving (122 °C, 20 min) as control (A). As clearly shown in Fig.2(C), the integrated lignin was almost completely removed from cellulose structures by the 3x+swelling treatment so that cellulose fibrils appear to be separated and accessible to the enzymes. The 3x treatment partially removed lignin from the surface, but cellulose fibril bundles still remained, supposedly leaving internal lignin.

The efficiency of lignin removal on treated rice straw was evaluated in detail by XRD (Fig. 3) and FTIR analysis (Fig. 4). XRD analysis was performed using pure cellulose (microcrystalline cellulose) as control (Fig. 3A). The observed crystallinity index (CrI_{obs}) of the samples were 48%, 53%, 67% for autoclaved (without chemicals), acidified sodium chlorite (3x) treated, acidified sodium chlorite (3x) and sodium bicarbonate treated samples, respectively, while the CrI_{obs} of pure cellulose was 68%. Treatment with sodium bicarbonate after acidified sodium chlorite (3x) increased the CrI_{obs} of the samples to closer with that of pure cellulose. This result indicates that more non-cellulosic components (lignin included) removed from the cellulose structures. The increasing of CrI_{obs} of rice straw after treated with acidified sodium chlorite (3x) was not significant unlike those after treated with acidified sodium chlorite (3x) and sodium bicarbonate. It means that surface lignin was removed by acidified sodium chlorite (3x), but internal lignin could be removed only by treatment with sodium bicarbonate.

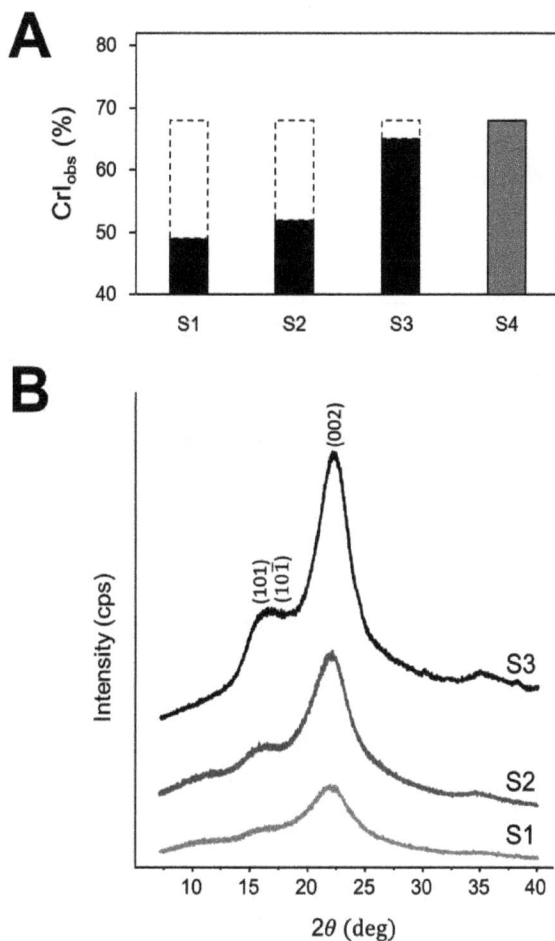

Figure 3. XRD analysis of pretreated rice straw. (A) The levels of crystalline index (CrI$_{obs}$) of rice straw compared to pure cellulose solids. (B) XRD spectra of rice straw samples. (S1) Autoclaved rice straw. (S2) Rice straw after treated with acidified sodium chlorite (three times). (S3) Rice straw after treated with acidified sodium chlorite (3x) and sodium bicarbonate. (S4) Microcrystalline cellulose (20 µm).

The XRD spectra of treated rice straws are shown in Fig. 3B. The intensities of peaks corresponding to (101), (10$\underline{1}$) and (002) lattice planes were increased by the treatment, indicating a significant increasing in crystallinity after acidified sodium chlorite (3x) and sodium bicarbonate treatments. The peaks corresponding to (101) and (10$\underline{1}$) were not clearly observed in the control sample, probably due to interference from other components in the samples (e. g.

lignin). After the treatment with acidified sodium chlorite (3x), these peaks could be significantly noticed, but not so obvious as those without extended treatment by sodium bicarbonate.

Based on FTIR analysis (Fig. 4), there are shown three representative chemical changes related to lignin removal, as assigned in Table 1.

Wavenumber (cm⁻¹)	Assignment	Reference(s)
3450-3350	O-H stretching	Nelson and O'Connor, 1964
2901-2892	C-H stretching	Schwanninger et al., 2004
1745	Carbonyl bonds (associated with lignin side chain removal)	Kumar et al., 2009
1732	Alkyl esther from cell wall hemicellulose C=O; strong carbonyl groups in branched hemicellulose	Liu et al., 2005; Pandey, 1999; Sene et al., 1994
1650-1640	C=O vibration; Amide I, aromatics	Haberhauer et al., 1998
1638-1604	Doublet phenolics of remained lignin	Sene et al., 1994
1517-1516	Aromatic C-O stretching mode for lignin; guayacyl ring of lignin	Liu et al., 2005
1512	Aromatic C-O stretching mode for lignin; guayacyl ring of lignin; lignocellulose	Ouatmane et al., 2000
1430	CO_2 stretching; carboxylic acids	Smith et al.,1999
1375-1370	C-H stretch of cellulose	Liu et al., 2005; Stewart et al., 1995
1247-1242	C-O-H deformation and C-O stretching of phenolics	Stewart et al., 1995; Sene et al., 1994
1162-1159	Antisymmetric stretching C-O-C glycoside; C-O-C β-1,4 glycosil linkage of cellulose.	Liu et al., 2005; Michell, 1990
1109-1098	C-O vibration of crystalline cellulose; glucose ring stretch from cellulose	Pandey, 1999; Stewart et al., 1995
1060, 1035	C-O vibration of cellulose	Stewart et al., 1995
900-897	Amorphous cellulose vibration; glucose ring stretch	Pandey, 1999; Stewart et al., 1995

Table 1. Assignment of the main bands in FTIR spectra for rice straw

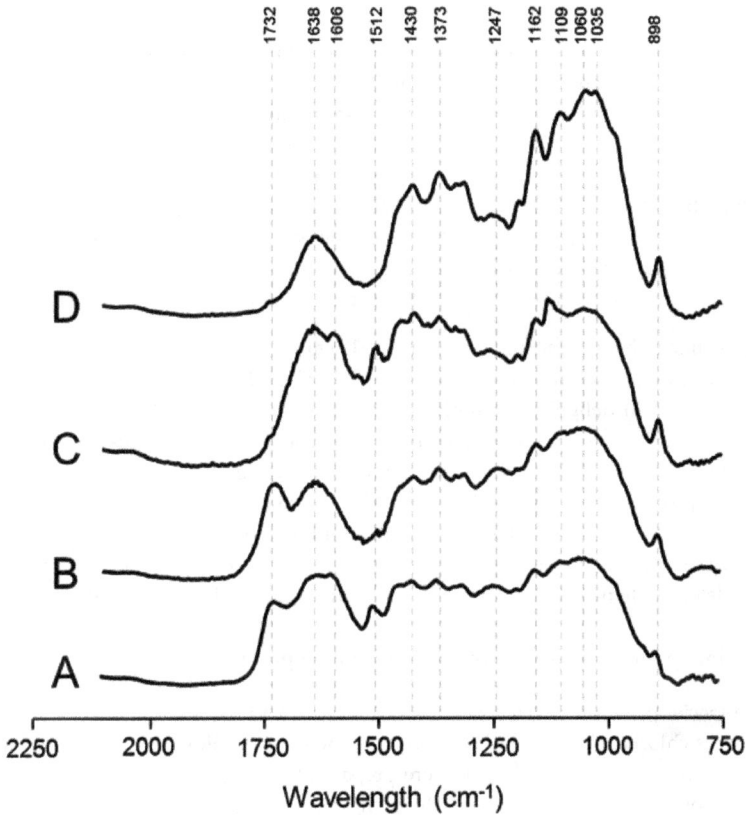

Figure 4. Chemical changes in rice straw solids determined by FTIR of wavelength ranged from 780 to 2200 (cm⁻¹). Symbols: (A) autoclaved rice straw, (B) rice straw after treated with acidified sodium chlorite (3x), (C) rice straw after treated with sodium bicarbonate, (D) rice straw after treated with acidified sodium chlorite (3x) and sodium bicarbonate.

1. The chemical changes in band position between 1745 and 1732 cm⁻¹. These are assigned to the carbonyl (C=O) stretching, attributed to aromatic skeletal vibrations in lignin structures (Kumar et al., 2009; Liu et al., 2005). Carbonyls mainly exist in the side chains of lignin structural units and are also an important functional group in the side chains. Particularly, the band position of 1732 cm⁻¹ is attributed to the linkage of lignin side chain

with branched hemicellulose. The significant change in these bands was almost unchanged by treatment with acidified sodium chlorite (3x) (B). After treated with sodium bicarbonate, these bands disappeared, indicating that the lignin linked to branched hemicellulose was removed from the cellulose structures (C, D), as also previously suggested (Liu and Wyman, 2004).

2. The chemical changes in band positions of 1606 cm^{-1} and 1638 cm^{-1}. The band at 1638 cm^{-1} is assigned to an aromatic stretch, and the band at 1606 cm^{-1} appears associated with the α-β double band of the propanoid side group in lignin-like structures (Sene et al., 1994). As shown in Fig. 4, The band at 1606 cm^{-1} was weak in delignified samples by acidified sodium chlorite treatment (3x) (B), and then disappeared after treated with sodium bicarbonate (C, D), indicating that lignin and its aggregates were also removed from the cellulose structure. Since the absorption band at 1638 cm^{-1} was often overlapped with the band assigned to absorbed water in cellulose (Chen et al., 1997; Gastaldi, et al., 1998), characteristic of the band at 1638 cm^{-1} may be similar to that at 1606 cm^{-1}.

3. The chemical changes in band position of 1512 cm^{-1}. This band is assigned to an aromatic C-O stretching mode in lignin (Ouatmane et al., 2000). The absorption band at 1512 cm^{-1} till remained in delignified samples by acidified sodium chlorite treatment (3x), and disappeared by sodium bicarbonate treatment. The chemical changes in band position between 1745 and 1732 cm^{-1} indicate the removal of integrated lignin, while the changes in band positions of 1606 and those of 1512 cm^{-1} indicate the removal of surface lignin. According to these results, it is clear that delignification by acidified sodium chlorite (3x) could efficiently remove surface lignin covered cellulose. However, to remove lignin completely, delignified rice straw must be treated with sodium bicarbonate.

3.2. Chlorine species produced during delignification process

Chlorine species produced in one time treating (1x) with acidified sodium chlorite for delignification were chlorite, chloride and chlorate ions and chlorine dioxide as shown in Fig. 5, in which rice straw and chemical lignin were prepared by equal weight and the control was treated without substrate. In general, sodium chlorite dissociates depending on pH in water and is converted to chlorite ion, in which it produces chlorine dioxide and chloride ion in acidic condition of pH 2 or less. In this study, the reaction was conducted in range of pH 4.5-4.8 under buffering by acetic acid and the main chemical species were chlorite with some chloride and little release of chlorine dioxide as observed in the control (Fig. 5.).

Chlorite is a strong oxidant and acts selectively on lignin (Ahlgren and Goring, 1971). In the delignification process of rice straw, lignin was clearly oxidized and removed by chlorite and then chlorite was reduced to chloride as confirmed by changes of chemical species in chemical lignin and rice straw (Fig. 5). Some chlorate was also produced. Meanwhile, about 20% of chlorite remained after one time treating in rice straw, even though 38% of lignin left as shown in (1x) of Fig. 1(A). This is because one time treating was not enough to remove the internal lignin of rice straw.

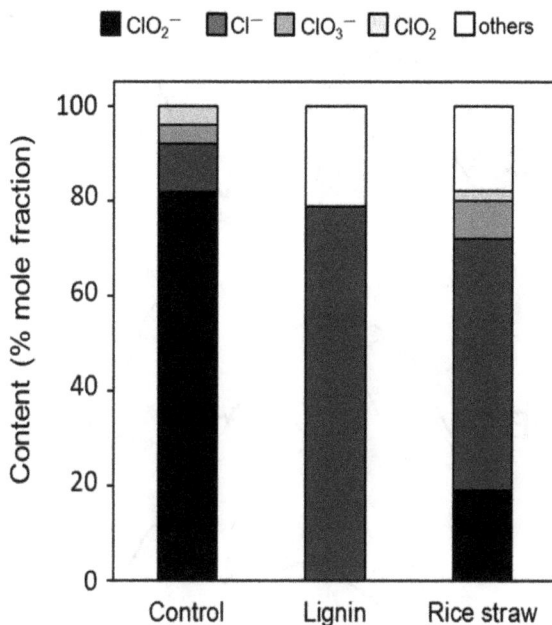

Figure 5. Formation of chlorite ion (ClO_2^-), chloride ion (Cl^-), chlorate ion (ClO_3^-) and chlorine dioxide (ClO_2) by oxidation of substrate using sodium chlorite as an oxidant. Rice straw and chemical lignin solids were used as substrates. Control means the reaction without any substrates.

3.3. Alteration in crystalline and chemical structures of microfibrils by pretreatment

Treatment with sodium bicarbonate also changes the properties of cellulose surface (Kwasniakova et al., 1996). As shown in Fig. 4, band position of 1430 cm^{-1} is assigned to CO_2 stretching for carboxylic groups on the surface of cellulose (Smith et al., 1999). The absorbance at this band position was significantly increased by treatment with sodium bicarbonate, indicating that the surface of cellulose microfibrils was carboxylated due to the treatment. The absorbance at 898 cm^{-1} is associated with the anti-symmetric out-of-phase ring stretch of amorphous cellulose (Stewart et al., 1995; Michell, 1990), and the 1060 and 1035 cm^{-1} bonds are related to C-O vibration of crystalline cellulose (Stewart et al., 1995). Both the crystalline (1035-1109 cm^{-1}) and amorphous (898 cm^{-1}) bands increased in intensity after treatment with sodium bicarbonate, suggesting that the sample had a higher percentage of crystalline cellulose, which we predicted that it was difficult to further hydrolyze with cellulase enzymes, particularly Cel7A. However, the saccharification yield of rice straw treated with sodium bicarbonate was significantly enhanced even though possessing high crystallinity. In this case, the carboxylation of cellulose surface supported the disintegration of cellulose structures, and allowed the enzyme to degrade the cellulose microfibrils.

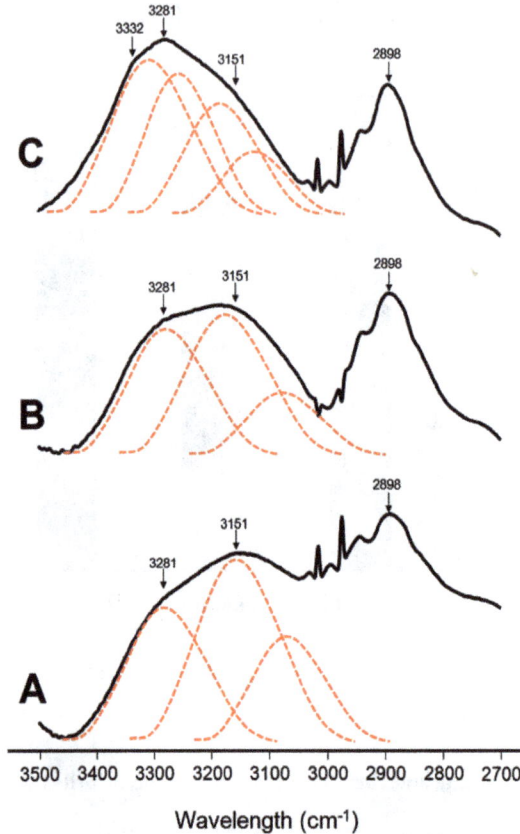

Figure 6. Chemical changes in rice straw solids determined by FTIR of wavelength ranged from 2700 to 3500 (cm⁻¹). Symbols: (A) autoclaved rice straw, (B) rice straw after treated with acidified sodium chlorite (3x), (C) rice straw after treated with acidified sodium chlorite (3x) and sodium bicarbonate.

Thermochemical treatments have a potential to change cellulose crystalline structure by disrupting inter/intra hydrogen bonding of cellulose chains (Zugenmaier, 2001). After treatment with sodium bicarbonate, the transformation of cellulose crystallinity from type I to type II was noticed, but it is not significant. As shown in Fig. 6, the typical changes in spectra length attributed to the transformation of crystal structure were, the decreasing of absorbance band at 3151 cm⁻¹, and the increasing of absorbance bands at 3281 and 3332 cm⁻¹, which were assigned to hydrogen-bonded OH stretching on the cellulose surface. Based on the previous study (Oh et al., 2005), there are three or four spectra overlapped in the weave length ranging from 2900 to 3400 cm⁻¹ giving the characteristic of bands with corresponding with the transformation of crystal type, as indicated by red dashed line in Fig. 6. The increasing of absorbance bands at 3281 and 3332 cm^{-1} was caused by the strong valence vibration of H bonded OH groups at O(2)H-O(6)H(intra) and O(3)H-O(5)H(intra), while the decreasing of absorbance band at 3151 cm^{-1} was caused by the weak valence vibration at O(6)H-O(3')H(inter) and O(6)H-O(2')H(inter), changing the stereochemistry at C2, C3, C5 and C6 inside the cellulose structure (Fig. 7). In case when the crystal type changed from type I to type II, the valence vibration at O(6)H-O(3')H(inter) and O(6)H-O(2')H(inter) became weak, whereas O(2)H-O(6)H(intra) and O(3)H-O(5)H(intra) became strong, due to the relocation of C-O changing the orientation of crystal. As shown in Fig. 6, the band at 3151 cm^{-1} was weak, but the bands at 3281 and 3332 cm^{-1} were strong, after treatment with sodium bicarbonate. By contrast, the treatment with acidified sodium chlorite (3x) only did not give the greatest changes at the length. Therefore, the treatment with acidified sodium chlorite (3x) could not change the crystal structure of rice straw cellulose to type II-like structure, unless combined with swelling by sodium bicarbonate.

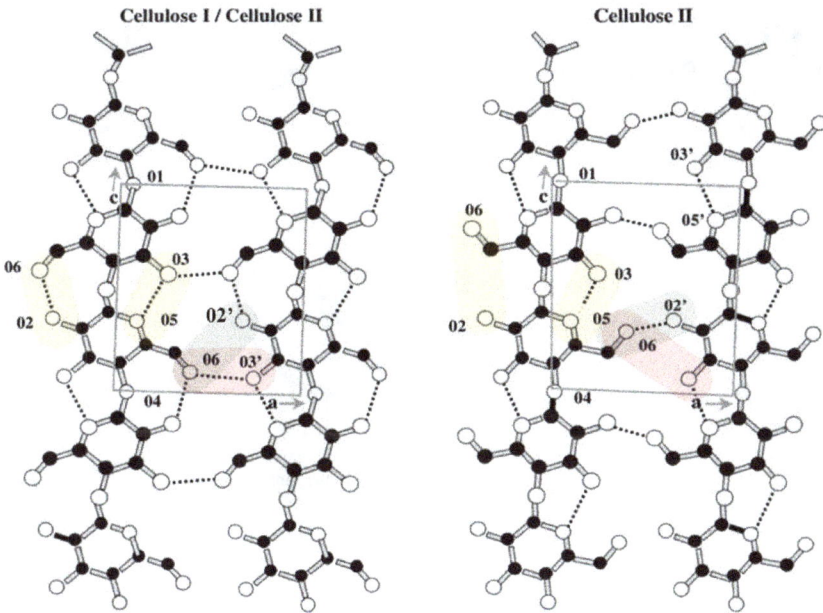

Figure 7. Proposed hydrogen-bonding patterns by Kolpak and Blackwell (1976): (a) 020 plane ('down' chains), (c) 020 plane ('up' chains).

3.4. Saccharification and fermentation of treated rice straw

Figure 8A shows the saccharification of rice straw, which was first delignified by acidified sodium chlorite (3x) only or alkaline treated by sodium bicarbonate only or treated with sodium bicarbonate after delignified by acidified sodium chlorite (3x). As a control, autoclaved rice straw without addition of any chemicals was used. The saccharification was performed under condition as mentioned in Materials and Methods section using commercially available enzymes mixture (Celluclast 1.5L and Novozym 188) at a low enzyme loading of 1/100 (g-enzymes/g-biomass solids). Single treatment with acidified sodium chlorite (3x) or with sodium bicarbonate could increase the saccharification rate of rice straw at 2.5 to 3-times higher than the control. However, high saccharification rate could be achieved from rice straw in case rice straw was treated with sodium bicarbonate after delignification by acidified sodium chlorite (3x). The rate was about 96% (wt.-reduced sugars/wt.-pretreated biomass), 4.2-times higher than the control, and even 1.5-times higher than the single treatments. This result indicates that the application of delignification and swelling processes on rice straw pretreatment can result in synergistic effect that enables broader success in achieving high enzymatic saccharification efficiency in processing rice straw to fermentable sugars, biofuels and value-added products.

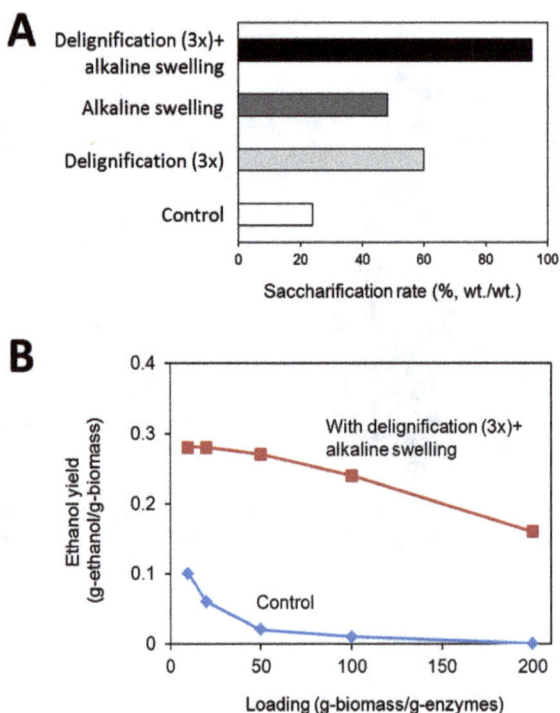

Figure 8. Saccharification (A) and fermentation (B) of treated rice straw. Saccharification was performed in 0.1M sodium acetate buffer (pH 4.8) for 3 days using commercial available cellulase enzyme mixtures (Novozym Celluclast 1.5L and 188) at enzyme loading of 1/100 (g-enzymes/g-biomass). Fermentation was performed anaerobically at 30 °C until no longer ethanol produced. Untreated rice straw was used as a substrate in all control experiments.

To test the feasibility of using the treated rice straw as a substrate in simultaneous saccharification and fermentation (SSF), the fermentations were performed at solid loadings of 50, 100, 150 and 200 (g-biomass/g-enzymes) under condition as mentioned in Materials and Methods section. The fermentations of untreated rice straw at similar solid loadings were used as controls. Figure 8B shows the yield of ethanol highly obtained from fermentation of treated (solid line) and untreated (dashed line) rice straws by using *S. cerevisiae* NBRC2114 as a strain. As clearly form the figure, high ethanol yield of 0.28 (g-ethanol/g-biomass) was obtained in case of treated rice straw. Since the fermentation of untreated rice straw could only produce ethanol at 0.12 (g-ethanol/g-biomass), the treatment with sodium bicarbonate after delignification by acidified sodium chlorite (3x) remarkably enhanced efficiency of fermentation by 2.3-times (in this study), as indicated by the ethanol yield. In spite of the fact that the ethanol yields gradually decreased as solid loading increased, SSF of treated rice straw was still enabled, as indicated by the ethanol yield of 0.16 (g-ethanol/g-biomass) achieved at solid loading of 200 (g-biomass/g-enzymes). By contrast, it was difficult to perform SSF of untreated rice straw even at solid loading of 50 (g-biomass/g-enzymes). According to these results, we conclude that the treatment with sodium bicarbonate after delignification by acidified sodium chlorite (3x) is effective in enhancing the enzymatic saccharification of rice straw in SSF process, particularly at high biomass solid with low enzyme loadings.

3.5. Inside the chemical reactions involved in the pretreatment process

Figure 9 shows a proposed conceptual model for the mechanism of enhanced enzymatic saccharification of rice straw by delignification with acidified sodium chlorite and by swelling and surface rearrangement with sodium bicarbonate treatment, as described in this study. During the delignification process by acidified sodium chlorite, the hydroxyl groups and reducing end groups of cellulose can also be oxidized (Fengel and Wegener, 1984).

Chemical oxidation of the reducing-ends of cellulose could negatively interact with Cel7A, a reducing-end targeting cellobiohydrolase (Barr et al., 1996). In fact, the treatment by acidified sodium chlorite could successfully improve the efficiency of saccharification on treated rice straw three-times compared with the control (Fig. 8A). As reported by Ishizawa et al. (2009), the acidified sodium chlorite treatment has no detectable effect on treated rice straw three-times compared with the control (Fig. 8A). As reported by Ishizawa et al. (2009), the acidified sodium chlorite treatment has no detectable effect on digestibility or accessibility of crystalline cellulose by Cel7A. However, amorphous cellulose was more susceptible to the oxidative treatment and showed a slight decreasing in the initial cellulose conversion and enzyme binding levels using Cel7A.

In this study, the digestion of delignified rice straw by cellulase enzyme mixture (more than 60% of total protein was Cel7A, according to Nummi et al., 1983) was remarkably enhanced, probably not only due to the removal of surface lignin, but also due to the protection of other amorphous parts of cellulose by its rigid structure. It could be proven by XRD analysis as shown in Fig. 3, the slightly increasing in its crystallinity after treated by acidified sodium chlorite (3x), indicating that most amorphous cellulose was placed inside the structure untouched by chlorite even after the treatment.

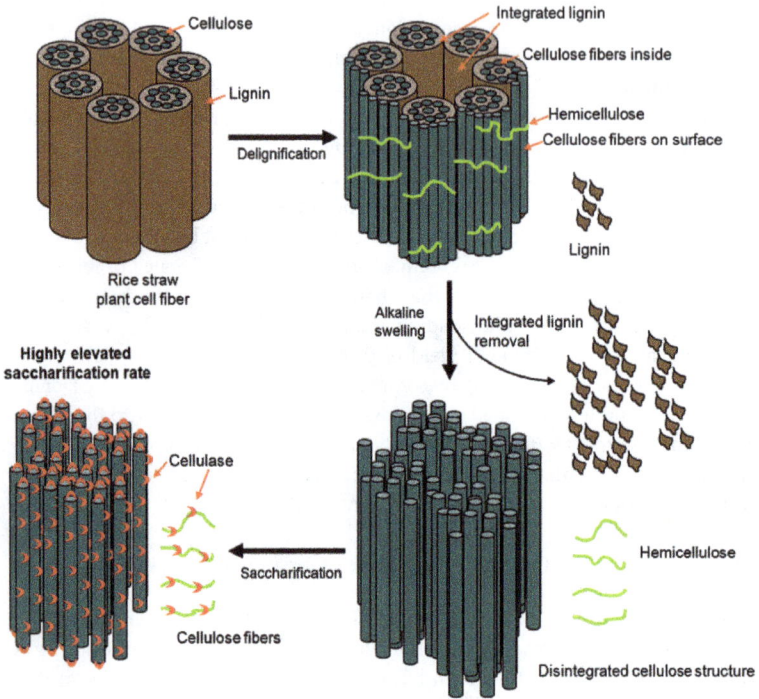

Figure 9. A proposed conceptual model for the mechanism of enhanced enzymatic saccharification of rice straw by delignification with acidified sodium chlorite and by swelling and surface rearrangement with sodium bicarbonate treatment.

Furthermore, the alkaline treatment (swelling) by sodium bicarbonate has a potential to dramatically enhance the saccharification of delignified rice straw, as shown in Fig. 9. Generally, the alkaline treatment resulted in remarkable decreasing of hemicellulose due to lignin removal (Liu and Wyman, 2004). The recovery of residual rice straw after sodium hydroxide treatment was low compared with sodium bicarbonate, even though alkaline treatment by sodium hydroxide is widely used to obtain fiber swelling upon mercerization. Taniguchi et al. (1982) reported that the saccharification of sodium hydroxide-treated rice straw was low compared to that treated with sodium bicarbonate. One possible explanation for this phenomenon is that the near complete removal of lignin allows adjacent hemicellulose-free cellulose microfibrils to aggregate upon elimination of the lignin spacer (Duchesne et al., 2001; Oksanen et al., 1997). Similar results have been reported in the literature (Fan et al., 1980), showing that the hydrolysis rate of wheat straw increases with delignification up to about 50%, after which cellulose hydrolysis increases only slightly. By contrast, the alkaline treatment by sodium bicarbonate allows the swelling similar to that with sodium hydroxide, but the difference is on the chemical modification of cellulose fiber surface by carboxylation with bicarbonate treatment (Kwasniakova et al., 1996). As a conclusion, the carboxylation on the cellulose surface could prevent the aggregation of cellulose fibrils.

4. Conclusion

The synergistic effect of delignification and swelling on the enzymatic saccharification of rice straw was evaluated by residual lignin estimation, x-ray diffraction, FTIR spectrophotometry, SEM as indirect methods, by enzymatic saccharification using purified or commercial available cellulase enzymes as a direct method. The removal of total lignin, including surface and internal lignin, is an indispensable pretreatment for achieving high efficient enzymatic saccharification of rice straw. Treatment with acidified sodium chlorite (3x) could remove surface lignin only, but internal lignin remained. By combination with the swelling using sodium bicarbonate, total lignin could be removed. Swelling with sodium bicarbonate did not result in the removal of total lignin only, yet it could breakdown the cellulose structure to release disintegrated cellulose fibrils and indeed change the chemical properties of cellulose surface by decarboxylation. By treated rice straw with acidified sodium chlorite (3x) and then with sodium bicarbonate, the saccharification yield was remarkable increased compared with that only by acidified sodium chlorite (3x) only. This indicates that delignification could not enhance the saccharification without the synergistic effect of the combination with swelling by alkaline treatment.

Acknowledgements

The author thanks Ms. Kiyoko Kawashima and Mr. Yuji Fukuda from the Collaborative Research Center of Meisei University for SEM, FTIR and XRD assistances. Also, the author thanks Mr. Kazuo Taku for his help in determination of chlorine species produced during delignification process and Dr. Shuzo Tanaka for valuable discussion. This research was financially supported by the grant-in-aid (K22027, K2339) from Ministry of the Environment, Japan.

Author details

Prihardi Kahar*

Address all correspondence to: pri@hino.meisei-u.ac.jp or prihardi@outlook.com

Program in Environment and Ecology, School of Science and Engineering, Meisei University, Hino-shi, Tokyo, Japan

References

[1] Ahlgren, P. A, & Goring, D. A. I. (1971). Removal of wood components during chlorite delignification of black spruce. Can. J. Chem. , 49, 1272-1275.

[2] Allen, S. E, Grimshaw, H. M, Parkinson, J, & Quermby, C. (1974). Chemical analysis of ecological materials. Blackwell Scientific, Oxford, , 386-389.

[3] Barzyk, D, Page, D, & Ragauskas, A. J. (1997). Carboxylic acid groups and fiber bonding; The fundamentals of papermaking materials, Transactions of the Fundamental Research Symposium IIth Sept, Cambridge, UK. , 2, 893-901.

[4] Bradford, M. M. A Rapid and sensitive method for the quantitation of microgram quantities of protein utilizing the principle of protein-dye binding. Anal. Biochem., , 72, 248-254.

[5] Chang, V. S, & Holtzapple, M. T. (2000). Fundamental factors affecting biomass enzymatic reactivity. Appl. Biochem. Biotechnol. , 84, 5-37.

[6] Chen, L. M, Wilson, R. H, & Mccann, M. C. (1997). Investigation of macromolecule orientation in dry and hydrated walls of single onion epidermal cells by FTIR microspectroscopy. J. Mol. Struct. , 408, 257-260.

[7] Corredor, D. Y, Salazar, J. M, Hohn, K. L, Bean, S, Bean, B, & Wang, D. (2009). Evaluation and characterization of forage sorghum as feedstock for fermentable sugar production. Appl. Biochem. Biotechnol. , 158, 164-179.

[8] Coughlan, M. P. (1985). The properties of fungal and bacterial cellulases with comment on their production and application, in: Russell, G.E. (Eds.), Biotechnology and genetic engineering reviews. 3, Newcastle-upon-Tyne, Interscience, , 37-109.

[9] Desai, S. G, & Converse, A. O. (1997). Substrate reactivity as a function of the extent of reaction in the enzymatic hydrolysis of lignocellulose. Biotechnol. Bioeng., , 56, 650-655.

[10] Din, N, Gilkes, N. R, Tekant, B, Miller Jr., R. C, Warren, R.A., & Kilburn, D.G. , (1991). Non-hydrolytic disruption of cellulose fibres by the binding domain of a bacterial cellulase. Bio/Technology , 9, 1096-1099.

[11] Ding, S. Y, & Himmel, M. E. (2006). The maize primary cell wall microfibril: A new model derived from direct visualization. J Agric. Food Chem. , 54, 597-606.

[12] Desai, S. G, & Converse, A. O. (1997). Substrate reactivity as a function of the extent of reaction in the enzymatic hydrolysis of lignocellulose. Biotechnol. Bioeng. , 56, 650-655.

[13] Duchesne, I, Hult, E-L, Molin, U, Daniel, G, Iversen, T, & Lennholm, H. (2001). The influence of hemicellulose on fibril aggregation of kraft pulp fibres as revealed by FE-SEM and CP/MAS 13C-NMR, Cellulose , 8, 103-111.

[14] Fan, L. T, Lee, Y-H, & Beardmore, D. H. (1980). Mechanism of the enzymatic hydrol-ysis of cellulose: Effects of major structural features of cellulose on enzymatic hydrol-ysis. Biotechnol. Bioeng., , 22, 177-199.

[15] Fang, J. M, Sun, R. C, Salisbury, D, Fowler, P, & Tomkinson, J. (1999). Comparative studyof hemicelluloses from wheat straw by alkali and hydrogen peroxide extrac-tions. Polym. Degrad. Stab. , 66, 423-432.

[16] Fengel, D, & Wegener, G. (1984). Wood: Chemistry, ultrastructure and reactions. Walter de Gruyter, Berlin.

[17] Frey-wyssling, A. (1954). The fine structure of cellulose microfibrils. Science , 119, 80-82.

[18] Gastaldi, G, Capretti, G, Focher, B, & Cosentino, C. (1998). Characterization and pro-prieties of cellulose isolated from the *Crambe abyssinica* Hull. Industrial Crops Prod., , 8, 205-218.

[19] Gusakov, A. V. (2011). Alternatives to *Trichoderma reesei* in biofuel production. Trends in Biotechnology , 29, 419-425.

[20] Haberhauer, G, Rafferty, B, Strebl, F, & Gerzabek, M. H. (1998). Comparison of the composition of forest soil litter derived from three different sites at various decompo-sitional stages using FTIR spectroscopy. Geoderma , 83, 331-342.

[21] Helle, S. S, Duff, S. J. B, & Cooper, D. G. (1993). Effect of surfactants on cellulose hy-drolysis. Biotechnol. Bioeng., , 42, 611-617.

[22] Hubbell, C. A, & Ragauskas, A. J. (2010). Effect of acid-chlorite delignification on cel-lulose degree of polymerization. Bioresour. Technol. , 101, 7410-7415.

[23] Ishizawa, C. I, Jeoh, T, Adney, W. S, Himmel, M. E, Johnson, D. K, & Davis, M. F. (2009). Can delignification decrease cellulose digestibility in acid pretreated corn stover? Cellulose , 16, 677-686.

[24] Jing, L, Qian, Y, Zhao, L. H, Zhang, S. M, Wang, Y. X, & Zhao, X. Y. (2009). Purifica-tion and characterization of a novel antifungal protein from *Bacillus subtilis* strain B29. J. Zhejiang. Univ. Sci. B. , 10, 264-272.

[25] Kahar, P, Taku, K, & Tanaka, S. (2010). Enzymatic digestion of corncobs pretreated with low strength of sulfuric acid for bioethanol production. J. Biosci. Bioeng. , 110, 453-458.

[26] Klein, G. L, & Snodgrass, W. R. (1993). Cellulose. In: Macrae, R., Robinson, R.K., Sad-dler, M.J., (eds.) Encyclopedia of food science, food technology and nutrition. Lon-don: Academic Press , 758-767.

[27] Klein-marcuschamer, D, Oleskowicz-popiel, P, Simmons, B. A, & Blanch, H. W. (2012). The challenge of enzyme cost in the production of lignocellulosic biofuels. Bi-otechnol. Bioeng., , 109, 1083-1087.

[28] Kolpak, F. J, & Blackwell, J. (1976). Determination of the structure of cellulose II. Macromolecules , 9, 273-278.

[29] Kumar, R, Mago, G, Balan, V, & Wyman, C. E. (2009). Physical and chemical characterizations of corn stover and poplar solids resulting from leading pretreatment technologies. Bioresour. Technol. , 100, 3948-3962.

[30] Kwasniakova, K, Kokta, B. V, & Koran, Z. (1996). Strength properties of black spruce wood under different treatment. Wood Sci. Technol. , 30, 463-475.

[31] Lin, Y, & Tanaka, S. (2006). Ethanol fermentation from biomass resources: current state and prospects. Appl. Microbiol. Biotechnol., , 69, 627-642.

[32] Liu, C. G, & Wyman, C. E. (2004). Impact of fluid velocity on hot water only pretreatment of corn stover in a flow through reactor. Appl. Biochem. Biotechnol. 113-116, 977-987.

[33] Liu, R, Yu, H, & Huang, Y. (2005). Cellulose (London, England), , 12, 25-34.

[34] Lynd, L. R, Weimer, P. J, Van Zyl, W. H, & Pretorius, I. S. (2002). Microbial cellulose utilization: fundamentals and biotechnology. Microbiol. Mol. Biol. Rev. , 66, 506-577.

[35] Michell, A. J. (1990). Second derivative FTIR spectra of native celluloses. Carbohydr. Res., , 197, 53-60.

[36] Mansfield, S. D, Mooney, C, & Saddler, J. N. (1999). Substrates and enzyme characteristics that limit cellulose hydrolysis. Biotechnol . Prog. , 15, 804-816.

[37] Miller, G. L. (1959). Use of dinitrosalicylic acid reagent for determination of reducing sugar. Anal. Chem., , 31, 426-428.

[38] Morehead, F. (1950). Ultrasonic disintegration of cellulose fibres before and after acid hydrolysis. Textile Res. J., , 20, 549-553.

[39] Nelson, M. L, & Connor, O. R.T., (1964). Relation of certain infrared bands to cellulose crystallinity and crystal lattice type. Part II. A new infrared ratio for estimation of crystallinity in celluloses I and II. J. Appl. Polymer Sci. , 8, 1325-1341.

[40] Nummi, M, Niku-paavola, M. L, Lappalainen, A, Enari, T. M, & Raunio, V. (1983). Cellobiohydrolase from Trichoderma reesei. Biochem. J., , 215, 677-683.

[41] Oh, S. Y, Yoo, D. I, Shin, Y. S, Kim, H. C, Kim, H. Y, Chung, Y. S, Park, W. H, & Youk, J. H. (2005). Crystalline structure analysis of cellulose treated with sodium hydroxide and carbon dioxide by means of X-ray diffraction and FTIR spectroscopy. Carbohyd. Res. , 340, 2376-2391.

[42] Oksanen, T, Buchert, J, & Viikari, L. (1997). The role of hemicelluloses in the hornification of bleached kraft pulps, Holzforchung , 51, 355-360.

[43] Ouatmane, A, Provengano, M. R, Hafidi, M, & Senesi, N. (2000). Compost maturity assessment using calorimetry, spectroscopy and chemical analysis. Compost Sci. Util. , 8, 124-134.

[44] Pandey, K. K. (1999). A study of chemical structure of soft and hardwood and wood polymers by FTIR spectroscopy. J. Appl. Polymer Sci. , 71, 1969-1975.

[45] Peterlin, A, & Ingram, P. (1970). Morphology of Secondary Wall Fibrils in Cotton. Textile Res. J., , 40, 345-354.

[46] Ragauskas, A. J, Williams, C. K, Davison, B. H, Britovsek, G, Cairney, J, & Eckert, C. A. Frederick Jr., W.J., Hallett, J.P., Leak, D.J., Liotta, C.L., Mielenz, J.R., Murphy, R., Templer, R., Tschaplinski, T., (2006). The path forward for biofuels and biomaterials. Science , 311, 484-489.

[47] Roche, C. M, Dibble, C. J, Knutsen, J. S, Stickel, J. J, & Liberatore, M. W. (2009). Particle Concentration and Yield Stress of Biomass Slurries during Enzymatic Hydrolysis at High-Solids Loadings. Biotechnology and Bioengineering , 104, 290-300.

[48] Schwanninger, M, Hinterstoisser, B, Gradinger, C, Messner, K, & Fackler, K. (2004). Examination of spruce wood biodegraded by *Ceriporiopsis subvermispora* using near and mid infrared spectroscopy. J. Near Infrared Spectrosc. , 12, 397-409.

[49] Segal, L, Creely, J. J, Martin, A. E, & Conrad, C. M. (1959). An empirical method for estimating the degree of crystallinity of native cellulose using the X-ray diffractometer. Text. Resear. J. , 29, 764-786.

[50] Sene, C. F. B, Mccann, M. C, Wilson, R. H, & Gdnter, R. FT-Raman and FT-Infrared spectroscopy: An investigation of five higher plant cell walls and their components. Plant Physiol. , 106, 1623-1633.

[51] Sluiter, A, Hames, B, Ruiz, R, Scarlata, C, Sluiter, J, Templeton, D, & Crocker, D. (2008). Determination of structural carbohydrates and lignin in biomass Laboratory Analytical Procedure (LAP). Technical Report NREL/TP-, 510-42618.

[52] Smith, B. (1999). Infrared Spectral Interpretation. CRC Press, Boca Raton, London, New York, Washington, DC.

[53] Stewart, D, Wilson, H. M, Hendra, P. J, & Morrison, I. M. (1995). Fourier-transform infrared and Raman-spectroscopic study of biochemical and chemical treatments of oak wood (*Quercus rubra*) and barley (*Hordeum vulgare*) straw. J. Agric. Food Chem. , 43, 2219-2225.

[54] Sun, J. X, Sun, X. F, Sun, R. C, & Su, Y. Q. (2004). Fractional extraction and structural characterization of sugarcane bagasse hemicelluloses. Carbohyd. Polym. , 56, 195-204.

[55] Wise, L. E, Murphy, M, & Addieco, A. A. (1946). Isolation of holocellulose from wood. Pap. Trade J., , 122, 35-43.

[56] Taniguchi, M, Tanaka, M, Matsuno, R, & Kamikubo, T. (1982). Evaluation of chemical pretreatment for enzymatic solubilization of rice straw. Eur. J. Appl. Microbiol. Biotechnol., , 14, 35-39.

[57] Teeri, T. T, Reinikainen, T, & Ruohonen, L. Alwyn Jones, T., Knowles, J.K.C., (1992). Domain function in *Trichoderma reesei* cellobiohydrolases. J. Biotechnol. , 21, 169-176.

[58] Zhang, Y-H. P, & Lynd, L. R. (2004). Toward an aggregated understanding of enzymatic hydrolysis of cellulose: Noncomplexed cellulase systems. Biotechnol. Bioeng., , 88, 797-824.

[59] Zhang, Y-H. P, Himmel, M, & Mielenz, J. R. (2006). Outlook for cellulase improvement: Screening and selection strategies. Biotechnol. Adv., , 24, 452-481.

[60] Zugenmaier, P. (2001). Conformation and packing of various crystalline cellulose fibers. Prog. Polym. Sci. , 26, 1341-1417.

Chapter 4

Comparison of the Performance of the Laccase Bioconversion of Sodium Lignosulfonates in Batch, Continuous and Fed Batch Reactors

Nidal Madad, Latifa Chebil, Hugues Canteri,
Céline Charbonnel and Mohamed Ghoul

Additional information is available at the end of the chapter

1. Introduction

Wood and food processes generate high quantities of by-products such as lignin, lignosulfonates and free phenols(Rodrigues et al., 2008). These compounds are natural molecules and renewable resources, but they constitute an important source of pollution. However, they can undergo several transformations and processes (hydrolysis, bioconversion and fractionation) to provide fractions with useful properties such as antioxidants, dispersing agent and plasticizer (Benavente-Garcia et al., 2000; Madad et al., 2011; Ouyang et al., 2006; Yang et al., 2008; Zhou et al., 2006). The recovery and development of these by-products are mainly carried out by chemical or physical processes such as thermal decomposition (Jiang et al., 2003), liquid (Correia et al., 2007) or membrane fractionation (Bhattacharya et al., 2005; Ferreira et al., 2005; Venkateswaran and Palanivelu, 2006). The chemical process is often not environmentally friendly and may be expensive. To overcome some drawbacks of the above mentioned processes, enzyme hydrolysis or bioconversion of these raw materials is presented as a promising way (Kobayashi et al., 2001). In fact, the enzymatic processes can be conducted under mild reaction conditions and without using toxic reagents. Moreover, in some cases, they lead to a homogeneous molecular distribution of obtained products and enhanced properties (Gross et al., 1998; Joo et al., 1998; Kobayashi, 1999; Kobayashi et al., 1995; Kobayashi and Uyama, 1998; Kobayashi et al., 2001).

The use of enzymes is firstly applied to the delignification and the removal of free phenols from wastewaters (Dasgupta et al., 2007; Husain, 2010; Nazari et al., 2007; Riva, 2006; Widsten and Kandelbauer, 2008). Recently, the ability of some oxidoreductases and laccases to polymerize phenols have received great attention and applied with success in the field of wood by-products (Ikeda et al., 2001; Jeon et al., 2010; Mita et al., 2003; Reihmann and Ritter, 2006).

Depending on enzyme nature, enzymatic bioconversion of phenols requires either oxygen or hydrogen peroxide. The availability and the concentration of these substrates are essential to these reactions. Ghosh et al. (Ghosh et al., 2008) studied the effect of dissolved oxygen concentration on laccase efficiency during the removal of 2,4-dimethylphenol. These authors experimented several techniques such as dissolution by stirring or bubbling or a high initial saturation of the medium by oxygen. They reported that, whatever the technique used, as long as dissolved oxygen inside the reactor remains high, initial rates of reactions were similar and high compared to a reaction control with a low concentration of oxygen.

The main investigations in the field of enzymatic bioconversion were carried out in batch mode (Ghosh et al., 2008; Kim et al., 2009; Nugroho Prasetyo et al., 2010). However, in this mode, the degree of polydispersity remains high and hydroxyl phenolic groups are often only partially oxidized. This behavior, according to Areskogh et al.(Areskogh et al., 2010a) would be due to the ability of the lignosulfonates to form spherical microgels makes the phenolic groups buried in the core of the gel inaccessible. It could also be explained the inhibition of laccase by formed polymers (Kurniawati and Nicell, 2009). Another explanation is that the bioconversion by laccase is carried out in two ways leading either to C-O-C or to C-C linkages. The last way generates phenolic groups by ionic tautomerisation (Areskogh et al., 2010b). The concentration of the lignosulfonates also seems to influence the conversion rate of the phenolic groups, the polydispersity and the average molecular weight of polymers formed. High M_w were reached with high lignosulfonate concentrations (Areskogh et al., 2010a).

Fed batch and continuous modes are used in chemical bioconversion to control average M_w evolution and polydispersity and could also overcome some drawbacks of batch reactions; because fed batch allows controlling the enzyme and the substrate concentrations in the medium while the continuous system avoids the accumulation of the formed polymers in the medium. In spite of the potential of these two modes of reaction few data are available on their performance in the field of laccase bioconversion of phenols. Wu et al. (Wu et al., 1999) compared phenols removal efficiency by horseradish peroxidase in batch, continuous stirred tank, fed batch and a plug flow reactors. They reported that the plug flow reactor was the most appropriate for this reaction. Areskogh et al. (Areskogh et al., 2010a) compared also the effect of a successive addition of laccase during the lignosulfonates (SLS) bioconversion. They observed only minor differences in the average molecular weight increase which is dependent on the amount of enzyme.

The aim of this paper is to compare the efficiency of lignosulfonate bioconversion by laccase in terms of phenolic OH group consumption, average molecular weight and degree of polydispersity evolution under three modes of reaction conductions: batch with different enzyme/substrate ratio, continuous feed of laccase and lignosulfonates and three alternatives of fed batch feeding. The oxygen consumption was also monitored.

Comparison of the Performance of the Laccase Bioconversion of Sodium Lignosulfonates in Batch, Continuous and Fed Batch Reactors

75

2. Materials and methods

2.1. Enzyme and chemicals

Sodium lignosulfonates (SLS) from (Aldrich, Sweden) : 90 wt. % of SLS, 4 wt. % of reducing sugars and 6 wt% of total impurities. The average molecular weight (M_w), the number molecular weight (M_n) and the polydispersity (Pdi) values are equal to 17800 Da ± 1500, 2900 Da ± 400, and 6.2 ± 0.3, respectively.

Laccase from *Trametes versicolor* (21.4 U/mg) was purchased from Fluka (Sweden).

2.2. Laccase activity assay

The activity of laccase was determined spectrophotometrically by monitoring the oxidation of 2,2'-azinobis-(3-ethylbenzthiazoline)-6-sulfonate (ABTS) to its cation radical as substrate at 436 nm in 50 mM sodium succinate buffer at pH 4.5 and 30 °C using quartz cuvette of path length 10 mm. Enzyme activity was expressed in units (1 U = 1 μmol ABTS oxidized per min at room temperature).

2.3. Batch operation

Batch operations were performed in a bioreactor with a working volume of 1 L equipped with dissolved oxygen, pH and temperature sensors. The reactor was stirred vigorously at 500 rpm to solubilise SLS at 20°C and throughout the reactions. The lignosulfonates were solubilised in phosphate buffer solution at pH 4.5 and laccase was added to initiate reactions. For the analyses, samples were drawn out from the reactor at different intervals of time and laccase activity was stopped by heating at 90°C for five minutes.

2.4. Fed batch operation

Fed batch reactions were carried out by progressive adding, at different time intervals (every 30 minutes during the first 5 hours), of enzyme alone, substrate alone or both enzyme and substrate. The total amounts of enzyme and substrate for the three fed batch operations were 10 g/L and 30 U/mL of SLS and laccase, respectively. Samples were taken at different time intervals and enzyme activity was stopped by heating to 90°C for five minutes.

2.5. Continuous stirred tank reactor operation

The continuous stirred tank reactor was similar to the one used in batch step. Lignosulfonates (32 g/L) and laccase (63 U/mL) were prepared in two flasks separately and 500 mL of each solution were added progressively at a constant flow-rate into the reactor initially filled with buffered solution (1 L). The reactor was aerated and stirred vigorously at 500 rpm. Samples were taken at different time intervals and the enzyme activity was stopped by heating to 90°C for five minutes.

2.6. Size exclusion chromatography analysis (SEC)

Samples were analysed by Size exclusion chromatography (SEC) (HPLC LaChrom Merck, Germany). The system consists of a pump L-2130, an autosampler L-2200, and a Superdex 200HR 10/30 column (24 mL, 13 μm, dextran/cross linked agarose matrix). Detection was performed using UV detector diode L-2455 at 280 nm. Before analysis, the samples were filtered using regenerated cellulose membrane (0.22 μm) and aliquots of 50 μl were injected into the SEC system. A Buffer Phosphate pH 7, 0.15 M NaCl solution was used as an eluent. The flow rate was 0.4 mL at 25°C and the pressure is maintained at 11 bars. The calibration was performed by using polystyrenes sulfonate (PSS) as a standard to define molecular weight distribution.

Chromatographs were integrated in segments of thirteen second intervals. The number-average molecular weight (M_n), the weight-average molecular weight (M_w), and the polydispersity (Pdi) were calculated as follows (Faix, 1981):

Number average molecular weight

$$M_n = \frac{\sum\limits_{i=1}^{n} Area_i}{\sum\limits_{i=1}^{n} \dfrac{Area_i}{M_i}} \tag{1}$$

Weight average molecular weight

$$M_w = \frac{\sum\limits_{i=1}^{n} Area_i \times M_i}{\sum\limits_{i=1}^{n} Area_i} \tag{2}$$

Polydispersity

$$D = \frac{M_w}{M_n} \tag{3}$$

where M_i is the molecular weight and $Area_i$ the area of each segment i.

2.7. Determination of phenolic content

Phenolic content was determined using the method described by Areskogh et al. (Areskogh et al., 2010a).

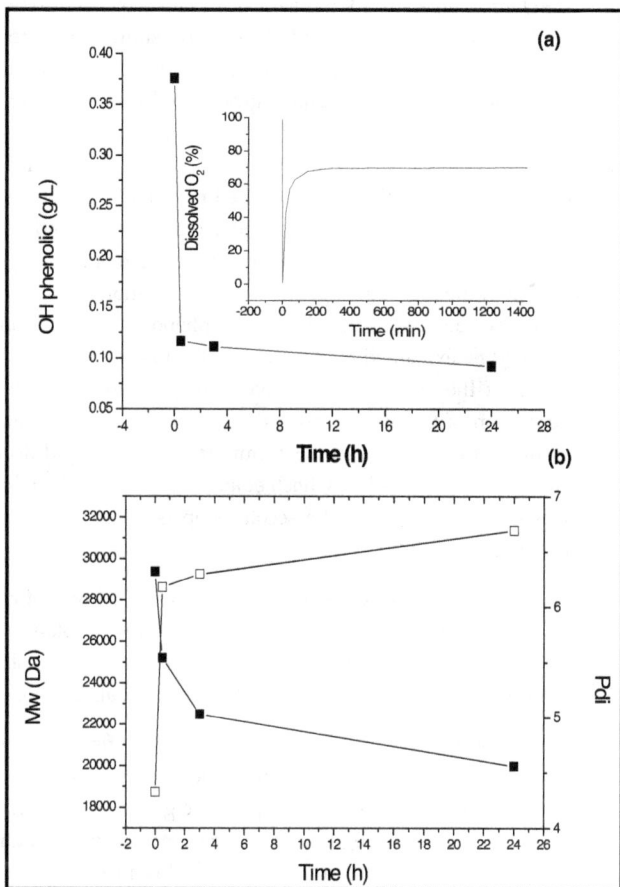

Figure 1. OH phenolic residual (a), dissolved O2 (a), Mw (b) and Pdi (b) variations in batchwise operation of reaction carried out with 10 g/L and 30 U/mL of SLS and laccase over time. (■) Pdi and (□) Mw.

3. Results and discussion

3.1. Kinetic study of enzymatic bioconversion in batch mode

The performance of the bioconversion reaction of lignosulfonates by laccase can be affected by the ratio of SLS/laccase. To verify this assumption, the reaction of bioconversion was carried out with different ratios SLS/laccase; (1 g/L)/ (3 U/mL), (1 g/L)/ (30 U/mL), (10 g/L)/ (3

U/mL) and (10 g/L)/ (30 U/mL); in a stirred and aerated reactor. For the different assays M_w average, Pdi, phenol OH group content, and oxygen consumption were determined throughout the reaction. The results obtained with the four studied ratios, indicated similar profiles for the consumption of hydroxyl phenolic groups and oxygen. As an illustration, Figure 1 represents the variation of M_w average, Pdi, hydroxyl phenolic groups and oxygen evolution for the reaction with a SLS/laccase ratio equal to (10 g/L) / (30 U/mL). It appears that this reaction is made up of a two distinct steps. The first one is characterized by a rapid decrease of phenol OH group amount, dissolved oxygen, and Pdi value and a high increase of M_w average. The second one shows an increase of the dissolution of the oxygen to reach a plateau near the saturation of the medium, a progressive deceleration in the decrease of Pdi, and in the increase of M_w average and, a stabilisation of phenol OH group content around 0.1 g/L. These profiles could be explained by the fact that the first step consists of the initiation and the propagation of the enzymatic bioconversion. The rapid consumption of the oxygen ensures the formation of the SLS phenoxy radicals via laccase reduction. Thus, the role of the oxygen is important and can become a limiting step. The rapid decrease of dissolved oxygen has already been reported by Ghosh et al. (Ghosh et al., 2008) during the 2,4-dimethylphenol bioconversion by laccase. The second step is rather a combination stage where the need for oxygen is negligible.

The observed increase of the dissolved oxygen while M_w is still growing confirms theses assumption. After 24 h of reaction the hydroxyl phenolic groups are not totally oxidized; this is due to the fact that when the reaction of bioconversion is finished, the final obtained structure of polymers contains hydroxyl groups (schema 1) (Areskogh et al., 2010b).

Table 1 reports the conversion rate of phenolic OH groups and the final M_w and Pdi values of batch reactions. These results showed also that regardless of the enzyme concentration, either 3 or 30 U/mL, the highest conversion rate of phenolic groups (73 % and 75 %) is observed at the highest SLS concentration (10g/L). For a given concentration of lignosulfonates, the enzyme concentration slighly affects the conversion rate; this means that a concentration of 3 U/mL of laccase is sufficient to polymerize the concentrations of the lignosulfonates tested in this work. It also appears that whatever the concentration of the enzyme M_w average is significantly improved at high concentrations of lignosulfonates (10 g/L). It increases from 17800 Da to 30600 Da and 31400 Da respectively for 3 U/mL and 30 U/mL of laccase. Pdi decrease approximately to a value of 4, independently of the enzyme and lignosulfonate concentrations. The high conversion yield of phenolic OH groups obtained at 10 g/L of lignosulfonates suggests that higher is generated phenoxy radicals in the reaction media, higher is the consumption of phenolic OH groups and M_w values. This may be due to the fact that the probability of establishing a contact between two phenoxy radicals is increased when their concentration in the medium is high and the C-O-C coupling is also favoured. So, this reaction is under a "kinetic control". The low M_w (26400 Da) observed with 1 g/L suggests that in the presence of a diluted solution and acid pH (4.5), the reaction is under a "thermodynamic control" which promotes C-C linkage.

Comparison of the Performance of the Laccase Bioconversion of Sodium Lignosulfonates in Batch, Continuous and Fed Batch Reactors

79

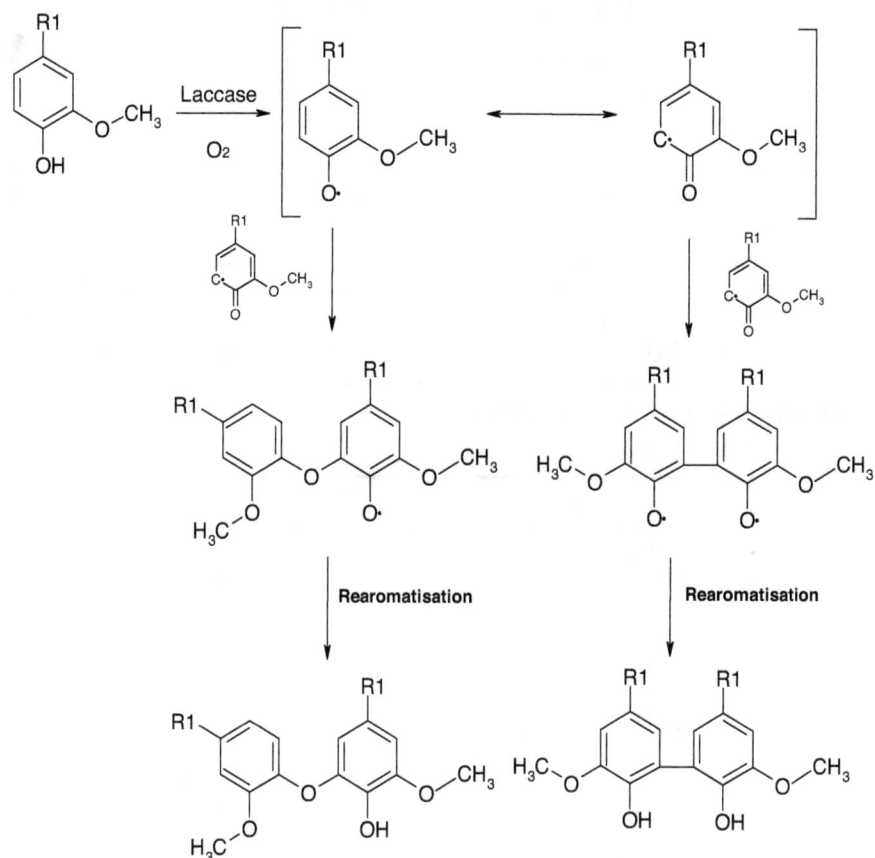

Scheme 1. Proposed reaction mechanism for the formation of C-O-C and C-C bonds when a lignosulfonates model is oxidized by laccase. (R1) lignin fragment

Reaction	Conversion rate (%)	Final Mw	Final Pdi
1 g/L of S and 3 U/mL of E	47 %	25700	4.6
1 g/L of S and 30 U/mL of E	52 %	26400	4.1
10 g/L of S and 3 U/mL of E	73 %	30600	4.4
10 g/L of S and 30 U/mL of E	75 %	31400	4.6

Table 1. The conversion rate, the specific conversion rate, the final M_w and the final Pdi of reactions carried out in batchwise operation. (S) Substrate; (E) Enzyme

3.2. Kinetic study of enzymatic bioconversion in continuous reactor

The operating conditions for the continuous feeding of the enzyme and lignosulfonates were chosen to add 16 g/L and 32 U/mL of lignosulfonates and laccase respectively and to have the same residence time (24 h) as that used in the batch mode.

The obtained results are summarized in Figure 2 a and 2 b. It appears that phenolic OH group content increases slightly in the medium to reach the same level as that observed at the end of the batch reaction (~0.1 g/L) ; while the conversion rate of phenolic OH groups remains constant near 85 % throughout the duration of the reaction. This conversion is higher than that obtained in the batch. Molecular weight average (M_w) of formed polymers increases gradually to 28400 Da during the first four hours and then, as in batch mode, this increase becomes less pronounced. Pdi values decrease quickly to reach a low value (3.7) and remain more or less constant along the time incubation (Figure 2b). The dissolved oxygen (Figure 2a) also decreases over time due to its continuous consumption by the added laccase.

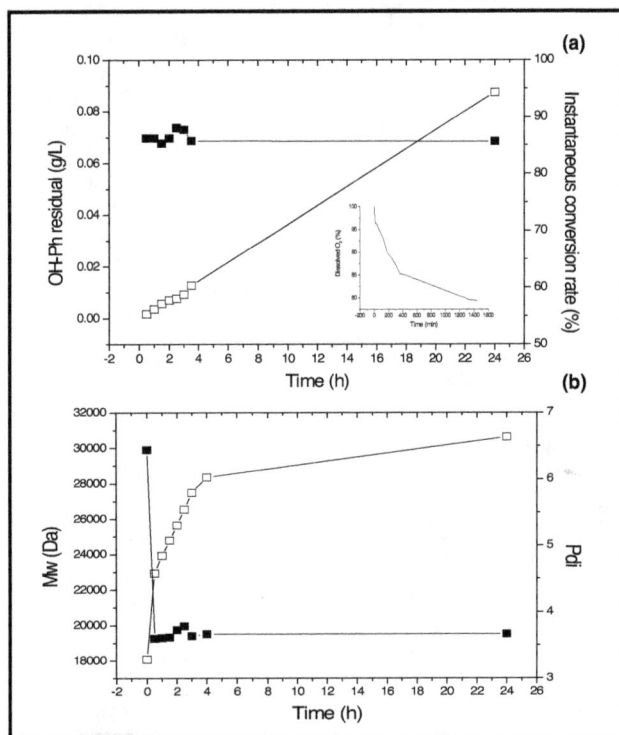

Figure 2. OH phenolic residual (a), dissolved O_2 (a), instantaneous conversion rate (a), M_w (b) and Pdi (b) variations in continuous operation over time.(a) (■) Instantaneous conversion rate (□) OH-Ph residual (b) (■) Pdi and (□) M_w

Comparison of the Performance of the Laccase Bioconversion of Sodium Lignosulfonates in Batch, Continuous and Fed Batch Reactors

81

Although 16 g/L of SLS were added during the 24h of the reaction, the final M_w average is of the same order of magnitude as the batch with 10 g/L of SLS. These results indicate that the increase of M_w average is rather favoured by the conditions allowing a high amount and instantaneous generation of free radicals rather than a progressive feeding of a high quantity of SLS. However, continuous adding of substrate and enzyme allows a low degree of polydispersity to be reached (3.7) compared to the batch (4.6). The low residual phenolic OH groups in the media and their high conversion rate suppose that the continuous mode promotes the C-O-C linkage.

3.3. Kinetic study of enzymatic bioconversion in fed batch operation

In fed batch mode three alternatives of feeding were tested i) with substrate alone, ii) with enzyme iii) or with both enzyme and substrate. For each assay the addition of substrate and enzyme was carried out in stepwise mode 10 times at a rate of 1g or 3000 U or both every 30 minutes during the five first hours of the reaction. Results are shown in figure 3, figure 4 and table 2.

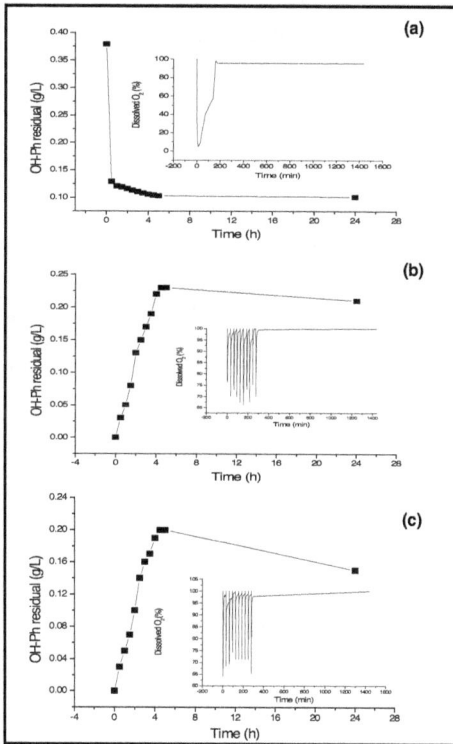

Figure 3. OH phenolic residual and dissolved O_2 variations in fed-batch operations over time. (a) Adding enzyme; (b) Adding substrate; (c) Adding both enzyme and substrate.

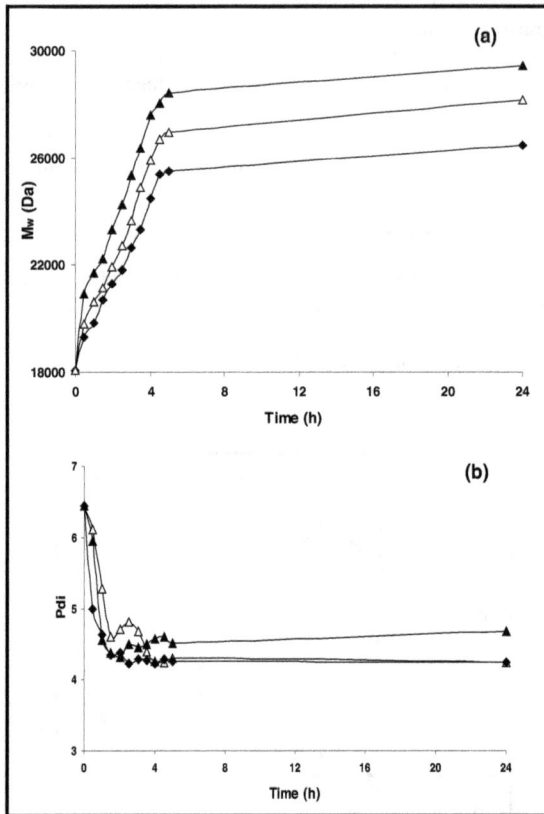

Figure 4. M_w (a) and Pdi (b) variations in fed-bach operations over time.(▲) Adding enzyme; (Δ) Adding both enzyme and substrate; (◆) Adding substrate.

Reaction	Conversion rate 5h (%)	Conversion rate 24 h (%)	Final M_w (Da)	Final Pdi
Fed-batch by adding enzyme	72 %	73 %	29400	4.7
Fed-batch by adding both enzyme and substrate	48 %	61 %	28200	4.2
Fed-batch by adding substrate	39 %	44 %	26500	4.2

Table 2. The conversion rate after 5 h and 24 h of reaction, the final M_w and the final Pdi of reactions carried out in fed-bach operations.

Comparison of the Performance of the Laccase Bioconversion of Sodium Lignosulfonates in Batch, Continuous and Fed Batch Reactors

83

Concerning the reaction carried out with enzyme feeding of the reactor (figure 3 a), similar results as batch mode operation were observed for both residual phenolic OH groups and oxygen consumption. In a first step a rapid oxidation of phenolic OH groups and oxygen uptake rate were observed, followed by an increase of the dissolved oxygen in the medium and a low oxidation of phenolic OH groups was observed during a second step. This behaviour confirms that only a low amount of enzyme is needed to oxidise the 10 g/L of SLS and the oxygen consumption occurs only during this first step of free radical generation.

After 24h of reaction, the conversion rate of phenolic OH groups (73 %) was in the same magnitude as that obtained for the batch mode (table 2). As it is indicated in figure 4 and table 2, the M_w average rose gradually during the period of enzyme addition (5h) and then stabilizes around 29000 Da along the remaining time of the reaction. Pdi value decreased rapidly and stabilised more or less at 4.7 until the end of the reaction (24 hours).

For reactions carried out with the addition of substrate or enzyme and substrate, a progressive increase of phenolic OH group content during the first 5 hours (0.2 g/L) then a slight decrease were observed. Moreover, dissolved oxygen decreases and then increases quickly after each addition, in a repetitive way (figure 3b and 3c). The conversion rate of phenolic OH groups after 24 hours is 61 % for enzyme and substrate addition and 44 % for substrate feeding.

As for continuous mode, a progressive increase of M_w was observed to reach 28200 Da and 26400 Da respectively for enzyme and substrate addition and substrate feeding. The Pdi dropped quickly to 4.2 and remained constant throughout the duration of the reaction (Figure 4 and Table 2).

The relatively low final Mw, the conversion rate and the accumulation of phenolic OH groups indicated that similar mechanisms, such as the one observed in batch mode with 1 g/L, occurre. This means that these two modes of reaction promote a "thermodynamic control" and then lead to C-C linkages instead of C-O-C coupling.

4. Conclusions

The obtained results in this work indicated that the increase of M_w average and the decrease of the polydispersity depend on the operating conditions. Batch mode with high concentration of SLS (10 g/L), promotes the increase of the M_w and probably the C-O-C coupling route. This seemesto be due to the high and instantaneous generation of free radicals, favouring the "kinetic control" of the reaction. The continuous mode also favours the formation of C-O-C bounds and indicates that the increase of M_w is strongly affected by the high amount of phenoxyl radicals generated than the quantity of added substrate. However, continuous feeding of enzyme and substrate leads to a low Pdi. Results for fed batch, carried out with enzyme feeding, is comparable to those obtained for batch with 10 g/L; the enzyme plays a minor role and a low amount is enough to oxidise the tested concentration of SLS.

Substrate adding and enzyme and substrate adding, as a dilute batch system, promotes C-C coupling ("thermodynamic control") and thus a low M_w increase. These results are likely to open new ways to control the enzymatic bioconversion of lignosulfonates. However these assumptions need to be verified by spectroscopy analyses of the formed polymers in order to have a better understanding of the mechanisms of allowing C-C or C-O-C coupling.

Author details

Nidal Madad, Latifa Chebil, Hugues Canteri, Céline Charbonnel and Mohamed Ghoul

*Address all correspondence to: latifa.chebil@ensaia.inpl-nancy.fr

Laboratoire d'Ingénierie des Biomolécules, ENSAIA-INPL, Vandœuvre-lès-Nancy, France

References

[1] Areskogh D, Li J, Gellerstedt G, Henriksson G. 2010. Investigation of the molecular weight increase of commercial lignosulfonates by laccase catalysis. Biomacromolecules 11: 904-910.

[2] Areskogh D, Li J, Nousiainen P, Gellerstedt G, Sipila J, Henriksson G. 2010. Oxidative polymerisation of models for phenolic lignin end-groups by laccase. Holzforschung 64: 21-34.

[3] Benavente-Garcia O, Castillo J, Lorente J, Ortuo A, Del Rio J. A. 2000. Antioxidant activity of phenolics extracted from Olea europaea L. leaves. Food Chemistry 68: 457-462.

[4] Bhattacharya P K, Todi R K, Tiwari M, Bhattacharjee C, Bhattacharjee S, Datta S. 2005. Studies on ultrafiltration of spent sulfite liquor using various membranes for the recovery of lignosulphonates. Desalination 174: 287-297.

[5] Correia P. F. M. M, Ferreira L. M, Reis M. T. A, Carvalho, J. M. R. 2007. A study on the selective recovery of phenol and formaldehyde from phenolic resin plant effluents by liquid-liquid extraction. Solvent Extraction and Ion Exchange 25:485-501.

[6] Dasgupta S, Taylor K. E., Bewtra J. K., Biswas N. 2007. Inactivation of enzyme laccase and role of cosubstrate oxygen in enzymatic removal of phenol from water. Water Environment Research 79: 858-867.

[7] Faix O, Lange W and E.C. Salud 1981. The use of hplc for the determination of average molecular weights and molecular weight distributions of milled wood lignins from Shorea polysperma. Holzforschung 35: 3-9.

[8] Ferreira F. C, Peeva L, Boam A, Zhang S, Livingston A. 2005. Pilot scale application of the Membrane Aromatic Recovery System (MARS) for recovery of phenol from resin production condensates. J Memb Sci 257: 120-133.

[9] Ghosh J. P, Taylor K. E, Bewtra J. K, and Biswas N. 2008. Laccase-catalyzed removal of 2,4-dimethylphenol from synthetic wastewater: Effect of polyethylene glycol and dissolved oxygen. Chemosphere 71: 1709-1717.

[10] Gross S. M, Givens R. D, Jikei M, Royer J. R, Khan S, DeSimone J. M, Odell P. G, and Hamer G. K. 1998. Synthesis and swelling of poly(bisphenol a carbonate) using supercritical CO2. Macromolecules 31: 9090-9092.

[11] Husain Q. 2010. Peroxidase mediated decolorization and remediation of wastewater containing industrial dyes: A review. Reviews Environ Sci Biotechnol 9: 117-140.

[12] Ikeda R, Uyama H, Kobayashi S. 2001. Laccase-catalyzed curing of vinyl polymers bearing a phenol moiety in the side chain. Polymer J 33: 540-542.

[13] Jeon J. R, Kim E. J, Murugesan K, Park H. K, Kim Y. M, Kwon J. H, Kim W. G, Lee J. Y, Chang Y. S. 2010. Laccase-catalysed polymeric dye synthesis from plant-derived phenols for potential application in hair dyeing: Enzymatic colourations driven by homo- or hetero-polymer synthesismbt. Microb Biotechnol 3: 324-335.

[14] Jiang H, Fang Y, Fu Y, Guo Q. X. 2003. Studies on the extraction of phenol in wastewater. J Hazardous Mater 101: 179-190.

[15] Joo H, Yoo Y. J, Dordick J. S. 1998. Polymers from biocatalysts. Korean J Chem Engineer 15: 362-374.

[16] Kim S, Silva C, Zille A, Lopez C, Evtuguin D. V, and Cavaco-Paulo A. 2009. Characterisation of enzymatically oxidised lignosulfonates and their application on lignocellulosic fabrics. Polymer Intern 58: 863-868.

[17] Kobayashi S. 1999. Enzymatic polymerization: a new method of polymer synthesis. Journal of Polymer Science, Part A: Polymer Chemistry 37: 3041-3056.

[18] Kobayashi S, Shoda S. I, Uyama H. 1995. Enzymatic polymerization and oligomerization. Advances in Polymer Sci 121: 1-30.

[19] Kobayashi S, Uyama H. 1998. Enzymatic Polymerization for Synthesis of Polyesters and Polyaromatics. in ACS Symposium Series, Vol. 684, pp. 58-73.

[20] Kobayashi S, Uyama H, Kimura S. 2001. Enzymatic polymerization. Chemical Reviews 101, 3793-3818.

[21] Kurniawati S, Nicell J. A. 2009. A comprehensive kinetic model of laccase-catalyzed oxidation of aqueous phenol. Biotechnol Progress 25: 763-773.

[22] Madad N, Chebil L, Sanchez C, Charbonnel C, and Ghoul M. 2011. Effect of molecular weight distribution on chemical, structural and physicochemical properties of sodium lignosulfonates Rasyan J chem 4: 189-202.

[23] Mita N, Tawaki S. I, Hiroshi U, and Kobayashi S. 2003. Laccase-catalyzed oxidative polymerization of phenols. Macromolecular Bioscience 3: 253-257.

[24] Nazari K, Esmaeili N, Mahmoudi A, Rahimi H, and Moosavi-Movahedi A. 2007. Peroxidative phenol removal from aqueous solutions using activated peroxidase biocatalyst. Enzyme Microb Technol 41: 221-233.

[25] Nugroho Prasetyo E, Kudanga T, Ostergaard L, Rencoret J, Gutierrez A, del Rio J. C, Ignacio Santos J, Nieto L, Jimenez-Barbero J, Martinez A. T, Li J, Gellerstedt G, Lepifre S, Silva C, Kim S Y, Cavaco-Paulo A, Seljebakken Klausen B, Lutnaes B. F, Nyanhongo G. S, and Guebitz G. M. 2010. Polymerization of lignosulfonates by the laccase-HBT (1-hydroxybenzotriazole) system improves dispersibility. Bioresource Technol 10: 5054-5062.

[26] Ouyang X, Qiu X, Chen P. 2006. Physicochemical characterization of calcium lignosulfonate-A potentially useful water reducer. Colloids Surf A 282-283: 489-497.

[27] Reihmann M, and Ritter H. 2006. Synthesis of phenol polymers using peroxidases in Advances in Polymer Science Vol. 194, pp. 1-49.

[28] Riva S. 2006. Laccases: blue enzymes for green chemistry. Trends in Biotechnology 24: 219-226.

[29] Rodrigues S, Pinto G. A. S, and Fernandes F. A. N. 2008. Optimization of ultrasound extraction of phenolic compounds from coconut (Cocos nucifera) shell powder by response surface methodology. Ultrasonics Sonochem 15: 95-100.

[30] Venkateswaran P, Palanivelu K. 2006. Recovery of phenol from aqueous solution by supported liquid membrane using vegetable oils as liquid membrane. J Hazardous Mater 131: 146-152.

[31] Widsten P, Kandelbauer A. 2008. Laccase applications in the forest products industry: A review. Enzyme Microb Technol 42: 293-307.

[32] Wu Y, Taylor K. E, Biswas N, Bewtra, J. K. 1999. Kinetic model-aided reactor design for peroxidase-catalyzed removal of phenol in the presence of polyethylene glycol. J Chem Technol Biotech 74: 519-526.

[33] Yang D, Qiu X, Pang Y, Zhou M. 2008. Physicochemical properties of calcium lignosulfonate with different molecular weights as dispersant in aqueous suspension J. Disper. Sci. Technol. 29: 1296-1303.

[34] Zhou M, Qiu X, Yang D, Lou H. 2006. Properties of different molecular weight sodium lignosulfonate fractions as dispersant of coal-water slurry. J. Dispersion Sci. Technol 27: 851-856.

Biodegradation of Hazardous Contaminants

Chapter 5

Biodegradation of Cyanobacterial Toxins

Sonja Nybom

Additional information is available at the end of the chapter

1. Introduction

Water is an essential natural resource, necessary for drinking, agriculture and industrial activities, and providing the human population with safe drinking water is one of the most important issues in public health. Cyanobacteria produce toxins that may present a hazard for drinking water safety. These toxins are structurally diverse and their effects range from liver damage, including liver cancer, to neurotoxicity. Toxic cyanobacteria have been reported in lakes and reservoirs around the world. The World Health Organization (WHO) has set a provisional drinking water guideline of 1 µg/L for microcystin-LR, one of the most commonly occurring cyanotoxin worldwide [1].

The occurrence of cyanobacteria and their toxins in water bodies used for the production of drinking water causes a technical challenge for water treatment and cleaning. Drinking water should be pure enough to be consumed or used with low risk of immediate or long term harm. The presence of toxins in drinking water creates a potential risk of toxin exposure for water consumers. Conventional water treatment procedures are in some cases insufficient in the removal of cyanobacterial toxins. Besides the chemical and physical methods used, biological degradation could be an efficient method of water detoxification. Therefore there is a need for simple, low-cost and effective water treatment procedures.

This review describes problems related to cyanobacterial toxins and safe drinking water, compares already existing methods of water treatment and cyanotoxin-removal and proposes novel methods of water decontamination. The majority of cyanotoxin-biodegradation studies so far have focused on bacteria isolated from water sources exposed to microcystin-containing blooms. The use of probiotic bacteria is proposed and discussed as a new and efficient means of cyanotoxin-degradation. The removal of cyanobacterial toxins and other environmental contaminants from drinking water is of great importance and probiotic bacteria show promising results in this respect. There is a high demand for effective and low-cost approaches for removing cyanotoxins from potable water due to the significant health risk and inadequate access to safe drinking water.

2. Cyanobacterial toxins

Cyanobacteria have a long evolutionary history and are among the oldest organisms in the world. There is evidence of the organisms even from around 3500 million years ago [2]. Cyanobacteria carry out oxygen-evolving photosynthesis. In eutrophic water, cyanobacteria recurrently form mass occurrences, so-called water blooms. Mass occurrences of cyanobacteria can be toxic. They have caused a number of animal poisonings and may also pose a threat to human health.

Cyanobacteria produce many different classes of biologically active compounds, including hepatotoxic cyclic peptides, microcystins and nodularins, cytotoxic cylindrospermopsins, neurotoxic anatoxin-a and -a(S), saxitoxins, neurotoxic amino acid β-N-methylamino-L-alanine (BMAA) and non-toxic irritating lipopolysaccharides [3]. Although both neurotoxins and hepatotoxins are distributed worldwide [4,5], it appears that hepatotoxic blooms of cyanobacteria are more commonly found than neurotoxic blooms, and neurotoxins are considered to be of lower risk as they are less stable [6]. In contrast, hepatotoxins are highly stable and exposure to these toxins has resulted in significant toxicity to both animals and humans.

Cyanobacteria are ubiquitous in their distribution in both fresh and marine waters. Toxic cyanobacterial blooms have been reported in most parts of the world, reviewed in [7]. Cyanobacterial blooms are a result of the increasing eutrophication in waterbodies [7]. Most of these cyanobacteria are harmful to animals and humans because of their production of toxins. Over the past several centuries, human nutrient over-enrichment in water, particularly nitrogen and phosphorus, associated with urban, agricultural and industrial development, has promoted eutrophication, which favours algal and cyanobacterial bloom formation. Decay of these excessive blooms results in decreased dissolved oxygen and the release of cyanotoxins in the water, which can result in mortality of animals and even humans [7].

2.1. Microcystins

Globally, the most frequently reported cyanobacterial toxins are cyclic heptapeptide hepato-toxins, microcystins (MC). These can be found primarily in some species of the freshwater genera *Microcystis, Anabaena, Planktothrix, Nostoc,* and *Anabaenopsis*. Microcystins are named after *Microcystis aeruginosa*, the cyanobacterium in which the toxin was first isolated and described [8].

Microcystins are cyclic heptapeptides with variable amino acids and a general structure of cyclo(-D-Ala(1)–L-X(2)–D-MeAsp(iso-linkage)(3)–L-Z(4)–Adda(5)–D-Glu(iso-linkage)(6)–Mdha(7), in which amino acid residues at 2 and 4 are variable L-amino acids, D-MeAsp is D-erythro-β-methylaspartic acid, and Mdha is N-methyldehydroalanine, while the amino acid

Adda is (2S,3S,8S,9S)-3-amino-9-methoxy-2,6,8-trimethyl-10-phenyldeca-4(E),6(E)-dienoic acid (Figure 1). The Adda component of microcystins is contributing to their toxicity [4,9]. There are around 100 structural variants of microcystins described in the literature (listed in [3,10,11]). The most widely-distributed [4] and studied microcystin variant is microcystin-LR (MC-LR), with the amino acid residues leucine and arginine in positions 2 and 4, respectively, and a molecular weight of 994. Production of MC-LR is dependent on various factors like strain specificity, genetic differences and metabolic processes required for toxin production [9]. A single bloom can have both toxigenic and non-toxigenic strains within it [7]. The toxins are generally bound to the cell membrane and are released as cells age and die, and under stress. They can also passively leak out of cells or be released by lytic bacteria [4].

MC-LR is hepatotoxic and a potent tumour promoter. The primary target organ of MC-LR is the liver [12,13] although it also affects the kidney, gastrointestinal tract and colon [14]. Microcystins are potent and specific inhibitors of serine/threonine-specific protein phosphatases 1 and 2A [15]. Microcystins are distributed in waterbodies worldwide, and the toxicity on exposure to microcystins has been reported worldwide in fish, animals and humans (reviewed in [16]). The World Health Organization has set a provisional drinking water guideline of 1 μg/L for MC-LR [1]; new edition in [17].

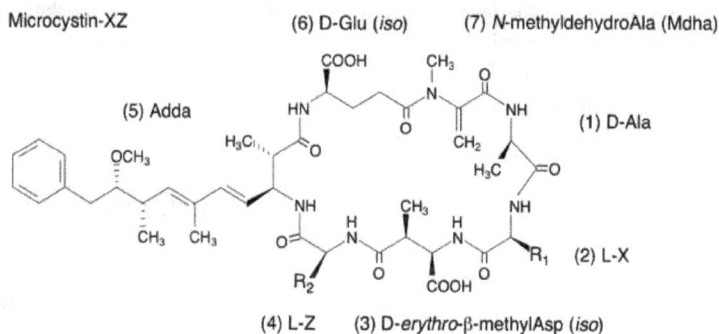

Figure 1. General structure of the hepatotoxic cyclic peptides, microcystins.

2.2. Other cyanobacterial toxins

The cyclic pentapeptide nodularin (NOD) is common in brackish water. It occurs in the Baltic Sea as well as in saline lakes and estuaries. In the Baltic Sea, marine blooms of *Nodularia spumigena* are among some of the largest cyanobacterial mass events in the world. Cylindrospermopsin (CYN), originally isolated from the cyanobacterium *Cylindrospermopsis raciborskii*, is an alkaloid cytotoxin with the structure of a tricyclic guanidine moiety attached to a hydroxymethyluracil [18] and a molecular weight of 415. Cylindrospermopsin inhibits protein synthesis and mainly affects the liver [19], but can also affect the kidney, spleen, thymus, and heart. It is a cyanotoxin occurring in tropical or subtropical regions that has recently been detected also in temperate regions.

Cyanobacterial neurotoxins belong to a diverse group of heterocyclic compounds called alkaloids. Three types of cyanobacterial neurotoxins, anatoxin-a, anatoxin-a(S), and saxitoxins, are known. A mild neurotoxin, BMAA, has been found in a variety of cyanobacteria [20,21]. Anatoxin-a is a small alkaloid with a molecular weight of 165, and it mimics the effect of acetylcholine and causes rapid death by respiratory arrest. Homoanatoxin-a (MW 179) is an anatoxin-a homologue. Anatoxin-a is perhaps the most common cyanobacterial neurotoxin, especially in North America and Europe, and has caused numerous animal poisonings. Anatoxin-a(S) is an irreversible acetylcholine esterase inhibitor and its characteristic signs of poisonings in mice include salivation. Anatoxin-a(S) was first reported in North America where it has caused animal poisonings and later also in Denmark [22].

Saxitoxins, also known as paralytic shellfish poisons (PSP toxins) were originally isolated and characterised from marine dinoflagellates [23]. Saxitoxins are sodium channel blocking agents causing paralysis and have caused human poisonings due to their ability to concentrate in shellfish [23].

Lipopolysaccharide endotoxins are generally found in the outer membrane of the cell wall of Gram-negative bacteria, also in cyanobacteria. Bacterial lipopolysaccharides are pyrogenic and toxic [24]. It is often the fatty acid component of lipopolysaccharides that elicits an irritant, pyrogenic or allergenic response in humans and mammals. Cyanobacterial lipopolysaccharides may contribute to human health problems via exposure to mass occurrences of cyanobacteria.

3. Occurrence and levels of cyanobacteria and hepatotoxins

Toxic cyanobacteria are found worldwide both in inland and coastal water environments. Cyanobacteria occur in various environments including water, such as fresh and brackish water, oceans, hot springs, moist terrestrial environments such as soil, and in symbioses with plants, lichens and primitive animals. Some environmental conditions, including sunlight, warm weather, low turbulence and high nutrient levels, can promote growth. A high density of suspended cells may lead to the formation of surface scums and high toxin concentrations.

The toxins are not actively secreted to the surrounding water; most of the toxin is intracellular in growing cells. The release of toxin occurs during senescence of the cultures and when cultures shift from growth phase to stationary and death phases. Under field conditions, the majority of microcystin is intracellular during active growth of the cells [25]. There are reports of hepatotoxic blooms from all continents around the world [7]. Some of the highest reported cyanotoxin concentrations in bloom samples (measured by HPLC) have been 7300 µg/g dry weight microcystin in a *Microcystis* bloom from China [26], 18000 µg/g dry weight nodularin in a *Nodularia* bloom from the Baltic sea [27] and 5500 µg/g dry weight cylindrospermopsin from Australia [3]. Toxic and non-toxic strains from the same cyanobacterial species cannot be separated by microscopic identification. To confirm that a particular cyanobacterial strain produces toxins, it is important to isolate a culture of that strain, and to detect and quantify toxin concentrations in the pure culture.

4. Human health effects caused by cyanobacterial toxins

Many cyanobacteria produce potent toxins. As reported in literature, problems caused by cyanobacteria are encountered around the world and problems related to safe drinking water production are common (reviewed in e.g. [7]). The human health effects caused by cyanobacterial toxins vary in severity from mild gastroenteritis to severe and sometimes fatal diarrhoea, dysentery and hepatitis. Microcystins, including the most common variant MC-LR, are hepatotoxic and potent tumour promoters. Acute symptoms reported after exposure to microcystin-containing cyanobacteria include gastrointestinal disorders, nausea, vomiting, fever and irritation of the skin, ears, eyes, throat and respiratory tract, abdominal pain, kidney and liver damage. There are several reports of human health effects associated with ingestion of water containing microcystins, with effects ranging from gastroenteritis [28] to liver damage [12] and even death [29,30].

Humans can be exposed to a range of cyanotoxins contained either in cyanobacterial cells or released into the water. The dissolved toxins are stable against low pH and enzymatic degradation and will therefore remain intact within the digestive tract. As microcystins do not readily penetrate the cell membrane [31], they enter the body from the intestine via the organic anion transporting polypeptides [32]. From the blood microcystins are then concentrated in the liver as a result of active uptake by hepatocytes [33]. The toxins are covalently bound to protein phosphatases in the hepatocyte cytosol [34]. Human health problems are often associated with chronic exposure to low microcystin concentrations in inadequately treated drinking water, contaminated food (such as fish, mussels and prawns) or with the consumption of algal supplements contaminated with cyanotoxins. Exposure routes include the oral route, through inhalation, through dermal exposure or the nasal mucosa [35,36].

4.1. Risk assessment

Poisonings caused by cyanotoxins produced during heavy blooms have affected both humans and wild and domestic animals. Both hepatotoxic and neurotoxic poisonings have been associated with mass occurrences of cyanobacteria [7]. Many reported incidents of human health effects have involved inappropriate treatment of water supplies. The health risk caused by cyanotoxin exposure is difficult to quantify, since the actual exposure and resulting effects have not been conclusively determined. The most likely route for human exposure is the oral route via drinking water [37], and from recreational use of lakes and rivers [36].

Due to the growing concern about health effects of cyanotoxins especially via drinking water, WHO has adopted a provisional guideline value of 1.0 µg/L for MC-LR in 1998 [1]. The newest 4th edition to the drinking water guideline was published in 2011 [17]. Assessment of different water treatment procedures has shown that many of the treatment methods result in a reduction of cyanotoxin concentrations to below acutely toxic levels and below the WHO guideline value of 1.0 μg/L MC-LR in drinking water. During a cyanobacterial bloom the treatment procedures may however be insufficient, and also when different water treatment procedures are not used in combination. Therefore it is important to observe the water treatment efficiency during cyanobacterial blooms.

5. Treatment of drinking water containing cyanotoxins

Water is an essential natural resource, necessary for drinking, agriculture and industrial activities. Contamination of water can therefore influence humans, agricultural livestock and irrigated field crops, as well as wildlife drinking the water or living in the aquatic environment. Drinking water should be pure enough to be consumed or used with low risk of immediate or long term harm. In large parts of the world, the population has inadequate access to safe potable water and use sources contaminated with disease vectors, pathogens or unacceptable levels of toxins and other harmful substances.

Prevention of bloom formation is naturally the most efficient method for avoiding cyanobacterial toxin contamination of drinking water. Cyanotoxins are produced within the cyanobacterial cells and thus toxin removal involves procedures to destroy or avoid the cells. Cyanotoxins are also water soluble and therefore chemical or biological procedures reducing the toxicity or completely removing the toxins from the drinking water are needed. If high extracellular toxin concentrations are present in the raw water, problems will occur for drinking water treatment plants. Under natural circumstances high toxin concentrations appear during the breakdown of a cyanobacterial bloom. Cyanobacterial cells are also lysed in the presence of chemicals, such as potassium permanganate or chlorine [38].

In cyanotoxin-removal from drinking water there is a need for knowledge of the physical and chemical properties of the toxin, such as the hydrophobicity, molecular size, and functional groups, the nature of the toxin, i.e., intracellular or extracellular, cyanobacterial growth and bloom patterns, and effective treatment processes [7]. However, these treatments may not be sufficient during cyanobacterial blooms or when a high organic load is present, and toxin levels should therefore be monitored during all steps of water treatment processes. Some of the existing methods of drinking water treatment are shortly described in the following section.

5.1. Water treatment processes

Most drinking water plants use conventional treatment methods that are unable to yield complete removal of microcystins or are too expensive [39]. Conventional surface drinking water treatment utilises coagulation, flocculation, sedimentation, filtration and disinfection as basic methods. However, conventional treatment may need to be optimised for cyanotoxin-removal, relating to the form of the toxin to be removed (intra- or extracellular), the background water matrix, and possible dissolved toxin release during the treatment process [40]. Alternative processes, such as granular activated carbon, powdered activated carbon, and membrane filtration have been proven efficient for the removal of microcystins [41]. However, these methods are sometimes considered too expensive to exclusively remove a contaminant that is irregularly occurring.

Coagulation or flocculation involves the aggregation of smaller particles into larger particles using chemicals, such as ferric chloride or aluminium sulphate. Coagulation can be an efficient method for eliminating cyanobacterial cells from water, but soluble cyanotoxins are not very efficiently removed by this method [42]. Coagulation may also cause additional problems such as lysis of cyanobacterial cells leading to release of toxins. The activated carbon approach uses either powdered activated carbon, which can be added occasionally when there is a need, or granular activated carbon adsorbers, which are used continuously [43]. Both microcystins and cylindrospermopsin can be absorbed by activated carbon [43]. The disposal of the carbon containing cyanobacterial toxins may present a challenge for this type of treatment.

Rapid filtration is a method usually used after a coagulation step in conventional water treatment, but does not effectively remove cyanobacterial cells from water. Conventional water treatment requires regular backwashing of the filters, but if the washing process is inadequately performed, lysis of cyanobacterial cells on the filters can lead to release of toxins into the water [7]. Two types of membrane filtration, microfiltration and ultrafiltration, are commonly used to remove contaminants from drinking water. Both microfiltration and ultrafiltration have been shown to be effective in removal of intact cyanobacterial cells [44].

The most common chemical oxidants used in drinking water treatment are ozone, hydroxyl radicals, chlorine, chlorine dioxide, chloramine and permanganate. Chlorination and ozonation are effective for the removal of microcystins [43]. However, there are concerns regarding the release of toxin when cyanobacteria are chlorinated and with the formation of undesirable chlorination by-products [45]. Ozonation has been shown to be a very effective method for destroying microcystins and nodularins. In recent years, many water treatment plants have included a two-stage ozonation treatment [46].

Removal and inactivation of cyanobacteria and intracellular and extracellular cyanotoxins most often requires a combination of treatment processes or a multiple barrier approach. Furthermore, biological treatment of water is a method used for cyanotoxin-removal from drinking water. Biologically active filtration in the form of river bank filtration and both slow and rapid filtration have been reported to remove or to inactivate microcystins in drinking water (e.g. [47,48]) and are discussed more in detail in the following section.

6. Biodegradation of cyanotoxins

Biodegradation is a chemical disruption of organic materials by microorganisms or other biological agents. Microbial degradation of chemicals in the environment is an important route for the removal of these compounds. Biodegradation is also one of the essential processes for the reduction of microcystins in natural eutrophic lakes and reservoirs. Cyanotoxin-degrading bacteria are distributed all over the world. Of all the cyanotoxin-biodegradation studies, most have focused on microcystins as a consequence of their biodegradability in drinking water sources. This section mainly describes biodegradation studies of microcystins, but studies on nodularin, cylindrospermopsin, saxitoxins and anatoxin-a have also been performed to some extent.

People are frequently exposed to cyanobacterial toxins as well as other microbial contaminants through drinking water. Conventional water treatment procedures discussed in the previous section are in some cases insufficient in the removal of cyanobacterial toxins from drinking water, especially during cyanobacterial blooms. If the cyanobacterial cells are not removed by traditional water treatment methods, the cells and therefore the toxins remain in the drinking water and must be degraded to non-toxic compounds. Since microcystins have been released into the water body, the toxins can persist for weeks [20] before they are adequately degraded by for example bacteria.

6.1. Bacterial degradation of microcystins

Different biological methods have been applied to remove cyanobacteria and their toxins. One type of these methods is the use of microorganisms or biofilms capable of degrading microcystins. Biological treatment for removal of toxin contaminants is becoming more useful as toxins can be removed without the addition of chemicals that may have the potential to produce undesirable by-products. Biodegradation of microcystins in water has been proven to be very effective as they can be used a as carbon source by heterotrophic bacteria [25,38,49,50]. Methods utilizing microcystin-degrading microorganisms can be classified into two groups. One is the use of biofilms grown on the surface of substrates within bioreactors, such as biological sand [48,51,52], biofilm-reactors based on immobilised microorganisms [53], biological treatment facilities combined with conventional treatment processes [54], and granular activated carbon filters [55]. The other group depends on specific microorganisms efficient in microcystin-degradation, such as bacteria of the *Sphingomonas* sp. [56,57] and *Sphingopyxis* sp. [58].

Different variants of microcystins have been demonstrated to be degraded after incubation with water from a lake in Japan, which is frequently contaminated with cyanobacteria [59]. A more effective degradation was observed after adding bed sediment or mud from the lake. Christoffersen *et al.* found out that bacteria can efficiently degrade microcystins in natural waters with previous cyanobacterial contamination and that the degradation process is rapid and without lag phase [60].

Many other studies have also reported biological degradation of microcystin in natural waters from lakes and reservoirs, particularly those containing toxic cyanobacterial blooms [25,50,61,62]. Several strains of the genus *Sphingomonas* have been reported to degrade microcystins [49,57,63–67]. Table 1 lists strains reported to degrade different variants of microcystins and nodularin. A part of the recognised microcystin-degraders so far belonging to the family *Sphingomonadaceae* are closely related and possess homologues of the mlrA gene. Seventeen strains of Gram-negative bacteria with the ability to degrade microcystins were isolated by Lahti *et al.* [47]. Other reported microcystin-degrading bacteria include *Pseudomonas aeruginosa* [68], *Paucibacter toxinivorans* [69] and *Sphingosinicella microcystinivorans* [70]. In a study of Rapala *et al.* thirteen bacteria capable of degrading microcystins and nodularin were isolated from lake sediment [61]. Genomic characterisation of these strains indicated that they formed a single microdiverse species and a novel genus and species (*Paucibacter toxinivorans* gen. nov.,sp. nov.) was proposed. A bacterium isolated from water samples in Brazil showed high homology with the *Burkholderia* genus, belonging to the beta subdivision of proteobacteria [71], which was the first reported bacterium from the genus *Burkholderia* as a cyanobacterial toxin degrader.

Bacterial strain	Degradable toxins	Reference
Arthrobacter sp.	MC-LR	[72]
Bacillus sp. strain EMB	MC-LR, MC-RR	[74]
Brevibacterium sp.	MC-LR	[72]
Burkholderia sp.	MC-LR, [D-Leu¹]MC-LR	[71]
Lactobacillus rhamnosus GG and LC-705, *Bifidobacterium longum* 46	MC-LR, MC-RR, MC-YR, MC-LF, MC-LY, MC-LW	[79,80]
Methylobacillus sp. strain J10	MC-LR, MC-RR	[75]
Microbacterium sp.	MC-LR	[78]
Morganella morganii	MC-LR	[77]
Paucibacter toxinivorans sp. nov.	MC-LR, MC-YR, NOD	[69]
Poterioochromonas sp.	MC-LR	[81]
Pseudomonas aeruginosa	MC-LR	[68]
Rhizobium gallicum	MC-LR	[78]
Rhodococcus sp.	MC-LR	[72]
Sphingomona stygia	MC-LR, MC-RR, MC-YR	[65]
Sphingomonas sp. 7CY	MC-LR, MC-RR, MC-LY, MC-LW, MC-LF	[66]
Sphingomonas sp. ACM-3962	MC-LR, MC-RR	[25,63,82]
Sphingomonas sp. B9	MC-LR, MC-RR, 3-dmMC-LR, dhMC-LR, MC-LR-Cys, NOD	[67,83]
Sphingomonas sp. CBA4	MC-RR	[57]
Sphingomonas sp. MD-1	MC-LR, MC-RR, MC-YR	[56]
Sphingomonas sp. MDB2	MCs	[70]
Sphingomonas sp. MDB3	MCs	[70]
Sphingomonas sp. MJ-PV	MC-LR	[49]
Sphingomonas sp. Y2	MC-LR, MC-RR, MC-YR, 6(Z)-Adda-MC-LR	[64,70,84]
Sphingopyxis sp. C-1	MC-LR	[58]
Sphingopyxis sp. LH21	MC-LR, MC-LA	[52]
Sphingopyxis sp. USTB-05	MC-RR	[85]
Stenotrophomonas sp. strain EMS	MC-LR, MC-RR	[76]
17 different strains (Gram-negative, Proteobacteria)	MCs	[47]

Table 1. Reported microcystin-degrading bacteria

Recently, Gram-positive bacteria isolated from freshwater, belonging to *Actinobacteria* and identified as *Arthrobacter* sp., *Brevibacterium* sp. and *Rhodococcus* sp., were shown to remove MC-LR [72]. The mechanism of MC-LR removal for *Rhodococcus* sp. C1 [73] was shown to be similar to the previously reported degradation pathway for *Sphingomonas* by Bourne *et al.* [63]. A new strain AMRI-03 with close relationship to the genus *Bacillus* was isolated from a Saudi freshwater lake [74]. Another strain J10 isolated from Lake Taihu in China was identified as *Methylobacillus* sp. [75]. An EMS strain similar to *Stenotrophomonas maltophilia* was described by Chen *et al.* and was the first report of microcystin-degrading bacteria carrying the mlrA gene in the genus of the gamma division of proteobacteria [76]. Other reported examples of bacteria with such ability are *Morganella morganii* and *Pseudomonas* sp. [77]. Further recent findings include two isolates from Lake Okeechobee, Florida, capable of microcystin-degradation and classified as *Rhizobium gallicum* and *Microbacterium* sp. [78].

6.2. Enzymatic mechanisms of microcystin-biodegradation

The first proposal of microcystin-biodegradation suggested a proteolytic mechanism [63]. Within the genome of the first isolated microcystin-degrading bacterium, *Sphingomonas* sp. ACM-3962, Bourne *et al.* identified a gene cluster, mlrA, mlrB, mlrC and mlrD, responsible for the degradation of MC-LR [63,82]. Based on MS-analysis a linear MC-LR (protonated molecular ion at m/z 1013) and a tetrapeptide (protonated molecular ion at m/z 615) were recognised as the degradation products. The microcystin-degradation pathway was described as a linear, three-step process. It was suggested that the mlrA gene encoded an enzyme responsible for the hydrolytic cleaving of the cyclic structure of MC-LR (ring-opening at the Adda-Arg peptide bond). The resulting linear MC-LR molecule was then sequentially hydrolysed by peptidases encoded by the mlrB and mlrC genes to a tetrapeptide, and further to smaller peptides and amino acids (Figure 2). The final gene, mlrD, encoded for a possible transporter protein that may have allowed for active transport of microcystin or its degradation products. The genes mlrA, mlrB and mlrC encode a 336-residue metalloendopeptidase (responsible for linearization of microcystins), a serine protease and a metalloprotease, respectively. Further studies have confirmed the existence of the mlr cluster components also in other microcystin-degrading bacteria; Ho *et al.* identified homologues of four mlr genes in *Sphingopyxis* sp. LH21 [52]. Similarly, a homologous gene cluster was also detected in *Sphingopyxis* sp. C-1 [58].

However, as has been recently indicated, mlrC acts not only on the tetrapeptide but is also able to hydrolyze linear microcystin without earlier processing by mlrB [86]. Other products of microcystin-degradation have consequently been documented, but the complete fate of microcystin-derivatives is still unknown [83,87]. Additionally, enzymes other than proteases have been suggested to be involved in microcystin-utilisation, and besides typical proteolytic activity, also decarboxylation and demethylation have been proposed as alternative mechanisms [87].

Various studies have designed qualitative polymerase chain reaction assays for detection of mlrA [51,52,56]. Saito *et al.* reported gene homologues of mlrA in two microcystin-degrading bacteria, *Sphingomonas* sp. MD-1 and *Sphingomonas* sp. Y2, both of which were previously isolated from Japanese lakes [56]. More recently, Hoefel *et al.* designed and optimised a quantitative real-time polymerase chain reaction assay for the detection of the mlrA gene [88].

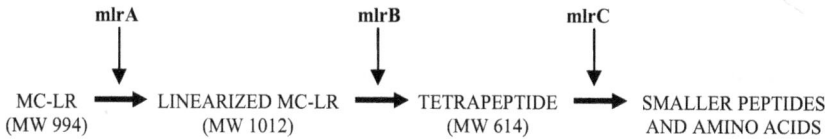

Figure 2. MC-LR degradation pathway by *Sphingomonas* sp. ACM-3962; mlrA-C: microcystinases A-C (modified from [63]).

6.3. Further aspects of microcystin-biodegradation

Before biological treatment can be considered a feasible option for effective removal of microcystins, there is a need to determine if any toxic biodegradation by-products are generated. Different studies have demonstrated that the biodegradation of microcystins does not yield toxic by-products. Bourne *et al.* [63] and Harada *et al.* [67] identified two intermediate products from the bacterial degradation of MC-LR by *Sphingomonas* sp. ACM-3962 and *Sphingomonas* sp. B9, respectively. Both studies identified linearized MC-LR and a tetrapeptide as the intermediate products, and isolated Adda as one of the final degradation products (Figure 2). Both these intermediate products were less active than the parent MC-LR. Studies with *Sphingpoyxis* sp. LH21 in treated reservoir water concluded that the decrease in cytotoxicity indicated that no cytotoxic by-products of microcystins were being generated [52].

Different factors may influence the biodegradation efficiency, such as water temperature. Published results suggest that the temperature range for the effective biodegradation of microcystins is between 11 and 37 °C, with more rapid degradation at the higher temperatures in most cases [52,64,79,88]. In addition, the bacterial composition and cell density within the water body also affects degradation; both the types of organisms present and their concentration.

Only few studies with respect to the biodegradation of a range of cyanobacterial metabolites in water bodies have been performed. This is relevant since multiple classes of cyanobacterial metabolites are often simultaneously present in water bodies. The following sections regarding the removal of cyanotoxins by probiotic bacteria will assess this issue, and results regarding the removal of a range of cyanotoxins are presented. The results on a range of bacterial species demonstrate the feasibility of biodegradation as a possible removal option for microcystins. The most important practical use of microbial aggregates, such as biological filters and biofilm, is in biological wastewater treatment, and some new technologies already utilize bacterial aggregates for degradation [89].

7. Probiotic bacteria involved in cyanotoxin-removal

Probiotics were earlier defined as "live microbial food supplements which beneficially affect the host either directly or indirectly by improving its intestinal microbial balance" [90]. Today, the most commonly accepted definition by WHO states that probiotics are "live microbial food supplements which, when given in adequate amounts have a demonstrated beneficial effect on human health" [91]. In order to be effective the probiotic micro-organisms must be able to survive the digestive conditions, including bile acids, and they must be able to colonise the gastrointestinal tract at least temporarily without any harm to the host [92]. Only certain strains of micro-organisms have these properties. Most probiotic micro-organisms are grouped in two bacterial genera, *Lactobacillus* (*L.*) and *Bifidobacterium* (*B.*).

The main site of action for the health benefits of probiotic bacteria is the gut. The intestinal mucosa forms a barrier between the external and internal environment of the human body. There are several important modes of action for probiotic bacteria, including modification of gut pH, colonisation ability, inhibition of the colonisation, adhesion and invasion of pathogens, direct antimicrobial effect, replacement of already adhered pathogens, competing for available nutrients and growth factors, regulation of the immune system of the host, normalisation of the gut microbiota, and different metabolic effects (reviewed in [93,94]). It is therefore believed that by adding these bacteria as probiotics to the diet, the normal microbiota can be altered. Many probiotic organisms originate in fermented foods, and they have a long history of safe use in human consumption. Lactobacilli and bifidobacteria common in the food industry belong to the European Qualified Presumption of Safety (QPS) status organisms, which can be used in foods and feeds [95].

7.1. Efficiency of probiotic strains in microcystin-removal

Recently published studies have reported efficient cyanotoxin-removal by several strains of probiotic bacteria [79,80,96,97]. The aim of these studies was to characterise the potential of probiotic lactic acid bacteria and bifidobacteria in removal of microcystins and cylindrospermopsin from aqueous solutions. Different physiological conditions possibly affecting the removal efficiency were studied and the mechanism of toxin removal was investigated.

In an initial screening study, 15 different strains of probiotic lactic acid bacteria and bifidobacteria were tested for their MC-LR removal capacities and evaluated for their potential in water decontamination [79]. The results showed a reproducible reduction of MC-LR in solution by the majority of the tested bacterial strains; the most efficient removal was achieved with *L. rhamnosus* strains GG and LC-705, *B. lactis* strains 420 and Bb12 and *B. longum* 46 [79]. The removal of MC-LR continued during the entire 24-hour incubation, which indicates that the removal process is quite slow. The effect of pH during incubation was also studied. pH was found to have an influence on toxin removal, with a higher removal percentage observed at neutral pH than at pH 3 [79]. It was also shown that viable bacteria were more efficient in microcystin-removal than non-viable bacteria [80]. Further studies showed that several strains were efficient in microcystin-removal and that different physiological conditions, including the effect of pH, temperature, toxin concentration, bacterial cell density and cell viability, had an effect on the removal efficiency [80].

The removal of MC-LR was shown to be temperature dependent, with the highest removal observed at 37 °C for all studied strains. At 4 °C, practically no removal of MC-LR could be observed and the removal percentages increased with increasing temperature [79]. This can be explained by the fact that at 4 °C, the bacterial cells are metabolically inactive, but at 22 and 37 °C, the bacteria become metabolically active, which is required for enzymatic activity. In addition, the role of glucose in activating the metabolism of the probiotic bacteria was assessed [96]. Since it was shown that viability is a requirement for efficient toxin removal, glucose was added as a source of nutrient to the bacterial solutions to enhance the bacterial viability. Glucose addition improved the removal efficiencies of all tested strains by enhancing both the removal rate and the amount of MC-LR removed after 24 hours of incubation [96]. Supplementation of glucose provides energy to the bacteria, and thereby, the microcystin-removal efficiencies also increase.

To investigate the role of the probiotic bacterial cell density, a range of bacterial cell densities were screened and tested for their microcystin-removal efficiencies [79]. The removal of MC-LR was shown to be dependent on the bacterial cell density, with a minimum of approximately 10^9 CFU/mL required for significant MC-LR removal [79]. The removal of MC-LR was further enhanced with increasing bacterial cell density. To assess whether a combination of several probiotic strains could enhance their microcystin-removal efficiencies the microcystin-removal of three probiotic strains (*L. rhamnosus* GG, *L. rhamnosus* LC-705 and *B. longum* 46) separately and in combination was studied [80]. With the probiotic mixture, microcystin-removal percentages of up to 90% could be observed and the results showed that the removal efficiency was improved with a mixture of the strains and compared to the individual strains [80].

In addition to MC-LR, probiotic bacterial strains were also incubated with other microcystins, including MC-RR, -YR, -LY, -LW and -LF. The results of the study show that probiotic strains were effective in the elimination of several different microcystins from solution [79]. Simultaneous removal of several toxins present in cyanobacterial extracts was also investigated. The time course for the removal of microcystins present in the cyanobacterial extracts *Microcystis* NIES-107 and *Microcystis* PCC 7820 by the probiotic strain *L. rhamnosus* GG is shown in Figures 3 a and b, respectively. The removal of all studied microcystins increased over time. The removal of the microcystins present in *Microcystis* NIES-107 after 24 hours of incubation was around 65–85% of total microcystin and for microcystins present in *Microcystis* PCC 7820 around 60–80%. The toxin-removal was thus shown to be efficient also when several different microcystins were present in the solution. This indicates that there is no competition taking place among the toxins during incubation with probiotic bacteria. In addition, the strains were shown to remove the cytotoxin cylindrospermopsin from aqueous solutions; the removal was somewhat less efficient, around 30% for all tested strains [80].

Probiotic bacteria have several advantages in comparison with the previously reported microcystin-degrading bacteria, as they have been classified as food grade, safe bacteria by the European Food Safety Authority (EFSA) [95]. Therefore probiotic bacteria can safely be included in both food and water. Previous studies have also shown the effect of probiotic bacteria in the removal of other environmental contaminants, such as heavy metals [98] and mycotoxins including aflatoxins and ochratoxins [99].

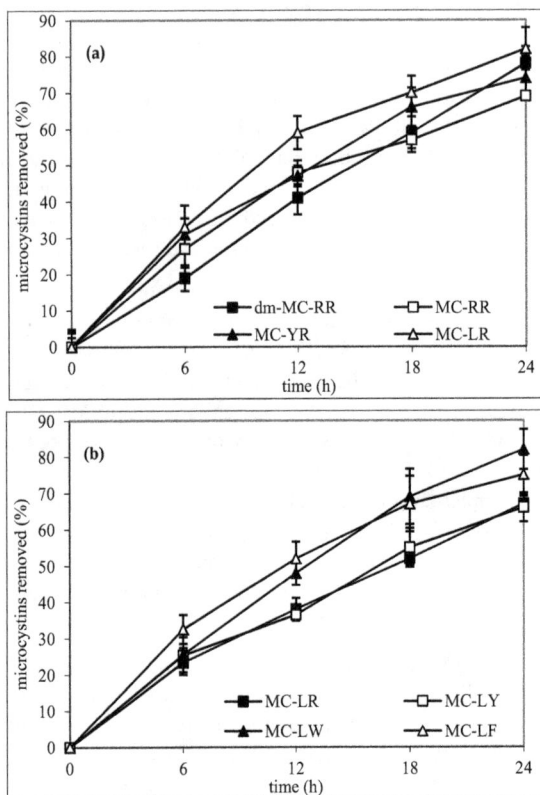

Figure 3. Removal of microcystins in cyanobacterial extracts by probiotic strain *Lactobacillus rhamnosus* GG (a) *Microcystis* NIES-107 and (b) *Microcystis* PCC 7820. Initial concentration of microcystins in extracts: 20-100 μg/L, bacterial concentration 10^{10} CFU/mL, temperature 37 °C, average ± SD, n = 3 (modified from [80]).

7.2. Mechanisms of microcystin-degradation by probiotic bacteria

As specific probiotic bacterial strains were shown to be efficient in microcystin-removal, the subsequent aim was to identify and specify the removal mechanisms. The location and mechanism of microcystin-removal were investigated by studying a possible extracellular enzymatic degradation of microcystins [97]. Furthermore, a comparison of the degradation pathways of previously identified microcystin-degrading bacteria with probiotic bacteria was performed.

The participation of cell-envelope proteinases in microcystin-removal was investigated. Following standard peptidase assay no proteolytic activity was found in the supernatants of the bacterial cell cultures of the investigated strains; enzymatic activity was found only in the cell suspensions. The activity of cell-associated proteinases of probiotic strain *L. rhamnosus* GG was measured after incubation with protease inhibitors. The protein inhibitor EDTA was shown to inhibit MC-LR removal [97]. The results suggest that the main proteolytic activity observed for the strain was due to metallo-enzymes. A possible extracellular enzymatic degradation of microcystins by probiotic bacteria was therefore investigated and it was suggested that extracellularly located cell-envelope proteinases appear to be involved in the decomposition of MC-LR [97]. A correlation between proteinase activity and MC-LR removal was also found when these parameters were simultaneously measured. The correlation between the activity of cell-envelope proteinases and the decrease of MC-LR concentration suggests that enzymes are involved in microcystin-removal [97]. The findings support the theory that enzymatic degradation of microcystins occurs when the toxin is incubated with probiotic bacteria, but the exact mechanism still remains unidentified.

Bacterial degradation of microcystins has previously been reported for strains of *Sphingomonas* and the degradation products and patterns have been determined for strains ACM-3962 and B9 [63,66,67,82]. For possible identification of toxin removal by probiotic bacteria, the removal process of MC-LR for *L. rhamnosus* GG was compared with the two *Sphingomonas* strains, and the degradation products were identified [97]. Linearized MC-LR and the tetrapeptide were observed for the two *Sphingomonas* strains, but these degradation products were not obtained using the probiotic strain, suggesting that the removal mechanisms between the strains differ [97]. Furthermore, no additional degradation products could be identified from samples incubated with the probiotic strain, which suggested that microcystin is rapidly degraded to smaller peptides and amino acids. Further studies are needed to identify possible degradation products and the precise steps of the degradation mechanism by probiotic bacteria.

8. Discussion and conclusions

The majority of cyanotoxin-biodegradation studies have focused on bacteria isolated from water sources exposed to microcystin-containing blooms. As described in this review, it is clear that many of the cyanobacterial metabolites are susceptible to biodegradation in water supplies. Currently an increasing focus on bacterial degradation of hepatotoxic cyanobacterial peptides is being observed [56,59,64,66,76]. Previous studies have demonstrated that the ability of bacteria to degrade microcystins is related to the presence of the gene mlrA that encodes a hydrolytic enzyme with specificity to the toxins. The potency to utilize these bacteria in microcystin-degradation has also been demonstrated in laboratory scale [51,54, 100]. Recently, a new type of bacteria, specific probiotic bacterial strains, was presented to be efficient in cyanotoxin-removal. Probiotic bacteria have several advantages in comparison with the previously reported microcystin-degrading bacteria, as they have been classified as food grade, safe bacteria by the EFSA [95]. Therefore probiotic bacteria can safely be included in both food and water, and can also safely be used in food technology. Furthermore, the beneficial health effects of probiotic bacteria give them an advantage for the use in different applications. A potential area of use could be probiotic dietary supplements used as a personal defense mechanism against cyanotoxins in the gastrointestinal tract when ingested through contaminated drinking water and to reduce the health risks caused by microcystins, as well as applications in biological decontamination of microcystin-containing water.

Several reports have showed that biological degradation of cyanotoxins may be a feasible method of water treatment. The bacterial strains or possible enzymes identified in the removal process could be used in a degradation process to remove toxins from drinking water. Technologies using potential purified enzymes identified in the removal process of bacteria could be a future approach for efficient cyanotoxin-removal. Today, the best way for cyanotoxin biodegradation is the use of biofilters with immobilised micro-organisms, as most water treatment processes already employ a filtration step. Also the removal of other cyanobacterial toxins, such as anatoxins, saxitoxins, and cylindrospermopsin, should be taken into account. In conclusion, the development of new water treatment technologies using efficient bacteria that would be able to remove or inactivate cyanotoxins, as well as other types of environmental contaminants, such as heavy metals, viruses and pathogenic bacteria found in drinking water, is an important aspect to consider in the future.

Acknowledgements

Dr. Jussi Meriluoto and Prof. Seppo Salminen are gratefully acknowledged for excellent supervision and guidance during the research work. Svenska litteratursällskapet i Finland r.f. is acknowledged for financial support.

Author details

Sonja Nybom

Department of biosciences, Biochemistry, Åbo Akademi University, Finland

References

[1] WHO. Cyanobacterial toxins: Microcystin-LR. In: Guidelines for drinking-water quality, Addendum to Volume 2, World Health Organization, Geneva; 1998. p95–110.

[2] Schopf JW. Microfossils of the Early Archean Apex chert: new evidence of the antiquity of life. Science 1993;260: 640–646.

[3] Sivonen K., Jones G. Cyanobacterial toxins. In: Chorus I., Bartram J. (eds) Toxic Cyanobacteria in Water: a Guide to Public Health Significance, Monitoring and Management. London: F & FN Spon; 1999. p41–111.

[4] Carmichael WW. Cyanobacteria secondary metabolites—the cyanotoxins. Journal of Applied Bacteriology 1992;72: 445–459.

[5] Carmichael WW. The toxins of cyanobacteria. Scientific American 1994;270: 78–86.

[6] Fawell J., James C., James H. Toxins from blue-green algae: Toxicological assessment of microcystin-LR and a method for its determination in water. Medmenham: Water Research Centre; 1994. p1–46.

[7] Chorus I., Bartram J. Toxic Cyanobacteria in Water: A guide to their public health consequences, monitoring and management. London: F & FN Spon; 1999.

[8] Carmichael WW, Beasley VR, Bunner DL, Eloff JN, Falconer I, Gorham P, Harada KI, Krishnamurthy T, Yu MJ, Moore RE, Rinehart K, Runnegar M, Skulberg OM, Watanabe M. Naming of cyclic heptapeptide toxins of cyanobacteria (blue-green algae). Toxicon 1988;26: 971–973.

[9] Ressom R., Soong F.S., Fitzgerald J., Turczynowicz L., El Saadi O., Roder D., Maynard T., Falconer I. Health effects of toxic cyanobacteria (blue-green algae). Canberra: National Health and Medical Research Council, Australian Government Publishing Service; 1994.

[10] Spoof L.
High-performance liquid chromatography
of microcystins and nodularins, cyanobacterial
peptide toxins. PhD thesis. Åbo Akademi University; 2004.

[11] Neffling MR. Fast LC-MS detection of cyanobacterial peptide hepatotoxins – method development for determination of total contamination levels in biological materials. PhD thesis. Åbo Akademi University; 2010.

[12] Falconer IR, Beresford AM, Runnegar MT. Evidence of liver damage by toxin from a bloom of the blue-green alga, *Microcystis aeruginosa*. Medical Journal of Australia 1983;1: 511–514.

[13] Meriluoto JA, Nygård SE, Dahlem AM, Eriksson JE. Synthesis, organotropism and hepatocellular uptake of two tritium-labeled epimers of dihydromicrocystin-LR, a cyanobacterial peptide toxin analog. Toxicon 1990;28: 1439–1446.

[14] Bell SG, Codd GA. Cyanobacterial toxins and human health. Reviews in Medical Microbiology 1994;5: 256–264.

[15] MacKintosh C, Beattie KA, Klumpp S, Cohen, P Codd GA. Cyanobacterial microcystin-LR is a potent and specific inhibitor of protein phosphatases 1 and 2A from both mammals and higher plants. FEBS Letters 1990;264: 187–192.

[16] Chorus I. Cyanotoxins: occurrence, causes, consequences. Heidelberg: Springer Verlag; 2001.

[17] WHO. Guidelines for Drinking Water Quality, 4th edition. World Health Organization, Geneva; 2011.

[18] Ohtani I, Moore RE, Runnegar MTC. Cylindrospermopsin: A potent hepatotoxin from the blue-green alga *Cylindrospermopsis raciborskii*. Journal of the American Chemical Society 1992;114: 7941–7942.

[19] Terao K, Ohmori S, Igarashi K, Ohtani I, Watanabe MF, Harada KI, Ito E, Watanabe M. Electron microscopic studies on experimental poisoning in mice induced by cylindrospermopsin isolated from blue-green alga *Umezakia natans*. Toxicon 1994;32: 833–843.

[20] Cox PA, Banack SA, Murch SJ, Rasmussen U, Tien G, Bidigare RR, Metcalf JS, Morrison LF, Codd GA, Bergman B. Diverse taxa of cyanobacteria produce beta-N-methylamino-L-alanine, a neurotoxic amino acid. Proceedings of the National Academy of Sciences 2005;102: 5074–5078.

[21] Metcalf JS, Banack SA, Lindsay J, Morrison LF, Cox PA, Codd GA. Co-occurrence of beta-N-methylamino-L-alanine, a neurotoxic amino acid with other cyanobacterial toxins in British waterbodies, 1990-2004. Environmental Microbiology 2008;10: 702–708.

[22] Onodera H, Oshima Y, Henriksen P, Yasumoto T. Confirmation of anatoxin-a(S) in the cyanobacterium *Anabaena lemmermannii* as the cause of bird kills in Danish lakes. Toxicon 1997;35: 1645–1648.

[23] Anderson DM. Red tides. Scientific American 1994;August, 52–58.

[24] Weckesser J, Drews G. Lipopolysaccharides of photosynthetic prokaryotes. Annual Review of Microbiology 1979;33: 215–239.

[25] Jones GJ, Orr PT. Release and degradation of microcystin following algicide treatment of a *Microcystis aeruginosa* bloom in a recreational lake, as determined by HPLC and protein phosphatase inhibition assay. Water Research 1994;28: 871–876.

[26] Zhang QX, Carmichael WW, Yu MJ, Li SH. Cyclic peptide hepatotoxins from freshwater cyanobacterial (blue-green algae) waterblooms collected in Central China. Environmental Toxicology and Chemistry 1991;10: 313–321.

[27] Kononen K., Sivonen K., Lehtimäki J. Toxicity of the phytoplankton blooms in the Gulf of Finland and Gulf of Bothnia, Baltic Sea. In: Smayda T.J., Shimizu Y. (eds) Toxic Phytoplankton Blooms in the Sea. Amsterdam: Elsevier; 1993. p269–274.

[28] Annadotter H, Cronberg G, Lawton LA, Hansson HB, Göthe U, Skulberg O. An extensive outbreak of gastroenteritis associated with the toxic cyanobacterium *Planktothrix (Oscillatoria) agardhii* (Oscillatoriales, Cyanophyceae) in Scania, South Sweden. In: Chorus I. (ed) Cyanotoxins: occurrence, causes, consequences. Springer Verlag, Heidelberg, Germany; 2001. p200–208.

[29] Teixeira MG, Costa MC, de Carvalho VL, Pereira MS, Hage E.. Gastroenteritis epidemic in the area of the Itaparica Dam, Bahia, Brazil. Bulletin of the Pan American Health Organization 1993;27: 244–253.

[30] Jochimsen EM, Carmichael WW, An JS, Cardo DM, Cookson ST, Holmes CE, Antunes MB, de Melo Filho DA, Lyra TM, Barreto VS, Azevedo SM, Jarvis WR. Liver failure and death after exposure to microcystins at a hemodialysis center in Brazil. New England Journal of Medicine 1998;338: 873–878.

[31] Eriksson JE, Grönberg L, Nygård S, Slotte JP, Meriluoto JAO. Hepatocellular uptake of 3H-dihydromicrocystin-LR, a cyclic peptide toxin. Biochimica et Biophysica Acta 1990;1025: 60–66.

[32] Feurstein D, Holst K, Fischer A, Dietrich DR. Oatp-associated uptake and toxicity of microcystins in primary murine whole brain cells. Toxicology and Applied Pharmacology 2009;234: 247–255.

[33] Runnegar MTC, Falconer IR, Silver J. Deformation of isolated rat hepatocytes by a peptide hepatotoxin from the blue-green alga *Microcystis aeruginosa*. Naunyn-Schmiedeberg's Archives of Pharmacology 1981;317: 268–272.

[34] Bagu JR, Sykes BD, Craig MM, Holmes CF. A molecular basis for different interactions of marine toxins with protein phosphatase-1. Molecular models for bound motuporin, microcystins, okadaic acid, and calyculin A. Journal of Biological Chemistry 1997;272: 5087–5097.

[35] Gupta S. Cyanobacterial toxins: Microcystin-LR. In: Guidelines for Drinking-water Quality, Addendum to vol. 2, Health Criteria and Other Supporting Information. Geneva: World Health Organization; 1998. p95–110.

[36] Pilotto L, Douglas R, Burch M, Cameron S, Beers M, Rouch G, Robinson P, Kirk M, Cowie C, Hardiman S, Moore C, Attewell R. Health effects of exposure to cyanobacteria (blue-green algae) during recreational water-related activities. Australian and New Zealand Journal of Public Health 1997;21: 562–566.

[37] Falconer I. An overview of problems caused by toxic blue-green algae (cyanobacteria) in drinking water. Environmental Toxicology 1999;14: 5–12.

[38] Kenefick SL, Hrudey SE, Peterson HG, Prepas EE. Toxin release from *Microcystis aeruginosa* after chemical treatment. Water Science and Technology 1993;27: 433–440.

[39] Westrick JA, Szlag DC, Southwell BJ, Sinclair J. A review of cyanobacteria and cyano-toxins removal/inactivation in drinking water treatment. Analytical and Bioanalytical Chemistry 2010;397: 1705–1714.

[40] Falconer IR. Cyanobacterial toxins of drinking water supplies: cylindrospermopsins and microcystins. Boca Raton: CRC Press; 2005.

[41] Svrcek C, Smith DW. Cyanobacteria toxins and the current state of knowledge on water treatment options: a review. Journal of Environmental Engineering and Science 2004;3: 155–185.

[42] Rositano J., Nicholson B. Water treatment techniques for the removal of cyanobacterial toxins from water 2/94. Australian Centre for Water Quality Research; 1994.

[43] Newcombe G. Removal of algal toxins from drinking water using ozone and GAC. AWWA Research Foundation Report, American Water Works Association: Denver; 2002.

[44] Gijsbertsen-Abrahamse AJ, Schmidt W, Chorus I, Heijman SGJ. Removal of cyano-toxins by ultrafiltration and nanofiltration. Journal of Membrane Science 2006;276: 252–259.

[45] Tsuji K, Watanuki T, Kondo F, Watanabe MF, Nakazawa H, Suzuki M, Uchida H, Harada KI. Stability of microcystins from cyanobacteria – IV. Effect of chlorination on decomposition. Toxicon 1997;35, 1033–1041.

[46] Westrick JA. Cyanobacterial toxin removal in drinking water treatment processes and recreational waters. Advances in Experimental Medicine and Biology 2008;619: 275–290.

[47] Lahti K., Niemi M.R., Rapala J., Sivonen K. Biodegradation of cyanobacterial hepato-toxins-charaterisation of toxin degrading bacteria. In: Reguera B., Blanco J., Fernán-dez M.L., Wyatt T. (eds) Harmful Algae. Proceedings of the VIII International Conference of Harmful Algae. Vigo, Spain; 1998. p363–365.

[48] Bourne DG, Blakeley RL, Riddles P, Jones GJ. Biodegradation of the cyanobacterial toxin microcystin LR in natural water and biologically active slow sand filters. Water Research 2006;40: 1294–1302.

[49] Jones GJ, Bourne DG, Blakeley RL, Doelle H. Degradation of the cyanobacterial hepa-totoxin microcystin by aquatic bacteria. Natural Toxins 1994;2: 228–235.

[50] Cousins IT, Bealing DJ, James HA, Sutton A. Biodegradation of microcystin-LR by in-digenous mixed bacterial populations. Water Research 1996;30: 481–485.

[51] Ho L, Meyn T, Keegan A, Hoefel D, Brookes J, Saint CP, Newcombe G. Bacterial deg-radation of microcystin toxins within a biologically active sand filter. Water Research 2006;40: 768–774.

[52] Ho L, Hoefel D, Saint CP, Newcombe G. Isolation and identification of a novel micro-cystin-degrading bacterium from a biological sand filter. Water Research 2007;41: 4685–4695.

[53] Tsuji K, Asakawa M, Anzai Y, Sumino T, Harada K. Degradation of microcystins us-ing immobilized microorganism isolated in an eutrophic lake. Chemosphere 2006;65: 117–124.

[54] Saitou T, Sugiura N, Itayama T, Inamori Y, Matsumura M. Degradation of microcys-tin by biofilm in practical treatment facility. Water Science and Technology 2002;46: 237–244.

[55] Wang H, Ho L, Lewis D, Brookes JD, Newcombe G. Discriminating and assessing ad-sorption and biodegradation removal mechanisms during granular activated carbon filtration of microcystin toxins. Water Research 2007;41: 4262–4270.

[56] Saito T, Okano K, Park HD, Itayama T, Inamori Y, Neilan BA, Burns BP, Sugiura N. Detection and sequencing of the microcystin LR-degrading gene, mlrA, from new bacteria isolated from Japanese lakes. FEMS Microbiology Letters 2003;229: 271–276.

[57] Valeria AM, Ricardo EJ, Stephan P, Alberto WD. Degradation of microcystin-RR by Sphingomonas sp., CBA4 isolated from San Roque reservoir (Córdoba–Argentina). Bi-odegradation 2006;17: 447–455.

[58] Okano K, Shimizu K, Kawauchi Y, Maseda H, Utsumi M, Zhang Z, Neilan BA, Su-giura N. Characteristics of a microcystin-degrading bacterium under alkaline envi-ronmental conditions. Journal of Toxicology 2009;954291, 8 pp.

[59] Ishii H, Abe T. Release and biodegradation of microcystins in blue-green algae, Mi-crocystis PCC7820. Journal of the School of Marine Science and Technology Tokai University 2000;49: 143–157.

[60] Christoffersen K, Lyck S, Winding A. Microbial activity and bacterial community structure during degradation of microcystins. Aquatic Microbial Ecology 2002;27: 125–136.

[61] Rapala J, Lahti K, Sivonen K, Niemelä SI. Biodegradability and adsorption on lake sediments of cyanobacterial hepatotoxins and anatoxin-a. Letters in Applied Microbi-ology 1994;19: 423–428.

[62] Tsuji K, Setsuda S, Watanuki T, Kondo F, Nakazawa H, Suzuki M, Harada KI. Micro-cystin levels during 1992–95 for lakes Sagami and Tsukui Japan. Natural Toxins 1996;4: 189–194.

[63] Bourne DG, Jones GJ, Blakeley RL, Jones A, Negri AP, Riddles P. Enzymatic pathway for the bacterial degradation of the cyanobacterial cyclic peptide toxin microcystin LR. Applied and Environmental Microbiology 1996;62: 4086–4094.

[64] Park HD, Sasaki Y, Maruyama T, Yanagisawa E, Hiraishi A, Kato K. Degradation of the cyanobacterial hepatotoxin microcystin by a new bacterium isolated from a hypertrophic lake. Environmental Toxicology 2001;16: 337–343.

[65] Saitou T, Sugiura N, Itayama T, Inamori Y, Matsumura M. Degradation charateristics of microcystins by isolated bacteria from Lake Kasumigaura. Journal of Water Supply: Research and Technology-AQUA 2003;52: 13–18.

[66] Ishii H, Nishijima M, Abe T. Characterization of degradation process of cyanobacterial hepatotoxins by a gram-negative aerobic bacterium. Water Research 2004;11: 2667–2676.

[67] Harada K, Imanishi S, Kato H, Mizuno M, Ito E, Tsuji K. Isolation of Adda from microystin-LR by microbial degradation. Toxicon 2004;44: 107–109.

[68] Takenaka S, Watanabe MF. Microcystin LR degradation by *Pseudomonas aeruginosa* alkaline protease. Chemosphere 1997;34: 749–757.

[69] Rapala J, Berg KA, Lyra C, Niemi RM, Manz W, Suomalainen S, Paulin L, Lahti K. *Paucibacter toxinivorans* gen. nov., sp. nov., a bacterium that degrades cyclic cyanobacterial hepatotoxins microcystins and nodularin. International Journal of Systematic and Evolutionary Microbiology 2005;55: 1563–1568.

[70] Maruyama T, Park HD, Ozawa K, Tanaka Y, Sumino T, Hamana K, Hiraishi A, Kato K. *Sphingosinicella microcystinivorans* gen. nov., sp. nov., a microcystin-degrading bacterium. International Journal of Systematic and Evolutionary Microbiology 2006;56: 85–89.

[71] Lemes GA, Kersanach R, Pinto Lda S, Dellagostin OA, Yunes JS, Matthiensen A. Biodegradation of microcystins by aquatic *Burkholderia* sp. from a South Brazilian coastal lagoon. Ecotoxicology and Environmental Safety 2008; 69: 358–365.

[72] Manage PM, Edwards C, Singh BK, Lawton LA. Isolation and identification of novel microcystin-degrading bacteria. Applied Environmental Microbiology 2009;75: 6924–6928.

[73] Lawton LA, Welgamage A, Manage PM, Edwards C. Novel bacterial strains for the removal of microcystins from drinking water. Water Science and Technology 2011;63: 1137–1142.

[74] Hu L, Zhang F, Liu C, Wang M. Biodegradation of microcystins by *Bacillus* sp. strain EMB. Energy Procedia 2012;16: 2054-2059.

[75] Hu LB, Yang JD, Zhou W, Yin YF, Chen J, Shi ZQ. Isolation of a *Methylobacillus* sp. that degrades microcystin toxins associated with cyanobacteria. New Biotechnology 2009;26: 205–211.

[76] Chen J, Hu LB, Zhou W, Yan SH, Yang JD, Xue YF, Shi ZQ. Degradation of microcystin-LR and RR by a *Stenotrophomonas* sp. strain EMS isolated from Lake Taihu, China. International Journal of Molecular Sciences 2010;11: 896–911.

[77] Eleuterio L, Batista JR. Biodegradation studies and sequencing of microcystin-LR degrading bacteria isolated from a drinking water biofilter and a fresh water lake. Toxicon 2010;55: 1434-1442.

[78] Ramani A, Rein K, Shetty KG, Jayachandran K. Microbial degradation of microcystin in Florida's freshwaters. Biodegradation 2011;23: 35–45.

[79] Nybom SMK, Salminen SJ, Meriluoto JAO. Removal of microcystin-LR by strains of metabolically active probiotic bacteria. FEMS Microbiology Letters 2007;270: 27–33.

[80] Nybom SMK, Salminen SJ, Meriluoto JAO. Specific strains of probiotic bacteria are efficient in removal of several different cyanobacterial toxins from solution. Toxicon 2008;52: 214–220.

[81] Zhang X, Hu HY, Hong Y, Yang J, Isolation of a *Poterioochromonas* capable of feeding on *Microcystis aeruginosa* and degrading microcystin-LR. FEMS Microbiology Letters 2008;288: 241–246.

[82] Bourne DG, Riddles P, Jones GJ, Smith W, Blakeley RL. Characterisation of a gene cluster involved in bacterial degradation of the cyanobacterial toxin microcystin LR. Environmental Toxicology 2001;16: 523–534.

[83] Imanishi S, Kato H, Mizuno M, Tsuji K, Harada K. Bacterial degradation of microcystins and nodularin. Chemical Research in Toxicology 2005;18: 591–598.

[84] Maruyama T, Kato K, Yokoyama A, Tanaka T, Hiraishi A, Park HD. Dynamics of microcystin-degrading bacteria in mucilage of *Microcystis*. Microbial Ecology 2003;46: 279–288.

[85] Yan H, Wang J, Chen J, Wei W, Wang H, Wang H. Characterization of the first step involved in enzymatic pathway for microcystin-RR biodegraded by *Sphingopyxis* sp. USTB-05. Chemosphere 2012;87: 12-18.

[86] Dziga D, Wasylewski M, Szetela A, Bocheńska O, Wladyka B. Verification of the role of MlrC in microcystin biodegradation by studies using a heterologously expressed enzyme. Chemical Research in Toxicology 2012;25: 1192–1194.

[87] Edwards C, Graham D, Fowler N, Lawton LA. Biodegradation of microcystins and nodularin in freshwaters. Chemosphere 2008;73: 1315–1321.

[88] Hoefel D, Adriansen CM, Bouyssou MA, Saint CP, Newcombe G, Ho L. Development of an mlrA gene-directed TaqMan PCR assay for quantitative assessment of microcystin-degrading bacteria within water treatment plant sand filter biofilms. Applied and Environmental Microbiology 2009;75: 5167–5

[89] Wu Y, Li T, Yang L. Mechanisms of removing pollutants from aqueous solutions by microorganisms and their aggregates: A review. Bioresource Technology 2012;107: 10–18.

[90] Fuller R. Probiotics in human medicine. Gut 1991;32: 439–442.

[91] WHO. Guidelines for the evaluation of probiotics in foods; 2002. http://www.who.int/foodsafety/fs_management/en/probiotic_guidelines.pdf

[92] FAO/WHO. Guidelines for the evaluation of probiotics in food. Working Group Report. Food and Health Agricultural Organization of the United Nations and World Health Organization, Washington; 2002.

[93] Salminen S, Bouley C, Boutron-Ruault MC, Cummings JH, Franck A, Gibson GR, Isolauri E, Moreau MC, Roberfroid MB, Rowland IR. Functional food science and gastrointestinal physiology and function. British Journal of Nutrition 1998;80: S147–S171.

[94] Andersson H, Asp NG, Bruce A, Roos S, Wadstrom T, Wold AE. Health effects of probiotics and prebiotics: A literature review on human studies. Scandinavian Journal of Nutrition 2001;45: 58–75.

[95] European Food Safety Authority. Scientific Opinion on the maintenance of the list of QPS biological agents intentionally added to food and feed (2010 update, published 14 December); 2010.

[96] Nybom SMK, Collado MC, Surono IS, Salminen SJ, Meriluoto JAO. Effect of glucose on the removal of microcystin-LR by viable commercial probiotic strains and strains isolated from dadih fermented milk. Journal of Agricultural and Food Chemistry 2008;56: 3714–3720.

[97] Nybom SMK Dziga D, Heikkilä JE, Kull TPJ, Salminen SJ, Meriluoto JAO. Characterization of microcystin-LR removal process in the presence of probiotic bacteria. Toxicon 2012;59: 171–181.

[98] Halttunen T, Salminen S, Tahvonen R. Rapid removal of lead and cadmium from water by specific lactic acid bacteria. International Journal of Food Microbiology 2007;114: 30–35.

[99] El-Nezami H, Kankaanpää P, Salminen S, Ahokas J Ability of dairy strains of lactic acid bacteria to bind a common food carcinogen, aflatoxin B1. Food Chemistry and Toxicology 1998;36: 321–326.

[100] Ho L, Hoefel D, Palazot S, Sawade E, Newcombe G, Saint CP, Brookes JD. Investigations into the biodegradation of microcystin-LR in wastewaters. Journal of Hazardous Materials 2010;15, 628–633.

Chapter 6

Microbial Degradation of Persistent Organophosphorus Flame Retardants

Shouji Takahashi, Katsumasa Abe and Yoshio Kera

Additional information is available at the end of the chapter

1. Introduction

1.1. Flame retardants

Flame retardants (FRs) are chemicals used in polymers to protect the public from accidental fires by preventing or retarding the initial phase of a developing fire (EFRA, 2007). These chemicals are now found in numerous consumer products, including construction materials, upholstery, carpets, electronic goods, furniture and also children's products such as car seats, strollers and baby clothing. FRs have become indispensable to modern life, and have saved numerous lives by preventing unexpected fires across the globe.

FRs are divided into two general classes based on their relation to host polymers: additive and reactive FRs (WHO, 1997). Additive FRs are simply mixed with host polymers. The lack of chemical bonding between the FRs and host polymers enables the FRs to leach out of or volatilize from host polymers over time into the ambient environment. Reactive FRs are incorporated into host polymers by covalent bonding into the polymer backbone, and are thus less likely to leach into the environment. Additive FRs are mainly used in thermoplastics, textiles and rubbers, whereas reactive FRs are usually used in thermoset plastics and resins (SFT, 2009a).

FRs are sub-divided into six groups characterized by their chemical composition: 1) aluminum hydroxide, 2) brominated, 3) organophosphorus, 4) antimony oxides, 5) chlorinated and 6) other FRs. These groups account for 40%, 23%, 11%, 8%, 7% and 11% of the annual FR global consumption in 2007, respectively (Beard & Reilly, 2009). The total market for FRs in the United States, Europe and Asia in 2007 amounted to about 1.8 million tons.

1.2. Organophosphorus flame retardants

Organophosphorus flame retardants (PFRs) are based primarily on phosphate esters, phosphonate esters and phosphite esters. The total consumption of FRs in Europe was an estimated 465,000 tons in 2006, of which 20% comprised PFRs (KLIF, 2010). Of the PFRs consumed, 55% were chlorinated. Halogenated PFRs are the preferred form of FRs because halogen inhibits flame formation in organic materials, and non-halogenated PFRs are typically used as flame-retardant plasticizers (KLIF, 2010).

1.3. Tris(1,3-dichloro-2-propyl) phosphate and tris(2-chloroethyl) phosphate

Tris(1,3-dichloro-2-propyl) phosphate (TDCPP) and tris(2-chloroethyl) phosphate (TCEP) are typical examples of additive chlorinated PFR (Fig. 1 and Table 1).

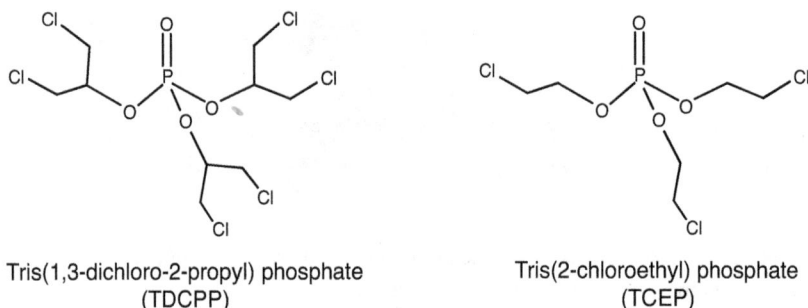

Tris(1,3-dichloro-2-propyl) phosphate
(TDCPP)

Tris(2-chloroethyl) phosphate
(TCEP)

Figure 1. Chemical structure of tris(1,3-dichloro-2-propyl) phosphate (TDCPP) and tris(2-chloroethyl) phosphate (TCEP)

TDCPP is a viscous colorless to light yellow liquid and is produced by the epoxide opening of epichlorohydrin in the presence of phosphorus oxychlorine (ATSDR, 2009). TDCPP is used primarily in flexible polyurethane foams but also in rigid polyurethane foams, resins, plastics, textile coatings and rubbers (California EPA, 2011). TDCPP was a common ingredient of sleepwear for children in the 1970s, but was voluntarily withdrawn by manufactures in 1977 because of its proven mutagenicity (California EPA, 2011). However, the PFR can still be found in many baby products (Stapleton et al., 2011). Currently, TDCPP is used mostly in flexible polyurethane foams for upholstered furniture and automotive products. TDCPP consumption has increased following the ban on common FR polybrominated diphenyl ethers (PBDEs). Consequently, total TDCCP production has increased, being an estimated 4,500-22,700 tons in the United States in 2006 and <10,000 tons in Europe in 2000 (van der Veen & de Boer, 2012).

TCEP is colorless to pale yellow liquid and is highly soluble in water (Fig. 1 and Table 1). The compound is chemically synthesized via condensation of phosphorus oxychloride and chloroalkyl alcohol at low temperatures and pressures to avoid formation of alkyl chlorides (ATSDR, 2009). Previously, the main purpose of TCEP was to reduce the brittleness of flame-resistant rigid or semirigid polyurethane foams. More recently, it has been used as a flame-retarding plasticizer and viscosity regulator in unsaturated polyester resin (accounting for around 80% of current use) (EURAR, 2009). TCEP-containing polymers are commonly used in the furniture, textile and building industries (for example, more than 80% of the TCEP consumption in the EU is invested in roofing insulation). TCEP is also used in car, railway and aircraft materials, and in professional paints. Since the 1980s, TCEP has been progressively replaced by other flame retardants, primarily tris(1-chloro-2-propyl) phosphate (TCPP). Consequently, global consumption of TCEP in the EU, which exceeded 9,000 tons in 1989, declined to below 4,000 tons by 1997. TCEP is no longer produced in the EU (EURAR, 2009).

	tris(1,3-dichloro-2-propyl) phosphate (US EPA, 2005)	tris(2-chloropropyl) phosphate (EURAR, 2009)
Cas number:	13674-87-8	115-96-8
Synonym:	Tris(1,3-dichloro-2-propyl) phosphate Tris-(2-chloro-,1-chloromethyl-ethyl)-phosphate 1,3-dichloro-2-propanol phosphate Phosphoricacid, tris(1,3-dichloro-2-propylester) Tris(1,3-dichloroisopropyl) phosphate Tris(1-chloromethyl-2-chloroethyl) phosphate Tri(β, β'-dichloroisopropyl) phosphate	Tris(2-chloroethyl) phosphate Tris(β-chloroethyl) phosphate 2-chloroethanol phosphate Phosphoricacid,tris(2-chloroethyl) ester Tris(2-chloroethyl) orthophosphate Tris(chloroethyl) phosphate
Abbreviation:	TDCPP TDCP	TCEP TClEP
Molecular weight:	430.91	285.49
Physical state:	Viscous, clear liquid	Clear, transparent, Low viscosity liquid
Melting point:	-58°C	<-70°C
Boiling point:	236-237°C at 5 mm Hg	Decomposition at 320°C at 1013 hPa
Density:	1.52	1.4193 (25°C)
Vapor pressure:	0.01 mmHg (30°C)	43 Pa (136.9°C) 0.00114 Pa (20°C, extrapolated)
Water solubility:	42 mg/L	7.82 g/L (20°C)
n-Octanol/water partition coefficient:	2.4	1.78

Table 1. General aspect of Tris(1,3-dichloro-2-propyl) phosphate (TDCPP) and tris(2-chloroethyl) phosphate (TCEP)

1.4. Occurrence and behavior of TDCPP and TCEP in the environment

TCEP and TDCPP have been detected in various environments worldwide, including indoor and outdoor air, surface and ground waters, and even drinking water (Tables 2 and 3). It is unlikely that these compounds are produced naturally. Their environmental presence is thus considered to be the result of human activity. Because these PFRs are physicochemically and microbiologically stable in the environment and are also reportedly toxic, they are a serious threat to human and ecosystem health.

1.4.1. TDCPP

Detected air concentrations of TDCPP have attained up to150 ng m^{3-1} in Sweden houses, and in Belgium office and stores, they have reached 73 ng m^{3-1} (Table 2). In outdoor air, TDCPP levels near a main road in Sweden ranged from <0.04-0.072 m^{3-1}, and significant amounts have been detected globally in air borne particles over the Pacific, Indian, Arctic and Southern Oceans. TDCPP has been also found in indoor dust at relatively higher concentrations. Levels of TDCPP have tended to be higher in public buildings than in domestic buildings.

With respect to water environments, TDCPP concentrations have been detected at up to ~50 ng L^{-1} in German rivers and at 1,335 ng L^{-1} in Italian lakes. In these countries, it also occurs in rain and/or snow, as a result of volatilization from host materials. A much higher TDCPP concentration was detected in raw water at a disposal site in Japan, suggesting that the compound leaches and migrates to water sources. In the United States and Germany, TDCPP has even been detected in drinking water processed in treatment plants (DWTs). Relatively higher concentrations of TDCPP occur in landfill site sediments. Much higher concentrations still have been found in sediments near a car demolition site in Norway.

TDCPP has been also detected in the effluents of sewage treatment plants (STPs) and waste water treatment plants (WWTPs) in European countries and Japan, revealing that effluents are a source of aquatic TDCPP contamination. Comparable levels have been observed in the influents, indicating that the compound persists in the treatment plants. Degradation of TDCPP in the environment has been reported as low. Together, these observations suggest that TDCPP is likely to accumulate in the environment.

Environment	Concentration	Location	Country	Reference
Indoor air:	<0.04-18 ng m^{3-1}	office and store	Norway	SFT, 2008
	<0.2-150 ng m^{3-1}	home, cinema, university, hospital, hotel, prison, library, office shops	Sweden	Marklund et al., 2005a
	<0.3-7 ng m^{3-1}	lecture and computer hall, electronic dismantling facility recycling plant	Sweden	Staaf & Ostman, 2005
	<73 ng m^{3-1}	work place	Belgium	Bergh et al., 2011
	<61.4 ng m^{3-1}	house	Japan	Kanazawa et al., 2010

Environment	Concentration	Location	Country	Reference
	1.3 ng m$^{3\text{-}1}$	newly constructed house	Japan	Saito et al., 2007
	<0.6 ng m$^{3\text{-}1}$, <8.7 ng m$^{3\text{-}1}$	house and office	Japan	Saito et al., 2007
Indoor dust:	0.2-67 µg g^{-1}	home, cinema, university, hospital, hotel, prison, library, office shops	Sweden	Marklund et al., 2003
	<0.08-6.64 µg g^{-1}	house	Belgium	van den Eede et al., 2011
	<0.08-56.2 µg g^{-1}	store	Belgium	van den Eede et al., 2011
	2.2-27 µg g^{-1}	home	Belgium	Bergh et al., 2011
	3.9-150 µg g^{-1}	day care	Belgium	Bergh et al., 2011
	3.3-91 µg g^{-1}	work place	Belgium	Bergh et al., 2011
	<1.1 µg g^{-1}	house	Spain	Garcia et al., 2007
	<0.09-56.1 µg g^{-1}	house	United States	Stapleton et al., 2009
	0.069-18 µg g^{-1}	hotel	Japan	Takigami et al., 2009
	<127 µg kg^{-1}	house	Japan	Kanazawa et al., 2010
Outdoor air:	<0.04-0.072 ng m$^{3\text{-}1}$	nearby main road	Sweden	Marklund et al., 2003
	<0.04-0.14 ng m$^{3\text{-}1}$	remote area from main road	Sweden	Marklund et al., 2003
	n.d.-5 pg m$^{3\text{-}1}$	sea	Arctic ocean	Moller et al., 2012
	16-52 pg m$^{3\text{-}1}$	sea	Japan	Moller et al., 2012
	5-8 pg m$^{3\text{-}1}$	sea	Northern pacific ocean	Moller et al., 2012
	49-780 pg m$^{3\text{-}13}$	sea	East Indian archipelago, Philippine sea	Moller et al., 2012
	n.d.-220 pg m$^{3\text{-}13}$	sea	Indian ocean	Moller et al., 2012
	80 pg m$^{3\text{-}1}$	sea	Southern ocean	Moller et al., 2012
Surface water:	10-18 ng L^{-1}	river	Germany	Andresen & Bester, 2006
	~50 ng L^{-1}	river	Germany	Andresen et al., 2004
	2-24 ng L^{-1}	rain	Germany	Regnery & Püttmann, 2009
	5-40 ng L^{-1}	snow	Germany	Regnery & Püttmann, 2009
	<19 ng L^{-1}	river	Austria	Martinez-Carballo et al., 2007
	<3.0-19 ng L^{-1}	river	Austria	Martinez-Carballo et al., 2007
	<1,335 ng L^{-1}	lake	Italy	Bacaloni et al., 2008
	108-448 ng L^{-1}	rain	Italy	Bacaloni et al., 2008
	680-6,180 ng L^{-1}	raw water of waste disposal site	Japan	Kawagoshi et al., 1999
Drinking water:	1.2-2.4 ng L^{-1}	water after drinking water treatment	Germany	Andresen & Bester, 2006
	<250 ng L^{-1}	water after drinking water treatment	United States	Stackelberg et al., 2004

Environment	Concentration	Location	Country	Reference
Sediment:	<0.15-54 µg kg⁻¹	lake and fjord at vicinity of WWFP	Norway	KLIF, 2010
	1,500-4,100 µg kg⁻¹	landfill site	Norway	SFT, 2008
	<250-8,800 µg kg⁻¹	car demolition site	Norway	SFT, 2008
	<709 µg kg⁻¹	waste disposal site	Japan	Kawagoshi et al., 1999
Sludge:	110-330 µg kg⁻¹		Norway	SFT, 2008
	3.0-260 µg kg⁻¹		Sweden	Stackelberg et al., 2004
Influent:	630-820 ng L⁻¹	WWTP	Norway	SFT, 2008
	240-450 ng L⁻¹	STP	Sweden	Marklund et al., 2005b
	330-1,600 ng L⁻¹	STP	Japan	Ishikawa et al., 1985
Effluent:	86-740 ng L⁻¹	WWTP	Norway	SFT, 2008
	130-340 ng L⁻¹	STP	Sweden	Marklund et al., 2005b
	20-120 ng L⁻¹	STP	Germany	Andresen et al., 2004
	19-1,400 ng L⁻¹	WWTP	Austria	Martinez-Carballo et al., 2007
	280-1,400 ng L⁻¹	STP	Japan	Ishikawa et al., 1985
Biota:	<6.0 ng g⁻¹	fish liver	Norway	SFT, 2009b
	<0.3-6.7 ng g⁻¹	fish muscle	Norway	SFT, 2009b
	<0.72-1.9 ng g⁻¹	bird egg	Norway	KLIF, 2010
	<0.11-0.16 ng g⁻¹	bird blood and plasma	Norway	KLIF, 2010
	<0.6-8.1 ng g⁻¹	whole fish	Norway	SFT, 2009b
	<1.5 ng g⁻¹	seabird liver	Norway	SFT, 2009b
	<0.3-1.2 ng g⁻¹	whole fish liver	Norway	SFT, 2009b
	<5.0 ng g⁻¹, <10-<30 ng g⁻¹	cod liver and mussel	Norway	SFT, 2008
	49-140 ng g⁻¹	freshwater fishes close to sources	Norway	Sundkvist et al., 2010
	16.-5.3 ng g⁻¹	human milk	Sweden	Sundkvist et al., 2010

Table 2. Occurrence and behavior of TDCPP

TDCPP has also been detected in biological samples, including fishes, mussels and birds. In Norway, fishes and mussels were observed to contain up to 8.1 and 30 ng g⁻¹ of TDCPP, respectively. In bird blood/plasma and eggs respectively, TDCPP levels range from <0.11-0.16 and from <0.72-1.9 ng g⁻¹. In Sweden, freshwater fishes close to emission sources contained 49-140 ng g⁻¹ TDCPP. Worryingly, TDCPP has also been detected in the breast milk of Swedish women.

1.4.2. TCEP

In Sweden, the highest detected air concentration of TCEP was 730 ng m⁻³ inside an office furnished with linoleum floor and a new photocopier (Table 3). In outdoor air, it can reach 6.2 ng m⁻³ beside a main road, but remote areas harbor less than 0.2 ng m⁻³, implicating road traffic as an important source of TCEP emission. TCEP has also been detected globally in air borne particles over the Pacific, Indian, Arctic and Southern Ocean. In Belgium, indoor dust can contain up to 260 $\mu g \, g^{-1}$ TCEP. TCEP concentrations in dusts of public spaces tend to exceed those in domestic dusts.

TCEP ranges from <3.0-1,236 ng L^{-1} in German rivers, lakes and reservoirs. In this country and in Italy, it has also been detected in rain and/or snow, indicating that, like TDCPP, TCEP volatilizes from its host materials. Groundwater TCEP levels up to 754 ng L^{-1} have been reported in Germany, suggesting that TCEP primarily mobilizes into water rather than attaching to soil. TCEP also occurs in drinking water or finished water from DWTs; recorded concentrations are as high as 99, 25 and 1.7 ng L^{-1} in the United States, Korea and Germany, respectively. Much higher concentrations have been observed in raw water of waste disposal sites in Japan. Relatively higher concentrations of TCEP have been detected in landfill site sediments in Japan and Norway (up to 7,400 and 380 $\mu g \, kg^{-1}$, respectively). Especially high concentrations were found in the sediment nearby a car demolition site.

TCEP has been also detected in STP or WWTP effluents in many countries. Comparable levels of TCEP are observed in the influents. These observations demonstrate that, like TDCPP, TCEP persists in the treatment plants.

Also similarly to TDCPP, TCEP has been detected in biological samples, including fishes, crabs, mussels and birds. In Norway, fishes and mussels respectively contain up to 26 and 23 ng g^{-1} TCEP. In birds and their eggs, TCEP levels can reach up to 6.1 ng g^{-1}. In fishes residing near emission sources in Sweden, they reach up to 69 and 160 ng g^{-1} respectively. Furthermore, like TDCPP, TCEP has been detected in the breast milk of Swedish women.

Environment	Concentration	Location	Country	Reference
Indoor air:	<0.2-23 ng m^{3-1}	office and store	Norway	SFT, 2008
	3, 9 ng m^{3-1}	lecture room and kindergarten	Sweden	Tollback et al., 2006
	0.4-730 ng m^{3-1}	home, cinema, university, hospital, hotel, prison, library, office shops	Sweden	Marklund et al., 2005a
	<0.3-10 ng m^{3-1}	Lecture and computer hall, electronic dismantling facility recycling plant	Sweden	Staaf & Ostman, 2005
	<22 ng m^{3-1}	car, theater, furniture store, office and electronics store	Sweden	Hartmann et al., 2004
	3-15 ng m^{3-1}	lecture room and office room	Sweden	Bjorklund et al., 2004
	<297 ng m^{3-1}	house	Japan	Kanazawa et al., 2010
	<136 ng m^{3-1}, <42.1 ng m^{3-1}	house and office	Japan	Saito et al., 2007
	1.2 ng m^{3-1}	newly constructed house	Japan	Saito et al., 2007
	<28 ng m^{3-1}	home	Belgium	Bergh et al., 2011
	7.8-230 ng m^{3-1}	day care center	Belgium	Bergh et al., 2011

Environment	Concentration	Location	Country	Reference
	<140 ng m$^{3-1}$	work place	Belgium	Bergh et al., 2011
Indoor dust:	0.19-94 µg g^{-1}	home, cinema, university, hospital, hotel, prison, library, office shops	Sweden	Marklund et al., 2003
	<0.08-2.65 µg g^{-1}	house	Belgium	van den Eede et al., 2011
	<33 µg g^{-1}	house	Belgium	Bergh et al., 2011
	<0.08-5.46 µg g^{-1}	store	Belgium	van den Eede et al., 2011
	2.5-150 µg g^{-1}	day care center	Belgium	Bergh et al., 2011
	1.3-260 µg g^{-1}	work place	Belgium	Bergh et al., 2011
	0.25-1.56 µg g^{-1}	house	Spain	Garcia et al., 2007
	<308 µg g^{-1}	house	Japan	Kanazawa et al., 2010
	0.082-2.3 µg g^{-1}	hotel	Japan	Takigami et al., 2009
Outdoor air:	0.51-6.2 ng m$^{3-1}$	nearby main road	Sweden	Marklund et al., 2003
	<0.2 ng m$^{3-1}$	remote area from main road	Sweden	Marklund et al., 2003
	126-585 pg m$^{3-1}$	ocean	Arctic ocean	Moller et al., 2012
	273-1,961 pg m$^{3-1}$	sea	Japan	Moller et al., 2012
	159-282 pg m$^{3-1}$	sea	Northern pacific ocean	Moller et al., 2012
	19-156 pg m$^{3-1}$	sea	East Indian archipelago, Philippine sea	Moller et al., 2012
	46-570 pg m$^{3-1}$	sea	Indian ocean	Moller et al., 2012
	74 pg m$^{3-1}$	sea	Southern ocean	Moller et al., 2012
Surface water:	<3-184 ng L^{-1}	lake and reservoir	Germany	Regnery & Püttmann, 2010
	12-130 ng L^{-1}	river	Germany	Andresen & Bester, 2006
	13-130 ng L^{-1}	river	Germany	Andresen et al., 2004
	<1,236 ng L^{-1}	river	Germany	Fries & Püttmann, 2003
	11-196 ng L^{-1}	rain	Germany	Regnery & Püttmann, 2009
	121 ng L^{-1}	rain	Germany	Fries & Püttmann, 2003
	19-60 ng L^{-1}	snow	Germany	Regnery & Püttmann, 2009
	13-130 ng L^{-1}	river	Austria	Martinez-Carballo et al., 2007
	<33 ng L^{-1}	lakes	Italy	Bacaloni et al., 2008
	7 ng L^{-1}	river	Italy	Bacaloni et al., 2007
	19-161 ng L^{-1}	rain	Italy	Bacaloni et al., 2008
	4,230-87,400 ng L^{-1}	raw water of waste disposal site	Japan	Kawagoshi et al., 1999
	14-347 ng L^{-1}	river and sea water	Japan	Ishikawa et al., 1985
	14-81 ng L^{-1}	lake and river	Korea	Kim et al., 2007
Ground water:	3-9 ng L^{-1}		Germany	European Commission DG ENV, 2011
	<312 ng L^{-1}		Germany	Fries & Püttmann, 2003

Environment	Concentration	Location	Country	Reference
	<754 ng L^{-1}		Germany	Fries & Püttmann , 2001
Drinking water:	0.74-1.7 ng L^{-1}	water after drinking water treatment	Germany	Andresen & Bester, 2006
	4-99 ng L^{-1}	water after drinking water treatment	United States	Stackelberg et al., 2007
	<99 ng L^{-1}	water after drinking water treatment	United States	Stackelberg et al., 2004
	14, 25 ng L^{-1}	water after drinking water treatment	Korea	Kim et al., 2007
Sediment:	<0.16-8.5 µg kg^{-1}	lake and fjord at vicinity of WWFP	Norway	KLIF, 2010
	27-380 µg kg^{-1}	landfill site	Norway	SFT, 2008
	2,300-5,500 µg kg^{-1}	car demolition site	Norway	SFT, 2008
	<160 µg kg^{-1}	river	Austria	Martinez-Carballo et al., 2007
	<7,400 µg kg^{-1}	waste disposal site	Japan	Kawagoshi et al., 1999
Sludge:	<9-<19 µg kg^{-1}		Norway	SFT, 2008
	6.6-110 µg kg^{-1}		Sweden	Marklund et al., 2005b
Influent:	2,000-2,500 ng L^{-1}	STP	Norway	SFT, 2008
	90-1,000 ng L^{-1}	STP	Sweden	Marklund et al., 2005b
	290, 180 ng L^{-1}	STP	Germany	Meyer & Bester, 2004
	983-1,123 ng L^{-1}	municipal STWs	Germany	Fries & Püttmann , 2003
	<0.025-0.3 ng L^{-1}	STP	Spain	Rodriguez et al., 2006
	540-1,200 ng L^{-1}	STP	Japan	Ishikawa et al., 1985
Effluent:	1600-2,200 ng L^{-1}	STP	Norway	SFT, 2008
	350-890 ng L^{-1}	STP	Sweden	Marklund et al., 2005b
	350, 370 ng L^{-1}	STP	Germany	Meyer & Bester, 2004
	214-557 ng L^{-1}	municipal STWs	Germany	Fries & Püttmann , 2003
	<0.025-0.7 ng L^{-1}	STP	Spain	Rodriguez et al., 2006
	500-1,200 ng L^{-1}	STP	Japan	Ishikawa et al., 1985
Biota:	0.5-5.0 ng g^{-1} 13-26 ng g^{-1}	fish muscle and liver	Norway	SFT, 2009b
	1.8-3.2 ng kg^{-1}	whole fish	Norway	SFT, 2009b
	<5 ng g^{-1}, <10-23 ng g^{-1}	cod liver and mussel	Norway	SFT, 2008
	<0.6-4.7 ng g^{-1}	sea bird liver	Norway	SFT, 2009b
	<0.17-19 ng g^{-1}	beach crab	Norway	KLIF, 2010
	<0.06-0.11 ng g^{-1}	blue mussel	Norway	KLIF, 2010
	<1.7-8.6 ng g^{-1}	burbot liver	Norway	KLIF, 2010
	<0.08-0.21 ng g^{-1}	trout	Norway	KLIF, 2010
	<0.33-6.1 ng g^{-1}	bird egg	Norway	KLIF, 2010
	<0.17-6.0 ng g^{-1}	bird blood and plasma	Norway	KLIF, 2010

Environment	Concentration	Location	Country	Reference
	1.5-69 ng g^{-1}	marine fishes	Sweden	Sundkvist et al., 2010
	<160 ng g^{-1}	freshwater fishes close to sources	Sweden	Sundkvist et al., 2010
	201-8.2 ng g^{-1}	human milk	Sweden	Sundkvist et al., 2010

Table 3. Occurrence and behavior of TCEP

1.5. Toxicological information of TDCPP and TCEP

Since the toxic effects of TCEP and TDCPP have been regarded as marginal compared to those of PBDEs, they have been extensively used. However, their non-negligible toxicities have been revealed in a number of studies (Tables 4 and 5). Together with their persistence in the environment, the environmental contamination of both compounds has become of serious concern.

1.5.1. TDCPP

Rats given oral doses of TDCPP absorb more than 90% of the compound within 24 h, with the highest concentrations being observed in kidney, liver and lung (EURAR, 2008). The acute toxicity of oral TDCPP has been reported as low, with LD$_{50}$ values ranging from 2,250 mg kg^{-1} for female mice to 6,800 mg kg^{-1} for male rabbits (Table 4). In a 2-year chronic toxicity study in rats, the lowest observable adverse effect level (LOAEL) was 5 mg kg^{-1} day^{-1}. In that study, statistically significant relationships between TDCPP dose and tumor incidences were observed in both male and female rats. Consequently, TDCPP is today classified as Carc. Cat. 3; R40 and Cat. 2; H351, denoting "limited evidence of a carcinogenic effect" and "suspected of causing cancer", respectively.

A number of TDCPP genotoxicity studies have been conducted in whole mammals that have resulted in negative conclusions regarding genotoxicity (Albemarle Corp. & ICL North America Inc., 2011). However, *in vitro* studies using bacteria and mammalian cells have suggested that TDCPP exerts genotoxic effects, and an *in vivo* study showed its covalent binding to DNA (US EPA, 2005; Morales & Matthews, 1980).

Similarly, neurotoxicity studies of TDCPP involving hens and rats reveal no clear evidence that TDCPP is neurotoxic. However, a study based on undifferentiated and differentiating PC12 cells showed its potential neurotoxicity (Dishaw et al., 2011).

Whether, and to what extent, TDCPP is toxic to humans remains unknown. However, TDCPP has been shown to alter sex hormone balance in human cell lines, via alteration of steroidogenesis or estrogen metabolism (Liu et al., 2012). In addition, TDCPP concentrations in house dusts have been linked to altered hormone levels and decreased semen quality in men (Meeker & Stapleton, 2010).

TDCPP is regarded as toxic to aquatic organisms (EURAR, 2008). An acute toxicity study on fish trout yielded an LC$_{50}$ value of 1.1 mg L^{-1}. Acute and chronic toxicity studies conducted on the invertebrate *Daphnia* produced an EC$_{50}$ value of 3.8 mg L^{-1}. In a chronic study on the alga *Pseudokirchneriella*, ErC_{10} (10% growth-rate inhibition) was recorded as 2.3 mg L^{-1}. Thus, TDCPP is classified as N; R51/53, denoting "Toxic to aquatic organisms, may cause long-term adverse effects in the aquatic environment". In addition, an LC_{50} of 23 mg kg^{-1} has been reported for a terrestrial organism, the earthworm *Eisenia*.

Toxicity	Organism	Reference
Acute toxicity	LD_{50}=6,800 mg kg^{-1} male rabbit	US EPA, 2005
	LD_{50}=3,160 mg kg^{-1} male rat	EURAR, 2008
	LD_{50}=2,670 mg kg^{-1} male mice	
	LD_{50}=2,250 mg kg^{-1} female mice	
	LD_{50}=2,236 mg kg^{-1} male rat	
	LD_{50}=2,489 mg kg^{-1} female rat	
Chronic toxicity	LOAEL=5 mg kg^{-1} day^{-1} rat for hyperplasia and convoluted tubule epithelium	EURAR, 2008
Cytotoxicity	hepatocytes and neuronal cells	Crump et al., 2012
Neurotoxicity	*in vitro* PC12 cells	Dishaw et al., 2011
Carcinogenicity	rat	California EPA, 2011
Genotoxicity	*in vivo Salmonella typhimurium*	California EPA, 2011
	in vitro mouse, Chinese hamster and rat cells	
Toxic to aquatic organisms	fishes, invertebrates and algae	EURAR, 2008
	LC_{50}=1.1 mg L^{-1} rainbow trout (96 h)	
	EC_{50}=3.8 mg L^{-1} *Daphnia magana* (48 h)	
	LOEC=1.0 mg L^{-1} *Daphnia* for reproduction (21 days)	
	NOEC=0.5 mg L^{-1} *Daphnia* for reproduction (21 days)	
	ErC_{10}=2.3 mg L^{-1} algae	
	LC_{50}=23 mg kg^{-1} earthworm *Eisenia*	
	NOEC=2.9 mg kg^{-1} earthworm *Eisenia* for reproduction	
	NOEC=17 mg kg^{-1} plant *Mustard*	
Alter hormone levels	human and zebra fish cells	Liu et al., 2012
Decreased sperm quality	human	Meeker & Stapleton, 2010

Table 4. Toxicological information of TDCPP

1.5.2. TCEP

Rats given oral doses of TCEP absorb over 90% of the compound within 24 h, with marked accumulations in liver, kidney, fat and the gastrointestinal tract (EURAR, 2009). In animals,

TCEP appears to be mainly toxic to brain, kidney and liver. Toxicity studies have implicated TCEP as moderately toxic; in rats, oral administration yields an LD_{50} of 430-1,230 mg kg^{-1} and skin contact reveals a low acute dermal toxicity (LD_{50} >2,150 mg kg^{-1}) (Table 5). A 2-year chronic toxicity study of TCEP yielded LOAELs of 44 mg kg^{-1} day^{-1} in rats and 175 mg kg^{-1} day^{-1} in mice. The same study indicated that TCEP is potentially neurotoxic, with no observed adverse effect levels (NOAELs) in rats and mice being 88 mg kg^{-1} day^{-1} and 175 mg kg^{-1} day^{-1}, respectively.

Toxicity	Organism	Reference
Acute toxicity	LD_{50}=430-1,230 mg kg^{-1} rat	EURAR, 2009
	LD_{50}>2,150 mg kg^{-1} rat for dermal	EURAR, 2009
Chronic toxicity	LOAEL=44 mg kg^{-1} day^{-1} rat for kidney lesions (2 years)	EURAR, 2009
	LOAEL=175 mg kg^{-1} mouse for kidney morphology (2 years)	EURAR, 2009
Neurotoxicity	rat and mouse	EURAR, 2009
	NOAEL=88 mg kg^{-1} day^{-1} rats (16 weeks by gavage)	
	NOAEL=175 mg kg^{-1} day^{-1} mouse (16 weeks by gavage)	
Reproductive toxicity	rat and mouse	EURAR, 2009
	NOAEL=175 mg kg^{-1} day^{-1} mouse for fertility	
Carcinogenicity	rat and mouse	SCHER, 2012
Toxic to aquatic organisms	killifish, trout and goldfish	EURAR, 2009
Alter sex hormone balance	human cells and Zebra fish	Liu et al., 2012
Alter cell cycle regulatory protein expression	rabbit renal proximal tubule cells	Ren et al., 2008

Table 5. Toxicological information of TCEP

In the 2-year study, increased incidences of adenomas and carcinomas were linked to TCEP exposure, revealing TCEP as a potential carcinogen (EURAR, 2009). TCEP is thus classified as Carc. Cat. 3; R40. Because TCEP additionally exhibits reproductive toxicity in rats and mice, it is also classified as Repr. Cat. 2; R60, denoting "may impair fertility". TCEP at environmental concentrations has been reported to affect the expression of cell cycle regulatory genes in primary cultured rabbit renal proximal tubule cells (Ren et al., 2008).

TCEP is toxic to aquatic organisms, being classified as N; R51/53 (EURAR, 2009). Short term exposure to TCEP is mildly-moderately adverse to the aquatic invertebrate organisms *Daphnia* and *Planaria,* and TCEP presents low acute toxicity to killifish, trout and goldfishes.

The toxic effects of TCEP in humans are largely unknown. However, neurotoxic signs have been reported in a 5-year old child who slept in a room with wood paneling containing 3% TCEP (Ingerowski & Ingerowski, 1997). In addition, an epidemiological study of children in school environments found a potential association between the TCEP content in air-bone dusts and impaired cognitive ability (UBA, 2008). TCEP has been further reported to alter the sex hormone balance in human cells, as well as in fish cells.

1.6. Removal technique for TDCPP and TCEP

The persistence of chlorinated FRs TCEP and TDCPP in current waste water and drinking water treatment processes has accelerated the investigation of alternative water treatment techniques that will dispel these compounds.

Echigo *et al.* showed that TDCPP in distilled water and an effluent from a solid waste landfill site is effectively degraded by O_3/vacuum UV or O_3/H_2O_2 process, although degradation products were not determined in this study (Echigo et al., 1996). Westerhoff *et al.* reported that >20% of approximately 30 ng L^{-1} of TCEP in surface water samples can be removed with powdered activated carbon, but that other adsorptive processes, metal salt coagulation and lime softening, and oxidative processes (chlorination and ozonation) are ineffective (Westerhoff et al., 2005). Lee *et al.* showed that > 90% removal efficiency of 100 μg L^{-1} of TCEP in river and sea waters is possible using tight nanofiltration membranes with a low molecular weight cutoff of approximately 200 (Lee et al., 2008). Watts *et al.* demonstrated that the higher removing efficacy (> 95%) of 5 mg L^{-1} of TCEP in a water is achieved by a UV/H_2O_2 advanced oxidation process with the highest UV fluence at 6,000 mJ cm^{-2} (Watts & Linden, 2008). In this study, the generation of stoichiometric amount of chloride ion was observed. In addition, Benotti *et al.* reported that UV/TiO_2 supplemented with H_2O_2 can decrease the concentration of TCEP in a river water, although the degradation was not so effective and not completed (Benotti et al., 2009).

2. Microbial degradation and detoxification of TDCPP and TCEP

FRs have been widely distributed commercially and are necessary to prevent or reduce mortality from accidental fires. However, the leaching of additive FRs has led to global contamination of the environment. The chlorinated PFRs TCEP and TDCPP persist in the environment and exhibit varying toxic effects, raising concerns about their effects on human and ecological health. Although several physicochemical methods for removing TCEP and TDCPP have been reported (as described above), biotechnological techniques offer an attractive alternative, being potentially cost-effective, eco-friendly and enabling *in situ* remediation of contaminants. However, prior to recent isolation of TCEP- and TDCPP-degrading bacteria by our group, no biological degrading agent for such compounds was known.

2.1. Isolation and characterization of TDCPP- and TCEP-degrading bacteria

2.1.1. Enrichment of TCEP and TDCPP-degrading bacteria

2.1.1.1. Enrichment cultivation of TCEP and TDCPP-degrading bacteria

To obtain microorganisms that can degrade TDCPP and TCEP, we used an enrichment culture technique in which one of TDCPP or TCEP served as the sole phosphorus source (Takahashi et al., 2008). Forty six environmental samples (soils and sediments) in Japan were cultivated at 30°C in minimal medium containing approximately 20 μM of each compound. Significant degradation of TCEP and TDCPP was seen in ten and three of the samples, respectively. In the first cultivation round, each compound had disappeared within 2 to 5 days; successive sub-cultivations reduced the degradation time to within one day. The enrichment cultures displaying the highest degradation efficacy against TCEP and TDCPP were designated 67E and 45D, respectively. Culture 67E completely degraded 20 μM of TDCPP in 3 h and TCEP in 6 h (Fig. 2A and B), while culture 45D completely degraded the same concentration of TDCPP in 3 h and TCEP in 24 h. During the degradations, 2-CE was liberated from TCEP and 1,3-DCP from TDCPP, indicating that the degradation pathway involved hydrolysis of phosphoester bonds.

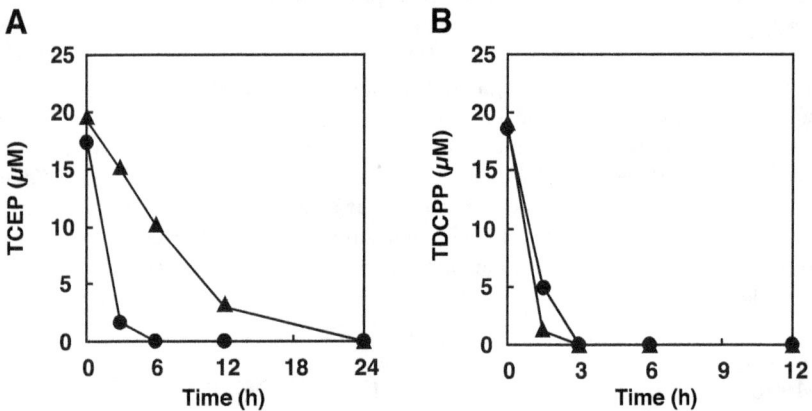

Figure 2. Degradation of TCEP (A) and TDCPP (B) by enrichment cultures. The enrichment cultures, 67E (circles) and 45D (triangles), were cultivated on 20 μM of TCEP or TDCPP as the sole phosphorus source.

2.1.1.2. 2-CE and 1,3-DCP degradation ability of enrichment cultures

The metabolites 2-CE and 1,3-DCP are also persistent and toxic: 1,3-DCP is a known genotoxin and carcinogen (NTP & NIEHS, 2005), while 2-CE exhibits genotoxicity, fetotoxicity and cardiotoxicity (National Toxicology, 1985). We analyzed whether the cultures can degrade the metabolites by measuring chloride ion formation. Cultures 67E and 45D liberated chloride ions from 2-CE and 1,3-DCP, respectively. After 120 h reaction, the proportion of chloride ion was approximately 100% and 68.5% of the total chlorine contained in the supplied 2-CE and 1,3-DCP, respectively. This shows that both cultures can dehalogenate their respective chloroalcohols and can therefore potentially detoxify chlorinated PFRs in the environment.

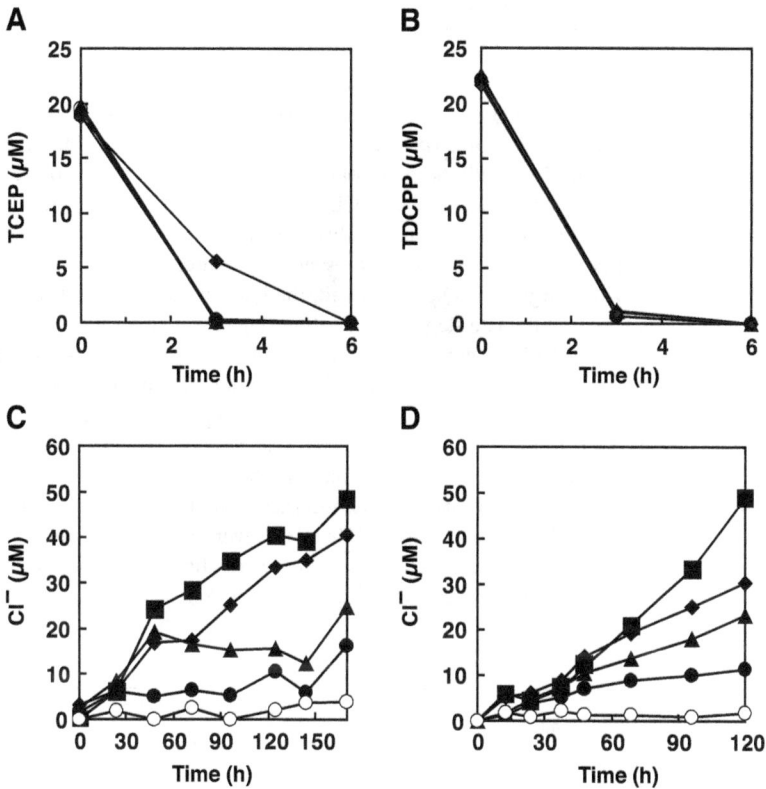

Figure 3. Effect of exogenous phosphate on the degradation of TCEP (A) and TDCPP (B) and the chloride ion formation from TCEP (C) and TDCPP (D). The enrichment cultures, 67E (A and C) and 45D (B and D), were cultivated on 20 µM of TCEP or TDCPP as the sole phosphorus source, respectively, with various concentrations of inorganic phosphate (NaH$_2$PO$_4$): 0 mM (closed circles), 0.02 mM (closed triangle), 0.2 mM (closed squares) and 2 mM (closed diamonds). Control culture without cell inoculation is indicated by open circles. Each data point represents the mean of at least two independent determinations.

2.1.1.3. Effect of exogenous phosphate on the degradation ability of enrichment cultures

Phosphate-sufficient conditions are well known to repress the expression of genes involved in phosphorus utilization. We thus examined the effect of exogenous inorganic phosphate on TDCPP and TCEP degradations and chloride ion formation (Fig. 3). At concentrations of 0.02, 0.2 and 2 mM, exogenous inorganic phosphate did not significantly inhibit TCEP and TDCPP degradation by the respective cultures (Fig. 3A and B), but chloride ion formation was enhanced at concentrations up to 0.2 mM (Fig. 3C and D). From these results, we concluded that efficient PFR detoxification could be achieved by optimizing the inorganic phosphate concentration.

2.1.1.4. Bacterial communities of enrichment cultures

To profile the bacterial communities in the cultures, we performed denaturing gradient gel electrophoresis (DGGE) analysis (Fig. 4). In the absence of inorganic phosphate, two bands (C1 and C2) were observed in the fingerprint of TCEP-supplemented 67E, which persisted throughout cultivation (Fig. 4A). With inorganic phosphate added, the intensity of C2 markedly decreased at later incubation stages (Fig.4A). In 45D supplemented with TDCPP, a single band (D3) was observed at the beginning of cultivation, but at later times two additional bands (D1 and D2) appeared, regardless of the presence or absence of inorganic phosphate (Fig. 4B). However, with inorganic phosphate added, the intensity of D2 and D3 decreased while that of D1 increased at the late stage of cultivation (Fig. 4B). The nucleotide sequence of C1 and D1 was affiliated with the genus *Acidovorax*, that of D2 with the genus *Aquabacterium*, and C2 and D3 were assigned to the genus *Sphingomonas* (Table 6). Together with the effect of exogenous inorganic phosphate on chlorinated PFRs degradation with liberation of chloride ions, these results imply that the *Sphingomonas*-related bacteria hydrolyze the PFRs, and that the *Acidovorax*-related bacteria dehalogenate the chloroalcohols. Among these bacterial genera, a strain of *Sphingomonas* sp. has been reported to hydrolyze some organophosphate pesticides, such as chlorpyrifos (Li et al., 2007). However, bacteria that are known to dehalogenate the chloroalcohols were not identified in the enrichment cultures, suggesting that a new member, possibly *Acidovorax* sp., is responsible for dehalogenating the chloroalcohols in the cultures.

Culture	Band	Phylogenetic affiliation
		Species
67E	C1	*Acidovorax* sp.
	C2	*Sphingomonas* sp.
45D	D1	*Acidovorax* sp.
	D2	*Aquabacterium* sp.
	D3	*Sphingomonas* sp.

Table 6. Phylogenetic affiliation of microorganisms represented by bands in DGGE profiles of the enrichment cultures 67E and 45D.

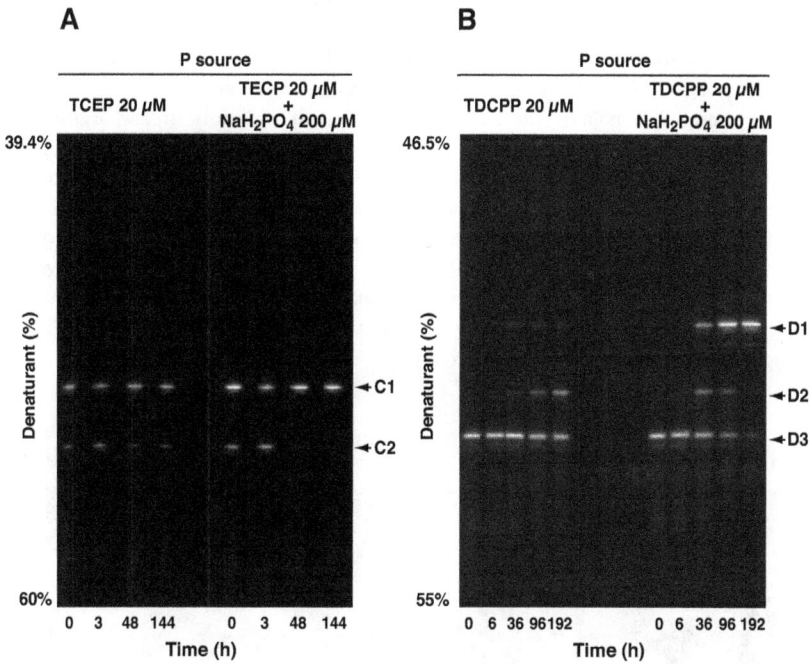

Figure 4. DGGE profile of the enrichment cultures 67E (A) and 45D (B) during cultivation in the presence of absence of inorganic phosphate. The arrowheads indicated the DNA fragments sequenced.

2.1.2. Isolation and characterization of TDCPP- and TCEP-degrading bacteria

2.1.2.1. Isolation of TDCPP- and TCEP-degrading bacteria

We attempted to isolate the bacteria responsible for degrading TDCPP and TCEP in the cultures 67E and 45D. (Takahashi et al., 2010). In the case of 45D, isolation was achieved by limiting dilution method. The culture was repeatedly serially diluted in a minimal medium containing 20 µM of TDCPP and cultivated at 30°C. Finally, the culture was spread onto a minimal agar plate containing 232 µM of TDCPP as the sole phosphorus source. A single colony grown on the plate was named strain TDK1 (Fig. 5A). In the case of 67E, the culture was spread onto a minimal agar plate containing 232 µM of TCEP as the sole phosphorus source and incubated at 30°C. Single colonies were then cultivated in a minimal medium containing 20 µM of TCEP as the sole phosphorus source. This isolation procedure was repeated three times, and a single colony was named strain TCM1 (Fig. 5B).

2.1.2.2. Identification of TDCPP- and TCEP-degrading bacteria

Both strains were short-rod-shaped bacteria (0.8-1.0 × 1.0-2.5 μm) and produced yellow, circular, convex colonies with smooth, glistening surfaces on a nutrient agar plate. As carbon sources, both strains assimilated glucose, maltose and L-arabinose; in addition, strain TCM1 assimilated potassium gluconate, while strain TDK1 assimilated D-mannose, N-acetyl-D-glucosamine, and D, L-malate. Both strains tested negative for indole, urease, arginine dihydrolase, nitrate reduction, gelatine hydrolysis, and glucose fermentation, and were positive for esculin hydrolysis. TCM1 and TDK1 tested negative and positive for cytochrome oxidase, respectively. The morphological and physiological characteristics of the strains were similar to those of *Sphingomonas* spp. Furthermore, the 16S rRNA gene sequence of the strains is closely related to those of sphingomonads, comprising the genera *Sphingomonas*, *Sphingobium*, *Novosphingobium* and *Sphingopyxis* (Takeuchi et al., 2001). The phylogenetic tree constructed from the sequences of these genera showed that strains TCM1 and TDK1 belong to *Sphingobium* and *Sphingomonas*, respectively

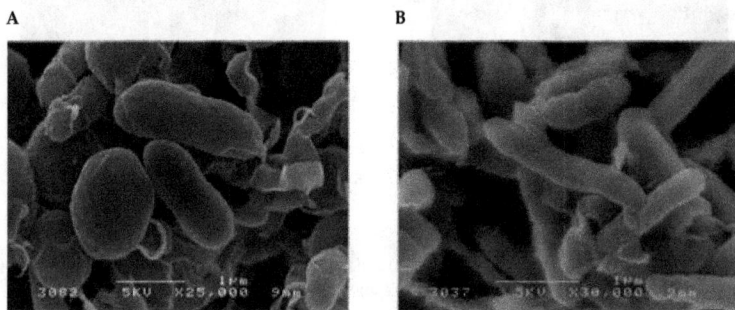

Figure 5. SEM micrographs of TCEP- and TDCPP-degrading bacteria *Sphingobium* sp. strain TCM1 (A) and *Sphingomonas* sp. stain TDK1 (B).

2.1.2.3. Degradation ability of TCEP and TDCPP-degrading bacteria

Both strains completely degraded 20 μM of TDCPP within 6 h (Fig. 6A and B). Strain TDK1, however, was 48 times less effective in degrading TCEP than TCM1 (TCEP degradation time was 144 h for TDK1, versus 3 h for TCM1) (Fig. 6A and B). During the degradations, 1,3-DCP and 2-CE were detected in the cultures of both strains and were not further degraded (Fig. 6C and D). These results showed that the strains degrade the compounds by hydrolyzing their phosphotriester bonds. To date, TCM1 and TDK1 are the only isolated microorganisms reported to degrade the persistent PFRs.

We then analyzed whether the strains can degrade other PFRs by utilizing them as sole phosphorus source. Both strains grew on tris(2,3-dibromopropyl) phosphate, tricresyl and triphenyl phosphates. Stain TDK1 did not grow on all trialkyl phosphates tested, whereas strain TCM1 grew moderately on tributyl phosphate and slightly on tris(2-butoxyethyl) phosphate, triethyl phosphate and trimethyl phosphate. These results demonstrate that the strains can degrade not only TDCPP and TCEP but also other PFRs, and that the strains have different substrate specificity for trialkyl phosphates.

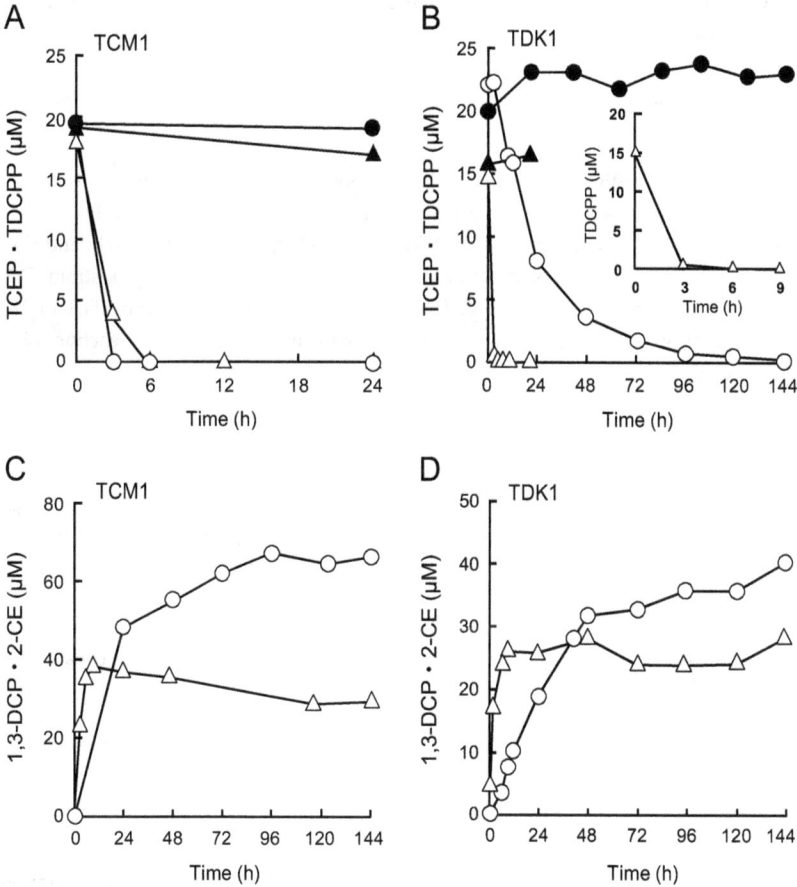

Figure 6. Degradation of TDCPP and TCEP by strains TCM1 (A) and TDK1 (B) and generation of 2-CE and 1,3-DCP (C and D). The cultivations were performed aerobically at 30°C in a minimal medium containing 20 µM of TCEP or TDCPP as the sole phosphorus source. (A and B) Open circles and triangles represent the concentrations of TCEP and TDCPP, respectively, and their filled forms represent concentrations for autoclaved control cells. (C and D) Open circles and triangles represent the concentrations of 2-CE and 1,3-DCP, respectively. Each data point represents the mean of at least two independent determinations.

2.2. Microbial detoxification of TDCPP and TCEP by two bacterial strains

We have successfully isolated TCEP- and TDCPP-degrading bacteria. However, neither strain can degrade the resulting toxic and persistent metabolites 2-CE and 1,3-DCP. Elimination of the metabolites is required before the strains can be used to degrade TDCPP and TCEP in practice. Fortunately, bacteria with chloroalcohol-degrading ability have been well-documented. We thus attempted to completely detoxify the PFRs by combining strain TCM1 with bacteria capable of degrading the chloroalcohols (Takahashi et al., 2012a; Takahashi, et al., 2012b).

2.2.1. Microbial detoxification of TDCPP using Sphingobium sp. strain TCM1 and Arthrobacter sp. strain PY1

Several 1,3-DCP-degrading bacteria have been reported, including *Arthrobacter* sp. strains PY1 (Yonetani et al., 2004) and AD2 (van den Wijngaard et al., 1991), *A. erithii* H10a (Assis et al., 1998), *Agrobacterium radiobacter* strain AD1 (van den Wijngaard et al., 1989), and *Corynebacterium* sp. strain N-1074 (Nakamura et al., 1991). Of these, *Arthrobacter* sp. strain PY1 exhibits high 1,3-DCP degradation ability. Therefore, we attempted to detoxify TDCPP by cohabitation of strain TCM1 and *Arthrobacter* sp. PY1 in a resting cell reaction (Fig. 7) (Takahashi et al., 2012a).

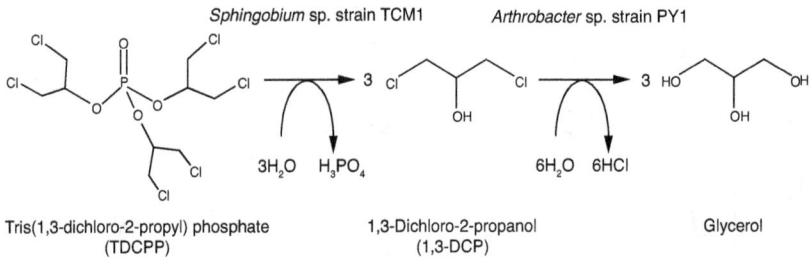

Figure 7. Complete detoxification of TDCPP by *Sphingobium* sp. strain TCM1 and 1,3-DCP-degrading bacterium *Arthrobacter* sp. strain PY1.

2.2.1.1. Freezing and lyophilization of strains TCM1 and PY1 cells

For resting cell preparation, we first examined the effect of freezing and lyophilization on the activity of strains TCM1 and PY1. The TDCPP-hydrolyzing activity of strain TCM1 intact cells was 1.07 μmol h^{-1} OD$_{660}$$^{-1}$, whereas respective activities of frozen and lyophilized cells were 0.90 and 0.84 μmol h^{-1} OD$_{660}$$^{-1}$. On the other hand, the 1,3-DCP-dehalogenating activity of strain PY1 intact cells was 0.22 μmol h^{-1} OD$_{660}$$^{-1}$, with respective frozen and lyophilized cell activities of 0.23 and 0.26 μmol h^{-1} OD$_{660}$$^{-1}$. These results reveal that freezing and lyophilization treatments cause no significant decline in degradation activities of the strains.

2.2.1.2. Optimum TDCPP and 1,3-DCP degradation conditions of strains TCM1 and PY1

We then determined the optimum temperature and pH for lyophilized cell activity (Fig. 8). At pH 9.0 for strain TCM1 and pH 8.5 for strain PY1, the highest activity of TCM1 and PY1 cells occurred at 30°C (2.53 μmol h^{-1} OD$_{660}$$^{-1}$) and 35°C (1.31 μmol h^{-1} OD$_{660}$$^{-1}$), respectively (Fig. 8A). At 30°C, the highest activity of TCM1 and PY1 cells occurred at pH 8.5 (2.48 μmol h^{-1} OD$_{660}$$^{-1}$) and pH 9.5 with 50 mM Tris-H$_2$SO$_4$ (0.95 μmol h^{-1} OD$_{660}$$^{-1}$), respectively (Fig. 8B). We thus established the optimum temperature as 30°C and 35°C and the optimum pH as 8.5 and 9.5 for strains TCM1 and PY1, respectively.

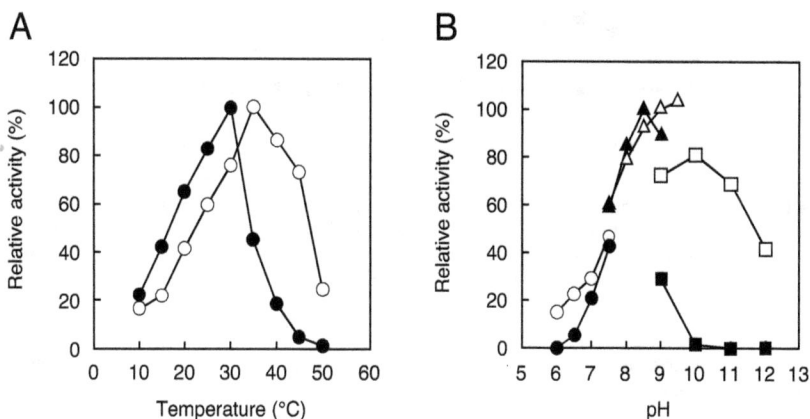

Figure 8. Effect of temperature and pH on the degradation activity of strains TCM1 and PY1. (A) effect of temperature: TDCPP hydrolyzation activity of strain TCM1 cells (closed circle) and 1,3-DCP dehalogenation activity of strain PY1 cells (open circle) were, respectively, assayed in 50 mM Tris-H$_2$SO$_4$ buffer (pH 9.0) and 50 mM Tris-H$_2$SO$_4$ buffer (pH 8.5). (B) effect of pH: TDCPP hydrolyzation activity of strain TCM1 cells (closed symbols) and 1,3-DCP dehalogenation activity of strain PY1 cells (open symbols) was assayed at 30°C in 50 mM MOPS-NaOH buffer (circle, pH 6.0-7.5), Tris-H$_2$SO$_4$ buffer (triangle, pH 7.5-9.5), and glycine-NaOH buffer (square, pH 9.0-12.0). Each datum represents means of two independent determinations.

2.2.1.3. Complete detoxification of TDCPP by mixed bacteria cells

Based on the optimum conditions, we set the reaction temperature to 30°C and pH to 9.0 (50 mM Tris-H$_2$SO$_4$) for TDCPP detoxification by mixed bacteria (Fig. 9). Under these conditions, the respective activities of strains TCM1 and PY1 were 2.21 and 0.92 μmol h^{-1} OD$_{660}$$^{-1}$. In the detoxification reaction using a mixture of TCM1 and PY1 cells (OD$_{660}$ 0.05 and 0.2, respectively), approximately 50 μM of TDCPP disappeared within 1 h, and 1,3-DCP and chloride ions were formed to levels of approximately 100 and 120 μM, respectively, after 2 h (Fig. 9A). This result suggests incomplete detoxification of TDCPP due to low 1,3-DCP dehalogenation activity. Increasing the strain PY1 population to an OD$_{660}$ of 4.0 decreased the TDCPP hydrolyzation rate of TCM1 cells, but completely eliminated the resulting 1,3-DCP after 10 h (Fig. 9B). At the same time, chloride ion concentration had reached its theoretical value expected from the initial TDCPP concentration, demonstrating that complete detoxification of TDCPP is achievable using strains TCM1 and PY1.

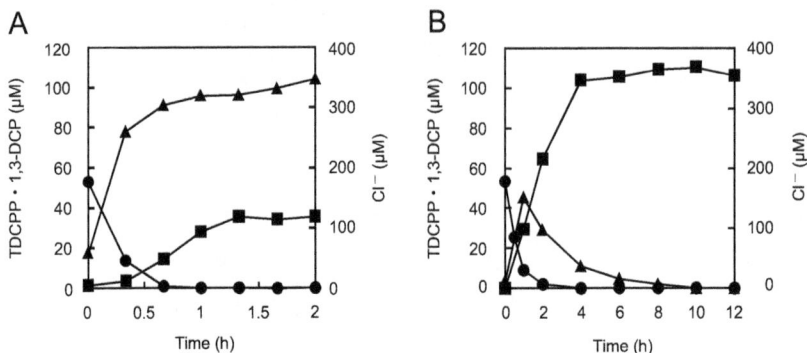

Figure 9. Complete detoxification of TDCPP by the mixed resting cells of strains TCM1 and PY1. The reactions were performed at 30°C with 50 μM TDCPP in 50 mM Tris - H_2SO_4 buffer (pH 9.0), and TDCPP (circles), 1,3-DCP (triangles) and chloride ion (squares) were determined. Cell concentrations of strains TCM1 and PY1 for each reaction were, respectively, OD_{660} of 0.05 and 0.2 (A) and 0.04 and 4.0 (B). Each datum represents means of two independent determinations.

2.2.2. Microbial detoxification of TCEP using Sphingobium sp. strain TCM1 and Xanthobacter autotrophicus strain GJ10

Several 2-CE-degrading bacteria have been reported, including *Xanthobacter autotrophicus* strain GJ10 (Janssen et al., 1985), *Pseudomonas putida* strain US2 (Strotmann et al., 1990) and *P. atutzeri* strain JJ (Dijk et al., 2003). Among these, the degradation of 2-CE by *X. autotrophicus* strain GJ10 has been well characterized. Therefore, we attempted to detoxify TCEP by co-habitation of strain TCM1 and *X. autotrophicus* strain GJ10 (Fig. 10) (Takahashi et al., 2012b).

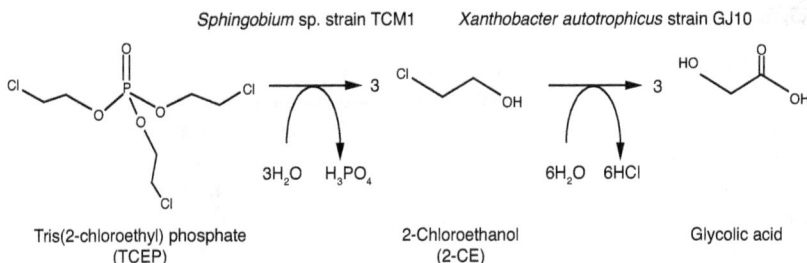

Figure 10. Complete detoxification of TCEP by *Sphingobium* sp. strain TCM1 and 2-CE-degrading bacterium *Xanthobacter autotrophicus* strain GJ10.

2.2.2.1. Optimum TCEP degradation condition of strain TCM1

We first determined the optimum temperature and pH for TCEP degradation by strain TCM1 in a resting reaction using lyophilized cells. At pH 7.4, the highest activity was obtained at 30°C (14.1 nmol min^{-1} OD$_{660}$$^{-1}$). Maintaining this temperature and varying the pH, the highest activity was recorded at pH 8.5 (14.6 nmol min^{-1} OD$_{660}$$^{-1}$). These optimum conditions were identical to those for TDCPP, suggesting that the same enzyme(s) might be involved in the degradation of both compounds.

Under the optimum conditions, TCM1 cells completely eliminated 10, 20 and 50 μM of TCEP within 3 h, but the generated 2-CE was approximately 50% of its theoretical value based on the initial TCEP concentrations (Fig. 11). Phosphotriesterase that can hydrolyze organophosphorus pesticides structurally similar to TCEP, such as chlorpyrifos, require two zinc ions for catalysis, and enzyme activity can be maximized by replacing Zn^{2+} with Co^{2+} (Omburo et al., 1992). A bacterial phosphodiesterase that can hydrolyze alkyl phosphodiesters similarly requires divalent metals (Gerlt & Wan, 1979). We therefore examined the effect of Co^{2+} as well as cell amount on TCEP hydrolysis (Fig. 11). In the reaction using approximately 10 μM of TCEP without Co^{2+}, 2-CE reached 21.2 μM (OD$_{660}$ of 0.8) after 3 h. Addition of 50 μM Co^{2+} resulted in an increase of 2-CE to 32.3 μM, equivalent to the theoretical value of 30 μM (Fig. 11B). These results showed that complete hydrolysis can be achieved at an OD$_{660}$ of 0.8 with 50 μM of Co^{2+}.

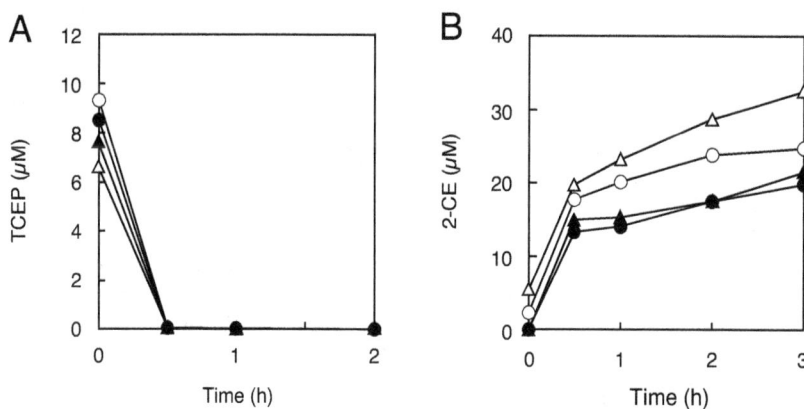

Figure 11. Effect of Co^{2+} and cell amount on TCEP hydrolysis by strain TCM1-resting cells. The reactions were performed at 30°C using the resting cells at OD$_{660}$ of 0.4 (circles) or 0.8 (triangles) with (open symbols) or without (closed symbols) 50 μM Co^{2+} in 50 mM Tris-H$_2$SO$_4$ buffer (pH 8.5) containing 10 μM TCEP, and TCEP (A) and 2-CE (B) were determined. Each datum represents the mean of two independent determinations. The inconsistency of the initial concentrations of TCEP at zero time with the set-up ones was mainly attributed to reaction progress in several minutes to stop the reaction.

2.2.2.2. Optimum 2-CE degradation condition of strain GJ10

We prepared resting cells of intact, frozen and lyophilized cells of X. *autotrophicus* strain GJ10 and examined their 2-CE degradation activity. Activity was detected only in frozen cells at 4.93 pmol min^{-1} OD$_{450}$$^{-1}$, four orders lower than the TCEP degradation activity of strain TCM1. This low 2-CE degradation activity might be attributable to the lack of coenzyme regeneration of enzymes involved in the degradation process. We next examined 2-CE degradation in a growing cell reaction. The growing cells completely degraded approximately 180 μM of 2-CE within 24 h. The degradation ability was estimated to be a minimum of 7.5 μM h^{-1}, comparable to the TCEP degradation ability of strain TCM1-resting cells (approximately 10 μM h^{-1}). This result shows that growing cells of strain GJ10 can degrade 2-CE effectively.

2.2.2.3. Complete detoxification of TCEP by two bacterial strains

Based on the results described above, we examined whether combining TCEP hydrolysis by TCM1 resting cells and 2-CE degradation by GJ10 growing cells would completely detoxify TCEP (Fig. 12). TCM1 resting cells abolished 9.6 μM of TCEP within 4 h, releasing 2-CE at 29.0 μM, equivalent to that estimated from the initial TCEP concentration, and consistent with complete TCEP hydrolysis (Fig. 12A and B). The generated 2-CE was abolished by GJ10 growing cells within 48 h, and chloride ion concentration reached 30.2 μM after 144 h, equivalent to that estimated from the generated 2-CE (Fig. 12C and D). Taken together, these results demonstrate that complete detoxification of TDCPP can be achieved using strains TCM1 and GJ10.

3. Concluding remarks

We have successfully isolated two novel bacterial strains capable of degrading the persistent and potential toxic PFRs, TCEP and TDCPP, which have become worldwide environmental contaminants. The two strains TCM1 and TDK1 belong to *Sphingobium* sp. and *Sphingomonas* sp. respectively. The strains are the first microorganisms reported to degrade the persistent PFRs. They degrade the compounds by hydrolyzing their phosphotriester bonds to produce metabolites 1,3-DCP from TDCPP and 2-CE from TCEP, which are themselves toxic and non-self-biodegradable. In a successful attempt to completely detoxify the FPRs, we combined TCM1 with the 1.3-DCP-degrading bacterium *Arthrobacter* sp. strain PY1 (for TDCPP degradation), and with the 2-CE-degrading bacterium X. *autotrophicus* strain GJ10 (for TCEP degradation). This is the first description of microbial FPR detoxification. The bacteria and the microbial detoxification techniques may prove useful for the bioremediation of sites contaminated with intractable compounds. Further studies on the PFRs-degrading bacteria as well as the chloroalcohols-degrading bacteria, and on the detoxification techniques, could help to establish more efficient detoxifications, and could also provide novel insights into microbial degradation of organophosphorus compounds. We are now working towards elucidating the enzymes and the genes involved in the degradation processes.

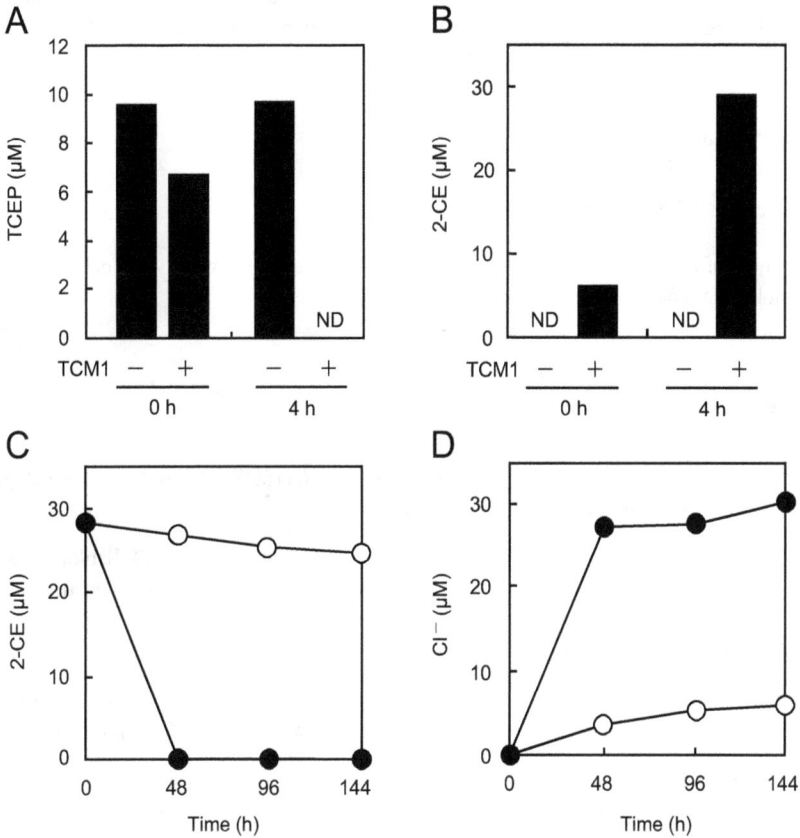

Figure 12. Complete detoxification of TCEP by *Sphingobium* sp. strain TCM1-resting cell reaction (A and B) and the following *X. autotrophicus* GJ10-growing cell reaction (C and D). The resting cell reaction was performed at 30°C with (+) or without (-) strain TCM1 cells at OD_{660} of 0.8 in 50 mM Tris-H_2SO_4 buffer (pH 8.5) containing 10 μM TCEP and 50 μM Co^{2+}, and TCEP (A) and 2-CE (B) were determined. The growing cell reaction was performed at 30°C with (closed symbols) or without (open symbols) strain GJ10 cells in a medium containing the generated 2-CE as the sole carbon source, and 2-CE (C) and chloride ion (D) was determined. ND means not detected. Each datum represents the mean of two independent determinations.

Acknowledgements

This research was supported in part by a Grant-in-Aid for Scientific Research (B) (to Y. K) from the Ministry of Education, Science, Sports, and Culture of Japan, by the River Environment Fund (REF) in charge of the Foundation of River and Watershed Environment Management (FOREM) (to Y. K.), by a grant from the Uchida Energy Science Promotion Foundation (to S. T.) and by the Kurita Water and Environment Foundation (to S. T).

Author details

Shouji Takahashi*, Katsumasa Abe and Yoshio Kera

Department of Environmental Systems Engineering, Nagaoka University of Technology, Kamitomioka, Nagaoka, Niigata, Japan

References

[1] Agency for Toxic ubstances & Disease Registry (ATSDR) (2009). Draft toxicological profile for phosphate ester flame retardants.

[2] Albemarle Corp. & ICL North America Inc. (2011). TDCPP Tris(1,3-dichloro-2-propyl) phosphate. Comments on July 2011 "Evidence of Carcinogenicity" Document Prepared by OEHHA and Arguments Opposing Listing TDCPP as a Carcinogen Under Proposition 65.

[3] Andresen, J. A., Grundmann, A., & Bester, K. (2004). Organophosphorus flame retardants and plasticisers in surface waters. *Sci. Total Environ.*, 332(1-3), 155-166.

[4] Andresen, J. A. & Bester, K. (2006). Elimination of organophosphate ester flame retardants and plasticizers in drinking water purification. *Water Res.*, 40(3), 621-629.

[5] Assis, H. M., Sallis, P. J., Bull, A. T., & Hardman, D. J. (1998). Biochemical characterization of a haloalcohol dehalogenase from *Arthrobacter erithii* H10a. *Enzyme Microb. Technol.*, 22(7), 568-574.

[6] Bacaloni, A., Cavaliere, C., Foglia, P., Nazzari, M., Samperi, R., & Lagana, A. (2007). Liquid chromatography/tandem mass spectrometry determination of organophosphorus flame retardants and plasticizers in drinking and surface waters. *Rapid Commun. Mass Spectrom.*, 21(7), 1123-1130.

[7] Bacaloni, A., Cucci, F., Guarino, C., Nazzari, M., Samperi, R., & Lagana, A. (2008). Occurrence of organophosphorus flame retardant and plasticizers in three volcanic lakes of central Italy. *Environ. Sci. Technol.*, 42(6), 1898-1903.

[8] Beard, A. & Reilly, T. (2009). Additives used in Flame Retardant Polymer Formulations: Current Practice & Trends "Fire Retardants and their Potential Impact on Fire Fighter Health". Workshop at NIST, Gaithersburg, MD USA, 30-Sep-2009.

[9] Benotti, M. J., Stanford, B. D., Wert, E. C., & Snyder, S. A. (2009). Evaluation of a photocatalytic reactor membrane pilot system for the removal of pharmaceuticals and endocrine disrupting compounds from water. *Water Res.*, 43(6), 1513-1522.

[10] Bergh, C., Torgrip, R., Emenius, G., & Ostman, C. (2011). Organophosphate and phthalate esters in air and settled dust - a multi-location indoor study. *Indoor Air*, 21(1), 67-76.

[11] Bjorklund, J., Isetun, S., & Nilsson, U. (2004). Selective determination of organophosphate flame retardants and plasticizers in indoor air by gas chromatography, positive-ion chemical ionization and collision-induced dissociation mass spectrometry. *Rapid Commun. Mass Spectrom.*, 18(24), 3079-3083.

[12] California Environmental Protection Agency (California EPA) (2011). Evidence on the carcinogenicity of Tris(1,3-dichloro-2-propyl) phosphate.

[13] Crump, D., Chiu, S., & Kennedy, S. W. (2012). Effects of tris(1,3-dichloro-2-propyl) phosphate and tris(1-chloropropyl) phosphate on cytotoxicity and mRNA expression in primary cultures of avian hepatocytes and neuronal cells. *Toxicol. Sci.*, 126(1), 140-148.

[14] Dijk, J. A., Stams, A. J., Schraa, G., Ballerstedt, H., de Bont, J. A., & Gerritse, J. (2003). Anaerobic oxidation of 2-chloroethanol under denitrifying conditions by *Pseudomonas stutzeri* strain JJ. *Appl. Microbiol. Biotechnol.*, 63(1), 68-74.

[15] Dishaw, L. V., Powers, C. M., Ryde, I. T., Roberts, S. C., Seidler, F. J., Slotkin, T. A., & Stapleton, H. M. (2011). Is the PentaBDE replacement, tris (1,3-dichloro-2-propyl) phosphate (TDCPP), a developmental neurotoxicant? Studies in PC12 cells. *Toxicol. Appl. Pharmacol.*, 256(3), 281-289.

[16] Echigo, S., Yamada, H., Matsui, S., Kawanishi, S., & Shishida, K. (1996). Comparison between O_3/VUV, O_3/H_2O_2, VUV and O_3 processes for the decomposition of organophosphoric acid triesters. *Water Sci. Technol.*, 34(9), 81-88.

[17] European Commission DG ENV (2011). Flame retardants found in groundwater.

[18] EU Scientific Committee on Health and Environmental Risks (SCHER) (2012). Opinion on tris(2-chloroethyl) phosphate (TCEP) in Toys.

[19] European Union Risk Assessment Report (EURAR) (2008). European Union Risk Assessment Report Tris (2-chloro-1-(chloromethyl)ethyl) phosphate (TDCP).

[20] European Union Risk Assessment Report (EURAR) (2009). European Union Risk Assessment Report Tris(2-chloroethyl) phosphate, TCEP.

[21] Fries, E. & Püttmann, W. (2001). Occurrence of organophosphate esters in surface water and ground water in Germany. *J. Environ. Monit.*, 3(6), 621-626.

[22] Fries, E. & Püttmann, W. (2003). Monitoring of the three organophosphate esters TBP, TCEP and TBEP in river water and ground water (Oder, Germany). *J. Environ. Monit.*, 5(2), 346-352.

[23] Garcia, M., Rodriguez, I., & Cela, R. (2007). Microwave-assisted extraction of organo-phosphate flame retardants and plasticizers from indoor dust samples. *J. Chromatogr A*, 1152(1-2), 280-286.

[24] Gerlt, J. A. & Wan, W. H. (1979). Stereochemistry of the hydrolysis of the endo iso-mer of uridine 2', 3'-cyclic phosphorothioate catalyzed by the nonspecific phospho-hydrolase from *Enterobacter aerogenes*. *Biochemistry*, 18(21), 4630-4638.

[25] Hartmann, P. C., Burgi, D., & Giger, W. (2004). Organophosphate flame retardants and plasticizers in indoor air. *Chemosphere*, 57(8), 781-787.

[26] Ingerowski, R. & Ingerowski, G. (1997). Moegliche neurotoxische Wirkung des chlor-ierten Posphorsaeureesters Tris(2-chlor-ethyl)-phosphat-eine umweltmedizinishe Ka-zuiseik. *Internist. Prax.*, 37, 229-230.

[27] Ishikawa, S., Shigezumi, K., Yasuda, K., & Shigemori, N. (1985). Determination of or-ganic phosphate esters in factory effluent and domestic effluent. *Suishituodakukenkyu*, 8(8), 529-535.

[28] Ishikawa, S., Taketomi, M., & Shinohara, R. (1985). Determination of trialkyl and tri-aryl phosphates in environmental samples. *Water Res.*, 19(1), 119-125.

[29] Janssen, D. B., Scheper, A., Dijkhuizen, L., & Witholt, B. (1985). Degradation of halo-genated aliphatic compounds by *Xanthobacter autotrophicus* GJ10. *Appl. Environ. Mi-crobiol.*, 49(3), 673-677.

[30] Kanazawa, A., Saito, I., Araki, A., Takeda, M., Ma, M., Saijo, Y., & Kishi, R. (2010). Association between indoor exposure to semi-volatile organic compounds and build-ing-related symptoms among the occupants of residential dwellings. *Indoor Air*, 20(1), 72-84.

[31] Kawagoshi, Y., Fukunaga, I., & Itoh, H. (1999). Distribution of organophosphoric acid triesters between water and sediment at a sea-based solid waste disposal site. *J. Mater. Cycles Waste Manag.*, 1(1), 53-61.

[32] Kim, S. D., Cho, J., Kim, I. S., Vanderford, B. J., & Snyder, S. A. (2007). Occurrence and removal of pharmaceuticals and endocrine disruptors in South Korean surface, drinking, and waste waters. *Water Res.*, 41(5), 1013-1021.

[33] Lee, S., Nguyen, Q., Lee, E., Kim, S., Lee, S., Jung, Y., Choi, S., & Cho, J. (2008). Effi-cient removals of tris(2-chloroethyl) phosphate (TCEP) and perchlorate using NF membrane filtrations. *Desalination*, 221(1-3), 234-237.

[34] Li, X., He, J., & Li, S. (2007). Isolation of a chlorpyrifos-degrading bacterium, *Sphingo-monas* sp. strain Dsp-2, and cloning of the mpd gene. *Res. Microbiol.*, 158(2), 143-149.

[35] Liu, X., Ji, K., & Choi, K. (2012). Endocrine disruption potentials of organophosphate flame retardants and related mechanisms in H295R and MVLN cell lines and in ze-brafish. *Aquat. Toxicol.*, 114-115, 173-181.

[36] Marklund, A., Andersson, B., & Haglund, P. (2005a). Organophosphorus flame retardants and plasticizers in air from various indoor environments. *J. Environ. Monit.*, 7(8), 814-819.

[37] Marklund, A., Andersson, B., & Haglund, P. (2005b). Organophosphorus flame retardants and plasticizers in Swedish sewage treatment plants. *Environ. Sci. Technol.*, 39(19), 7423-7429.

[38] Marklund, A., Andersson, B., & Haglund, P. (2003). Screening of organophosphorus compounds and their distribution in various indoor environments. *Chemosphere*, 53(9), 1137-1146.

[39] Martinez-Carballo, E., Gonzalez-Barreiro, C., Sitka, A., Scharf, S., & Gans, O. (2007). Determination of selected organophosphate esters in the aquatic environment of Austria. *Sci. Total Environ.*, 388(1-3), 290-299.

[40] Meeker, J. D. & Stapleton, H. M. (2010). House dust concentrations of organophosphate flame retardants in relation to hormone levels and semen quality parameters. *Environ. Health Perspect.*, 118(3), 318-323.

[41] Meyer, J. & Bester, K. (2004). Organophosphate flame retardants and plasticisers in wastewater treatment plants. *J. Environ. Monit.*, 6(7), 599-605.

[42] Moller, A., Sturm, R., Xie, Z., Cai, M., He, J., & Ebinghaus, R. (2012). Organophosphorus flame retardants and plasticizers in airborne particles over the northern pacific and Indian ocean toward the polar regions: evidence for global occurrence. *Environ. Sci. Technol.*, 46(6), 3127-3134.

[43] Morales, N. M. & Matthews, H. B. (1980). In vivo binding of the flame retardants tris-2,3-dibromopropyl phosphate and tris-1,3-dichloro-2-propyl phosphate to macromolecules of mouse liver, kidney and muscle. *Bull. Environ. Contam. Toxicol.*, 25(1), 34-38.

[44] Nakamura, T., Yu, F., Mizunashi, W., & Watanabe, I. (1991). Microbial transformation of prochiral 1,3-dichloro-2-propanol into optically active 3-chloro-1,2-propanediol. *Agric. Biol. Chem.* 55(7), 1931-1933.

[45] National toxicology program (NPT) (1985). NTP toxicology and carcinogenesis studies of 2-chloroethanol (ethylene chlorohydrin) (CAS No. 107-07-3) in F344/N rats and Swiss CD-1 mice (dermal studies). *Natl. Toxicol. Program Tech. Rep. Ser.*, 275, 1-194.

[46] National Toxicology Program (NPT) & National Institute of Environmental Health Sciences (NIEHS) (2005). 1,3-Dichloro-2-propanol [CAS No. 96-23-1] Review of Toxicological Literature.

[47] Omburo, G. A., Kuo, J. M., Mullins, L. S., & Raushel, F. M. (1992). Characterization of the zinc binding site of bacterial phosphotriesterase. *J. Biol. Chem.*, 267(19), 13278-13283.

[48] Regnery, J. & Püttmann, W. (2010). Occurrence and fate of organophosphorus flame retardants and plasticizers in urban and remote surface waters in Germany. *Water Res.*, 44(14), 4097-4104.

[49] Regnery, J. & Püttmann, W. (2009). Organophosphorus flame retardants and plasticizers in rain and snow from middle Germany. *CLEAN - Soil, Air, Water*, 37(4-5), 334-342.

[50] Ren, X., Lee, Y. J., Han, H. J., & Kim, I. S. (2008). Effect of tris-(2-chloroethyl)-phosphate (TCEP) at environmental concentration on the levels of cell cycle regulatory protein expression in primary cultured rabbit renal proximal tubule cells. *Chemosphere*, 74(1), 84-88.

[51] Rodriguez, I., Calvo, F., Quintana, J. B., Rubi, E., Rodil, R., & Cela, R. (2006). Suitability of solid-phase microextraction for the determination of organophosphate flame retardants and plasticizers in water samples. *J. Chromatogr A*, 1108(2), 158-165.

[52] Saito, I., Onuki, A. & Seto, H. (2007). Indoor organophosphate and polybrominated flame retardants in Tokyo. *Indoor Air*, 17(1), 28-36.

[53] Staaf, T. & Ostman C. (2005). Indoor air sampling of organophosphate triesters using solid phase extraction (SPE) adsorbents. *J. Environ. Monit.*, 7(4), 344-348.

[54] Stackelberg, P. E., Furlong, E. T., Meyer, M. T., Zaugg, S. D., Henderson, A. K., & Reissman, D. B. (2004). Persistence of pharmaceutical compounds and other organic wastewater contaminants in a conventional drinking-water-treatment plant. *Sci. Total Environ.*, 329(1-3), 99-113.

[55] Stackelberg, P. E., Gibs, J., Furlong, E. T., Meyer, M. T., Zaugg, S. D., & Lippincott, R. L. (2007). Efficiency of conventional drinking-water-treatment processes in removal of pharmaceuticals and other organic compounds. *Sci. Total Environ.*, 377(2-3), 255-272.

[56] Stapleton, H. M., Klosterhaus, S., Eagle, S., Fuh, J., Meeker, J. D., Blum, A., & Webster, T. F. (2009). Detection of organophosphate flame retardants in furniture foam and U.S. house dust. *Environ. Sci. Technol.*, 43(19), 7490-7495.

[57] Stapleton, H. M., Klosterhaus, S., Keller, A., Ferguson, P. L., van Bergen, S., Cooper, E., Webster, T. F., & Blum, A. (2011). Identification of flame retardants in polyurethane foam collected from baby products. *Environ. Sci. Technol.*, 45(12), 5323-5331.

[58] Strotmann, U. J., Pentenga, M., & Janssen, D. B. (1990). Degradation of 2-ehloroethanol by wild type and mutants of *Pseudomonas putida* US2. *Arch. Microbiol.*, 154(3), 294-300.

[59] Sundkvist, A. M., Olofsson, U., & Haglund, P. (2010). Organophosphorus flame retardants and plasticizers in marine and fresh water biota and in human milk. *J. Environ. Monit.*, 12(4), 943-951.

[60] Takahashi, S., Kawashima, K., Kawasaki, M., Kamito, J., Endo, Y., Akatsu, K., Horino, S., Yamada, R., & Kera, Y. (2008). Enrichment and characterization of chlorinated

organophosphate ester-degrading mixed bacterial cultures. *J. Biosci. Bioeng.*, 106(1), 27-32.

[61] Takahashi, S., Obana, Y., Okada, S., Abe, K., & Kera, Y. (2012a). Complete detoxification of tris(1,3-dichloro-2-propyl) phosphate by mixed two bacteria, *Sphingobium* sp. strain TCM1 and *Arthrobacter* sp. strain PY1. *J. Biosci. Bioeng.*, 113(1), 79-83.

[62] Takahashi, S., Miura, K., Abe, K., & Kera, Y. (2012b). Complete detoxification of tris(2-chloroethyl) phosphate by two bacterial strains: *Sphingobium* sp. strain TCM1 and *Xanthobacter autotrophicus* strain GJ10. *J. Biosci. Bioeng.*, 114(3), 306-311.

[63] Takahashi, S., Satake, I., Konuma, I., Kawashima, K., Kawasaki, M., Mori, S., Morino J, Mori J, Xu H, Abe K, Yamada R., & Kera Y. (2010). Isolation and identification of persistent chlorinated organophosphorus flame retardant-degrading bacteria. *Appl. Environ. Microbiol.*, 76(15), 5292-5296.

[64] Takeuchi, M., Hamana, K., & Hiraishi, A. (2001). Proposal of the genus *Sphingomonas sensu stricto* and three new genera, *Sphingobium*, *Novosphingobium* and *Sphingopyxis*, on the basis of phylogenetic and chemotaxonomic analyses. *Int. J. Syst. Evol. Microbiol.*, 51(4), 1405-1417.

[65] Takigami, H., Suzuki, G., Hirai, Y., Ishikawa, Y., Sunami, M., & Sakai, S. (2009). Flame retardants in indoor dust and air of a hotel in Japan. *Environ. Int.*, 35(4), 688-693.

[66] The European Flame Retardants Associatio (EFRA) (2007). Flame Retardants Frequently Asked Questions.

[67] The German Federal Environment Agency (UBA) (2008). Guidelines for Indoor Air Hygiene in School Buildings. Produced by the German Federal Environment Agency's Indoor Air Hygiene Commission.

[68] The Norwegian Climate and Pollution Agency (KLIF) (2011). Screening of organophosphor flame retardants 2010.

[69] The Norwegian Pollution Control Authority (SFT) (2009a). Guidance on alternative flame retardants to the use of commercial pentabromodiphenylether (c-PentaBDE).

[70] The Norwegian Pollution Control Authority (SFT) (2008). Screening of selected metals and new organic contaminants 2007.

[71] The Norwegian Pollution Control Authority (SFT) (2009b). Screening of new contaminants in samples from the norwegian arctic. Silver, Platinum, Sucralose, Bisphenol A, Tetrabrombisphenol A, Siloxanes, Phtalates (DEHP) and Phosphororganic flame retardants.

[72] Tollback, J., Tamburro, D., Crescenzi, C., & Carlsson, H. (2006). Air sampling with Empore solid phase extraction membranes and online single-channel desorption/ liquid chromatography/mass spectrometry analysis: determination of volatile and semi-volatile organophosphate esters. *J. Chromatogr A*, 1129(1), 1-8.

[73] US Environmental Protection Agency (US EPA) (2005). Flame retardant alternatives tris(1,3-dichloro-2-propyl) phosphate harzard review. Chemical Hazard Reviews. Furniture Flame Retardancy Partnership: Environmental Profiles of Chemical Flame-Retardant Alternatives for Low-Density Polyurethane Foam. Vol. 2, pp. 3-1-3-53.

[74] van den Eede, N., Dirtu, A. C., Neels, H., & Covaci, A. (2011). Analytical developments and preliminary assessment of human exposure to organophosphate flame retardants from indoor dust. *Environ. Int.*, 37(2), 454-461.

[75] van den Wijngaard, A. J., Janssen, D. B., & Witholt, B. (1989). Degradation of epichlorohydrin and halohydrins by bacterial cultures isolated from freshwater sediment. *Microbiology*, 135(8), 2199-2208.

[76] van den Wijngaard, A. J., Reuvekamp, P. T., & Janssen, D. B. (1991). Purification and characterization of haloalcohol dehalogenase from *Arthrobacter* sp. strain AD2. *J. Bacteriol.*, 173(1), 124-129.

[77] van der Veen, I. & de Boer, J. (2012). Phosphorus flame retardants: Properties, production, environmental occurrence, toxicity and analysis. *Chemosphere*, 88(10), 1119-1153.

[78] Watts, M. J. & Linden, K. G. (2008). Photooxidation and subsequent biodegradability of recalcitrant tri-alkyl phosphates TCEP and TBP in water. *Water Res.*, 42(20), 4949-4954.

[79] Westerhoff, P., Yoon, Y., Snyder, S., & Wert, E. (2005). Fate of endocrine-disruptor, pharmaceutical, and personal care product chemicals during simulated drinking water treatment processes. *Environ. Sci. Technol.*, 39(17), 6649-6663.

[80] Worls Health Organization (WHO) (1997). Environmental Health Criteria 192: Flame Retardants: A General Introduction: World Health Organization, Geneva.

[81] Yonetani, R., Ikatsu, H., Miyake-Nakayama., C., Fujiwara, E., Maehara, Y., Miyoshi., S., Matsuoka, H., & Shinoda, S. (2004). Isolation and characterization of a 1,3-dichloro-2-propanol-degrading bacterium. *J. Health Sci.*, 50(6), 605-612.

Chapter 7

Continuous Biotechnological Treatment of Cyanide Contaminated Waters by Using a Cyanide Resistant Species of *Aspergillus awamori*

Bruno Alexandre Quistorp Santos,
Seteno Karabo Obed Ntwampe and
James Hamuel Doughari

Additional information is available at the end of the chapter

1. Introduction

Various industries release a combination of free cyanide and cyanide complexes into the environment via a variety of disposal methods, particularly as wastewater. These industries utilise cyanide based compounds in various operations, including: the beneficiation of metals, electroplating, case hardening, automotive manufacturing, circuitry board manufacturing, and in chemical industries [23]. Cyanide is often found in organic, hydrocarbon chains or as inorganic, transition, alkali and alkali earth metal complexes [20]. Many cyanide complexes are highly unstable, thus temperature, pH and light can degrade the components to form free cyanide which is the most toxic form of cyanide [20, 26].

There is an overwhelming popularity in industry for the use of chemical treatment methods for the treatment of free cyanide and cyanide complexes compared to biochemical treatment methods. Chemical remediation methods like alkaline chlorine oxidation are commonly used to treat cyanide contaminated wastewater [23, 24]. Chemical oxidation is particularly ineffective in the treatment of cyanide-metal complexes containing heavy metals, such as copper, nickel and silver, due to the slow reaction rate [23]. The excess quantity of chlorine used in the treatment process increases the chemical oxygen demand (COD) of the wastewater thereby rendering the water undesirable for reuse, toxic to aquatic life and may produce organic substances. In order to reduce operational costs, some manufacturers partially treat the wastewater, resulting in untreated and/or partially decomposed cyanide being discharged. Other methods of treatment include copper catalysed hydrogen peroxide oxidation, ozonation and electrolytic decomposition [23]. However, these methods are unpopular due to the high capital costs, specialist equipment and maintenance requirements.

Several microorganisms, bacteria such as *Nocardia sp.* and *Rhodococcus sp.*, fungi such as *Aspergillus sp.* and *Fusarium sp.* and algae such as *Arthrospira maxima* and *Scenedesmus obliquus*, possess enzymatic mechanisms able to bioremediate free cyanide and cyanide complexes [1, 26]. However, limited studies have been conducted using organic waste and fungal strains in cyanide bioremediation. Several studies have been conducted using varying concentrations of free cyanide, with moderate success being achieved in some cases [1].

Since the early 1970's, progress has been made to develop economically viable continuous remediation processes such as membrane bioreactors (MBRs) [8]. A membrane is generally defined as being a selective barrier. The membrane utilised in a bioreactor can provide either a barrier to limit the transport of certain components, while being permeable to others, thus prevent certain components from contacting a biocatalyst, or contain reactive sites thus being a catalyst itself [5]. The application of MBRs for the production of enzymes has received considerable attention for their diverse industrial use. A number of microorganisms have been studied in MBR applications for wastewater using fungi, such as white-rot fungus, *Phanerochaete chrysosporium* (*P. chrysosporium*) [8].

Solid waste generation in South Africa is a problem growing at an exponential rate with the majority of landfill sites reaching maximum capacity. Approximately 427×10^6 tonnes of solid waste is generated in South Africa every year, of which 40% by mass is organic waste [10]. The average amount of waste generated per person in South Africa is 0.7 kg/annum, which is close to that of developed countries such as the United Kingdom (0.723 kg/annum) and Singapore (0.87 kg/annum), than for developing countries, such as Nepal (0.3 kg/annum) [10]. It is sensible to bioaugment biotechnological processes to utilise organic waste materials, particularly for industries which produce large quantities of it.

2. Overview: Free cyanide

Free cyanide is the simplest form of cyanide and has two forms, namely as a hydrocyanic molecule or hydrogen cyanide (HCN) which dissociates into an anionic cyanide molecule (CN⁻) in solution [20]. By definition, free cyanides are forms of molecular and ionic cyanides that are released in aqueous solution by the dissolution and dissociation of cyanide complexes. Simple and weak acid dissociable cyanides are the most unstable and most likely to form free cyanide in aqueous solution. Simple cyanide compounds are ionically bonded cyanide anions and alkali earth or alkali metals that are neutral, that exist in solid form and dissociate into alkali earth or alkali metals and free cyanide when in aqueous solutions [20].

Accordingly the environmental risk of cyanide wastewater is not limited to the effluent but also the possibility of emitting hydrocyanic gas. Hydrocyanic gas is toxic, colourless, distinctive almond smell at low concentrations, slightly soluble in water and readily dissociates into hydrogen and anionic cyanide at low pH in aqueous solution [20]. For safety reasons, it is advised to keep cyanide solutions at a high pH to prevent the evolvement of hydrocyanic gas since there is a direct relationship between the dissociation of hydrocyanic molecule and pH (Figure 1) including temperature [20].

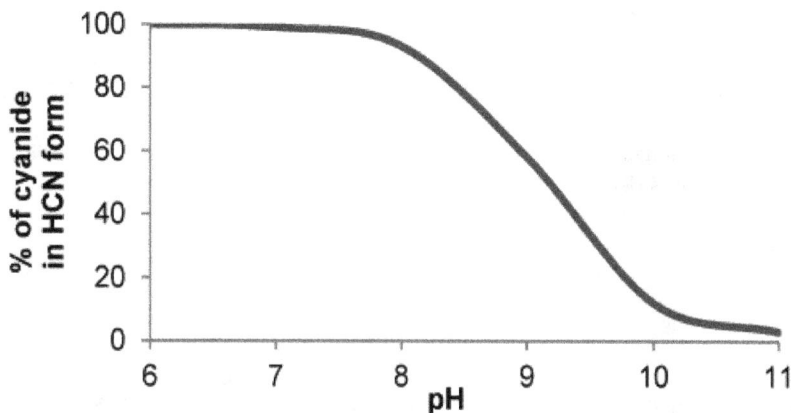

Figure 1. Relationship between HCN in solution and pH [20]

3. Differentiating filamentous fungi: *Aspergillus niger* and *Aspergillus awamori*

Black *Aspergilli* (*Aspergillus* section *Nigri*) species, are aerobic filamentous fungi often derived from soil. They have shown potential in biotechnology, food and medical applications. The trait of these species causing agricultural products to spoil had recently shown to be of benefit. Agricultural waste can be fermented to produce a variety of industrially important extracellular enzymes, such as cellulase, amylase, xylanase, pectinase, elastase, and organic acids, such as citric, galacturonic including gluconic acid. *Aspergillus niger* (*A. niger*) and *Aspergillus awamori* (*A. awamori*) are closely related species and *A. awamori* is often misidentified as *A. niger*. They share similar morphology and growth rates at various temperatures and produce several common enzymes [34].

3.1. Isolation and identification: *Aspergillus awamori*

A black filamentous mould was isolated from cyanide contaminated municipal wastewater discharge drain located in the Western Cape, South Africa. Swab samples were taken at various points along the drain and grown on 1 % (w/v) Citrus Pectin Agar (CPA) plates incubated at 37 °C for 5 days. After incubation, black mycelia of filamentous mould grew on the plate (Figure 2).

This was transferred to Potato Dextrose Agar (PDA) plates with 0.2% (v/v) Penicillin-Streptomycin (PEN-STREP; (10000 units/L of Penicillin and 10 mg of Streptomycin/ml) anti-biotic solution. The plates were again incubated at 37 °C for 5 days. One millilitre (1 ml) of 0.1% (w/v) Tween 80 solution was added to each plate and a spatula was used to harvest the spores and mycelium from the plate to form a spore-mycelium suspension. The suspension was then filtered through a glass wool using 20 ml syringes to entrap the mycelium onto the glass wool and produce a spore suspension which was stored at 4 °C. Afterwards, serial dilutions were made from the spore suspension and the number of spores in each 1 ml, were determined in duplicate using a Marienfeld Neubauer cell-counter and a Nikon Eclipse E2000 at a phase contrast of one and magnification of 100x. The absorbance of the diluted spore suspension was determined at 750 nm using a Jenway 6715 UV/Visible spectrophotometer with distilled water as a blank [31]. A calibration graph for the spore concentration was determined by plotting the absorbance against the spore concentration (spores/ml), to quantify spore concentration in the inoculum.

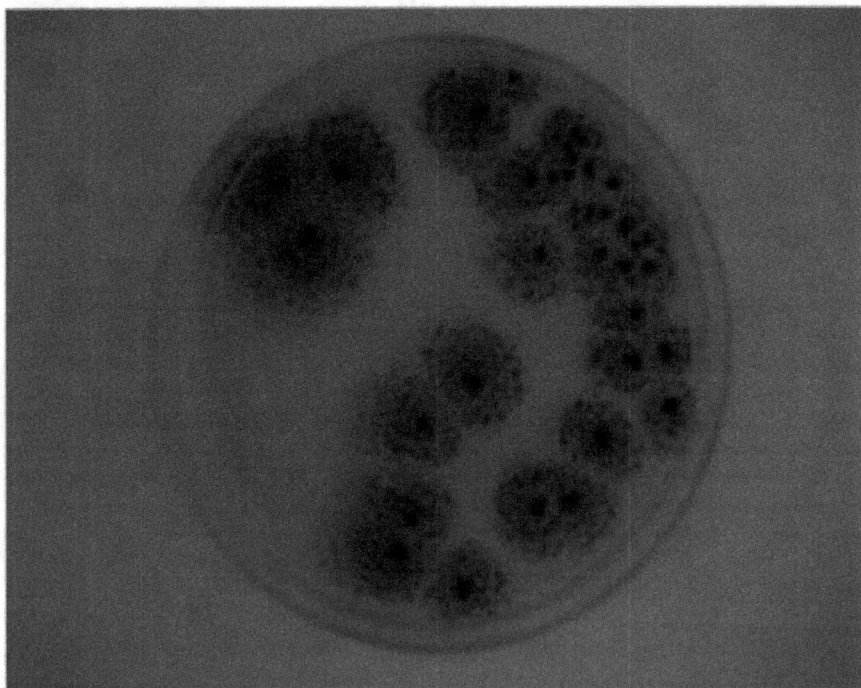

Figure 2. Growth of *A. awamori* on CPA after incubation at 37 °C for 5 days

To observe the morphological characteristics of the fungus, the isolates were inoculated on Malt Extract Agar (MEA) and Czapek Yeast Agar (CYA) and incubated at 26 °C for 7 days. Based on their growth rate, the fungus was presumptively identified as *A. awamori*. *A. awamori* was reported to show rapid growth on the CYA compared to *A. niger* which exhibited restricted growth. However, the growth and sporulation on MEA was better than CYA in the case of both *A. niger* and *A. awamori*. The fruiting bodies were mounted in lactic acid before they were observed under an oil immersion. The conidial heads for *A. awamori* were not well-defined columns in comparison to conidia heads observed for *A. niger*. The strain showed colony characteristics of both *A. niger* and *A. awamori*. On a general note, the conidiophores and conidia of *A. awamori* and *A. niger* are similar and morphologically indistinguishable, as shown in Figure 3 [34].

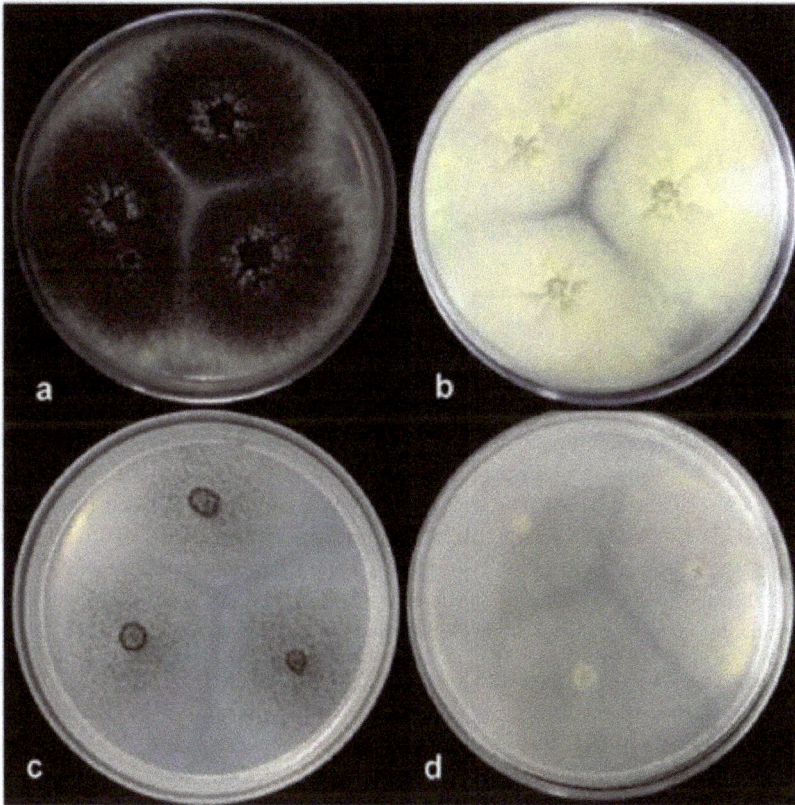

Figure 3. Growth of isolate on (a & b) MEA and (c & d) CYA

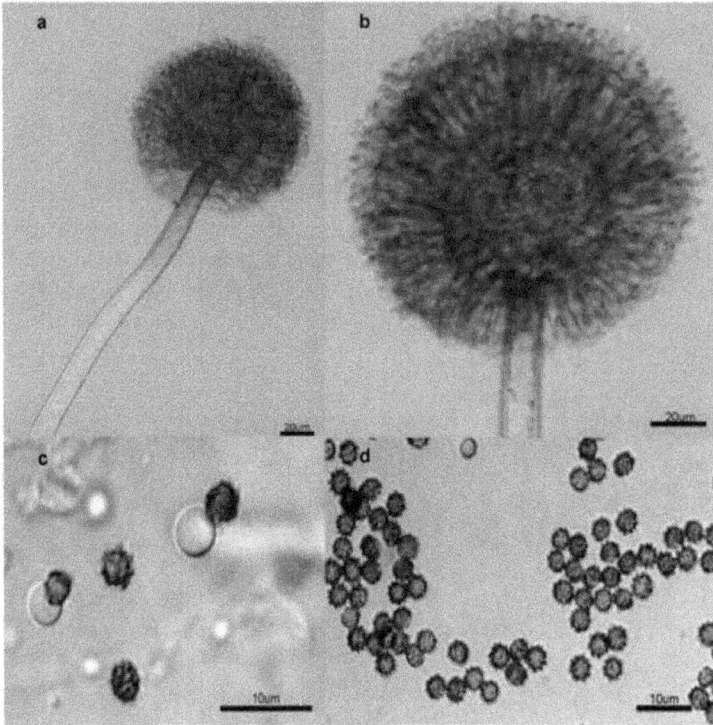

Figure 4. (a & b) Typical *Aspergillus* conidiophores with a radial head and (c & d) roughened, round conidia with regular low ridges and bars

Molecular characterization was carried out in order to confirm the identity of the fungal isolates. DNA was extracted from the pure isolates using the ZR Fungal/Bacterial DNA Kit (Zymo Research, California, USA). The subsequent Polymerase Chain Reaction (PCR) of the ITS1–5.8S–ITS2 rDNA region was prepared with primers ITS1 and ITS4 [36]. The β-tubulin gene was amplified using primers Bt2a and Bt2b [7] and the calmodulin gene with CL1 and CL2A [22], respectively. Sequencing reactions of the PCR products were set-up using a Big Dye terminator cycle sequencing premix kit (Applied Biosystems, CA).

Sequence reactions were analysed with an ABI PRISM 310 genetic analyser. Sequences were compared to those of a recent study by [34]. Datasets were aligned in Se-Al, including a sequence analysis in Se-Al. This was followed by a sequence analysis in PAUP* v4.0b10, using the BioNJ option for calculating a single tree for each dataset. Confidence in nodes was calculated using a bootstrap analysis of 1000 replicates. Only bootstrap values above 90% were indicated on the branches and *Aspergillus flavus* was chosen as the outgroup [34]. The isolated fungus was denoted as *Aspergillus* (CPUT) in Figure 5 and 6.

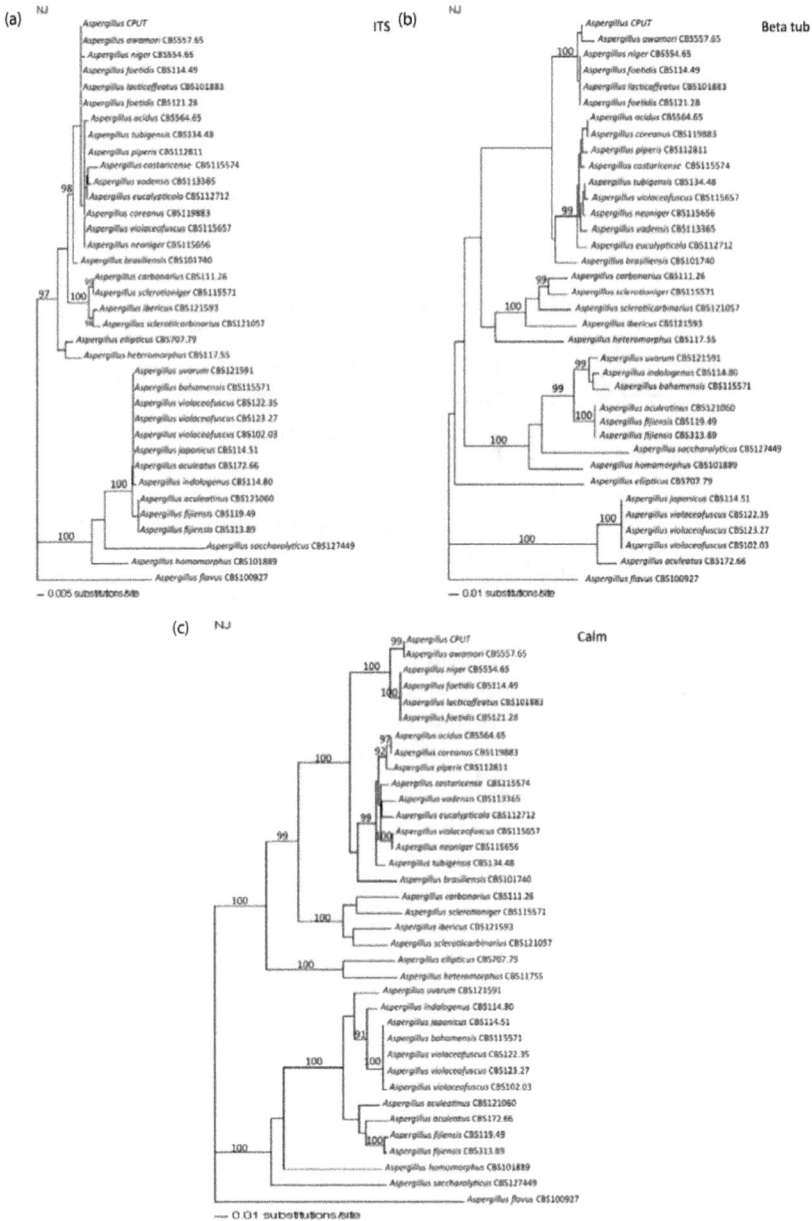

Figure 5. NJ tree based on the analysis of the (a) ITS, (b) β-tublin and (c) calmodulin gene regions

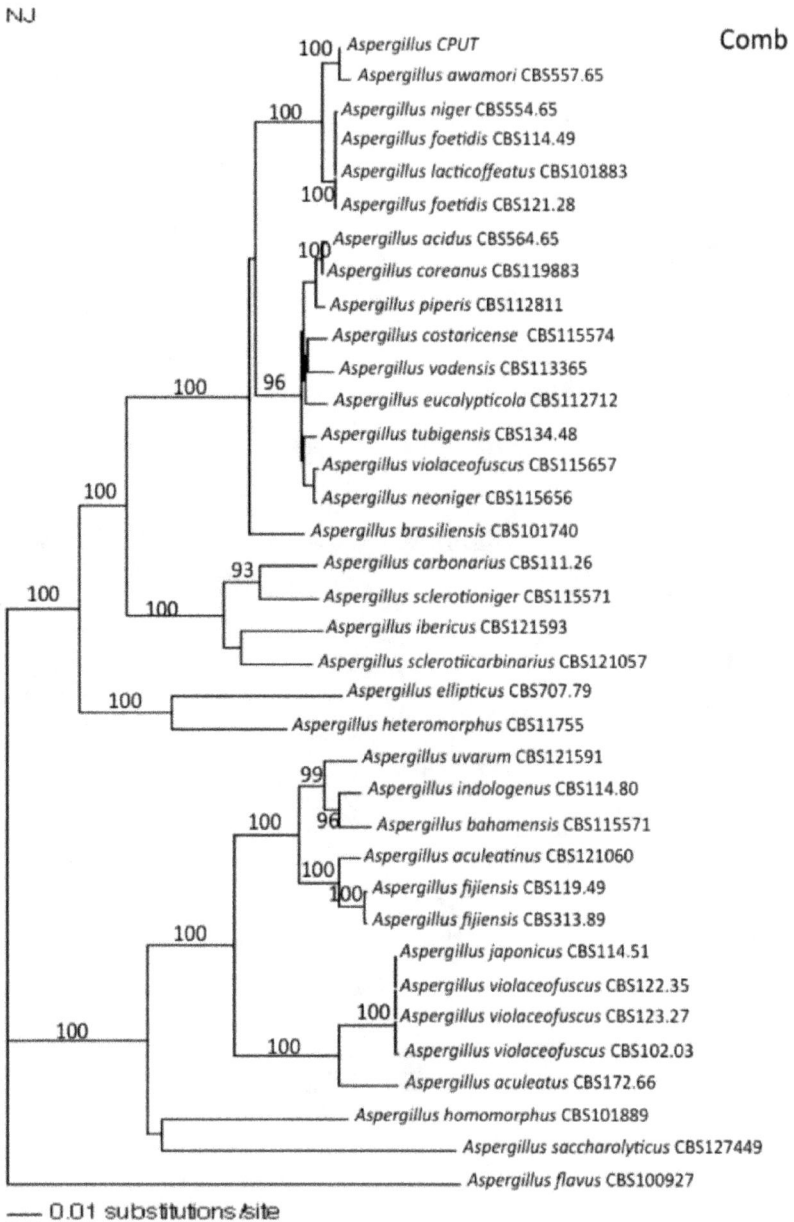

Figure 6. NJ tree based on the analysis of the combined gene regions

According to [34], the only way to separate *A. niger* and *A. awamori* is through a multi-gene phylogenetic analysis. This was done by using ITS, β-tubulin and calmodulin gene regions, as shown in Figure 5. The combined gene region analysis (Figure 6) indicates that the *Aspergillus* strain was similar and indeed identical to the sequence of the type strain of *A. awamori*, a fungus with diverse properties in biotechnology applications, including the production of nitrilase-a cyanide degrading enzyme.

3.2. Citrus peel supplemented growth medium for cyanide bioremediation using *Aspergillus awamori*

Citrus peels are composed of cellulose, pectin, hemi-cellulose, lignin, chlorophyll pigments, low relative-molecular-mass hydrocarbons including lipids, proteins, simple sugars, starches, water and ash [14, 29]. The major components in citrus peel are cellulose, hemi-cellulose, pectin and lignin which are inter-wound with each other to provide a rigid cell wall structure. *Aspergillus awamori* is able to produce enzymes for the breakdown of cellulose, hemi-cellulose and pectin and leaving lignin as the remaining structural component.

Cellulase, xylanase, pectinase are important enzymes in the hydrolysis of cellulose, hemi-cellulose and pectin into simpler sugars which can be utilised as a carbon source by the fungus. Hydrolysis of the orange peel has shown to yield significant quantities of neutral sugars: glucose, fructose and sucrose with low yields of xylose, arabinose, galactose and mannose. Hydrolysis of citrus peel also yield uronic acids, with galacturonic acid being the major of uronic acid liberated with trace quantities of other uronic acids. The optimum temperature and pH for these enzymes are in the range of 45 to 50 °C and 4.0 to 5.5, respectively [18, 19, 32]. Furthermore, *A. awamori* can produce nitrilase which hydrolyses the nitrile (cyanide) group (R-C≡N) into the corresponding carboxylic acid and ammonia, as shown in Figure 7 [26]. In aqueous solution, ammonia/ammonium equilibrium is observed.

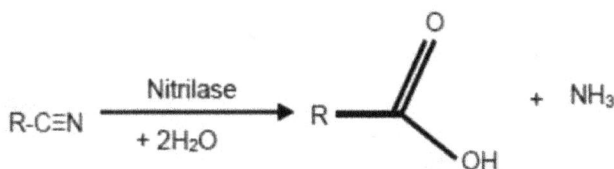

Figure 7. Nitrilase hydrolysis on cyanide group [26]

The majority of the citrus peel components are high carbon source materials and the ammonia produced from the cyanide degradation can be utilised as a nitrogen source by the microorganism. Studies have shown that the optimal pH and temperature for nitrilase production is 8 and 45 °C, respectively [12, 35].

The successful treatment of 400 ppm free cyanide supplemented with refined/readily metabolisable carbon sources has been reported [1]. However, current technology used can be capital intensive for large scale operations. Most studies on free cyanide bioremediation efficiency measured the free cyanide reduction periodically as opposed to product formation. This may be misleading since free cyanide is very volatile, even at room temperature, and the decline in free cyanide concentration observed may be a result of volatilisation into the atmosphere rather than actual biological remediation. The cyanide tolerance of the *A. awamori* (CPUT) isolate was initially assessed up to a 500 ppm CN⁻ (1.2515 g KCN/L) in PDA (Figure 8).

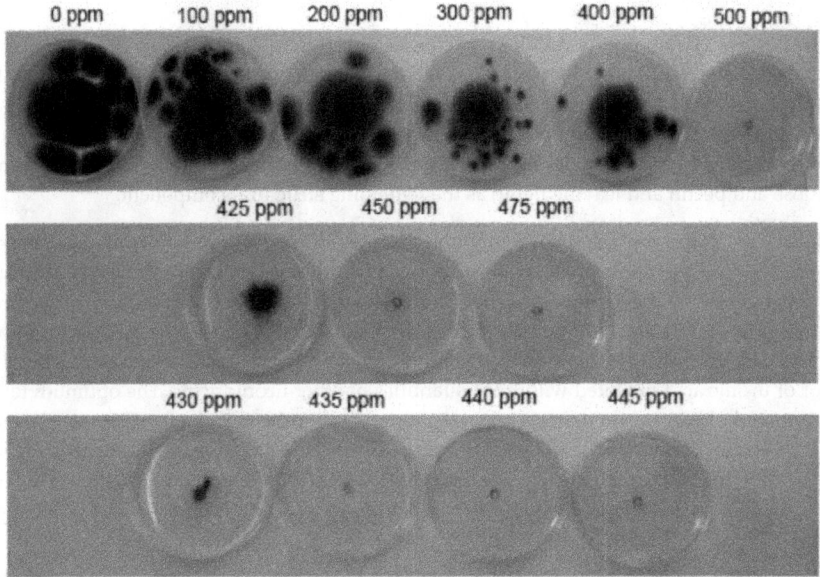

Figure 8. Cyanide tolerance analysis for *A. awamori* (CPUT) isolate

There was a clear decline in the growth of the fungus as the free cyanide concentration was increased. Appreciable growth occurred for the strain for free cyanide concentrations up to 200 ppm. A rapid decline in the growth was observed as the free cyanide concentration exceeded 300 ppm. The toxicity of cyanide reduces the functionality of the fungus metabolic processes, thus its growth. There was limited growth observed at cyanide concentrations above 430 ppm. Preliminary analysis on the effect of growth media on *A. awamori* (CPUT) for free cyanide bioremediation was performed in batch cultures, shaken at 180 rpm and 30 °C in a ZhiCheng (ZHWY-1102) shaking incubator. Media solutions of 42.5 ml of 1% (w/v) refined citrus pectin, 1 % (w/v) powered orange peel, Czapek yeast medium and sterile distilled water (standard) were added into 250 ml flasks. To each of the flasks, 1 ml of spore suspension (2x10⁶ spores) was added followed by 7.5 ml of a 1 g CN⁻/L cyanide solution.. The experiments were run in duplicate in which sampling was every 48 hours. The samples were centrifuged for 13000 rpm for 5 minutes before any analysis was conducted. Merck cyanide (CN⁻) (09701) and Merck ammonium (NH^{4+}) (00683) test kits were used to measure the free cyanide and ammonia/ammonium (NH_3/NH^{4+}) concentrations in solution.

The orange peel medium showed considerably higher cyanide reduction compared to the other nutrient media evaluated (Figure 9). The change in the cyanide concentration in the water medium (control) was due to volatilisation. At day 2, 7.5 ml of 1 g CN⁻/L free cyanide was again added to each flask to evaluate the robustness of the culture, in each growth media. The orange peel culture had the fastest recovery, even with a sudden increase in free cyanide concentration. The numerous enzymes released by the fungus, sufficiently hydrolysed the orange peel which resulted in better supplementation and maintenance of the fungus compared to the other media for cyanide bioremediation.

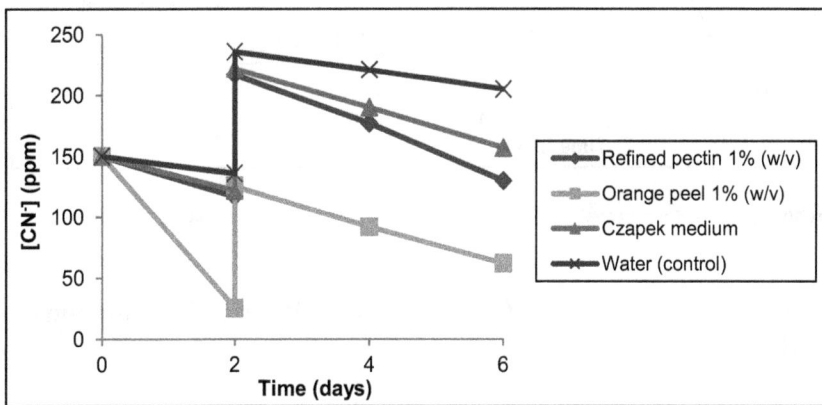

Figure 9. Cyanide bioremediation by *A. awamori* (CPUT) in batch cultures.

NH_3/NH^{4+} can be used as a nitrogen source for most fungi. The presence of quantifiable NH_3/NH^{4+} in solution during biodegradation is indicative of cyanide reduction. However, not all the NH_3/NH^{4+} was consumed by the fungus (Figure 10). Theoretical NH_3/NH^{4+} was calculated based on the quantity of cyanide degraded and the stoichiometry of the cyanide hydrolysis reaction (Figure 7). The experimental $[NH_3/NH^{4+}]$ were lower than that of the theoretical since the fungus metabolised some of the NH_3/NH^{4+} as a nitrogen source. However, the hydrolysis of the cyanide does not result in complete metabolism of NH_3/NH^{+}. The orange peel medium showed the highest concentration of the ammonia/ammonium in solution compared to the other media which is indicative of cyanide reduction. The rich carbon sources present in the orange peel and the further supplementation by NH_3/NH^{+}, had shown to be an added advantage of using waste orange peel as a potential nutrient source.

However, for an efficient bioremediation process, a continuous process must be developed to assess the applicability of the *A. awamori* (CPUT) isolate for continuous cyanide bioremediation processes. One of the effective technologies which have been determined to be effective on a large scale is the use of immobilised MBRs, for continuous remediation of contaminants [8]. The advantage of using MBR technology is that the biomass can be retained for elongated periods while the continuous remediation of the contaminated water is in progress.

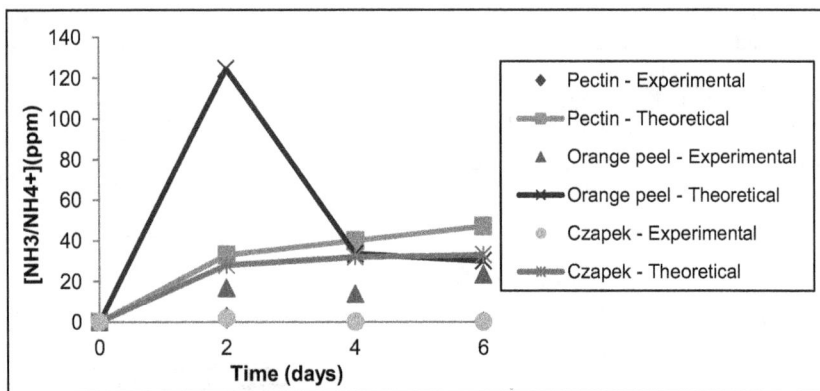

Figure 10. Experimental and theoretical ammonia/ammonium concentration in batch cultures

4. Membrane technology for fixed-film immobilisation in continuous remediation processes

Materials used in the construction of membranes include organic or non-organic (e.g. metal, ceramic), homogeneous (e.g. polymer, metal) and heterogeneous (e.g. polymer mixes, mixed glass) solids and solutions (mostly polymers) [25]. Polymeric membranes are commonly used because they are well developed, competitive in separation performances and economical [25]. Membrane processes are categorised according to the pore size, molecular cut-off and pressure at which they operate. These categories are inter-related, because as the pore size or the molecular cut-off size is decreased the pressure applied to the membrane increases [33]. Membrane separation processes can be broadly categorised into four groups, based on the pore size of the membranes (Figure 11).

Asymmetric membranes have shown to be effective in immobilising biofilms in MBRs since the membranes allow the transport of nutrients to the biomass immobilised on their external surface [27]. Asymmetric refers to the graded porosity of the membrane substructure and indicates that the membranes have an inside coating, a skin layer, and combines the high selectivity of a dense membrane with a high permeation rate of a thin membrane. The exposed microvoids on the externally unskinned surface enable a resilient attachment of the microorganism on the membrane [21]. The production of asymmetric membranes is by manipulating manufacturing parameters during the membrane formation process which results in a unique membrane morphology, characteristics and properties of the membrane [16].

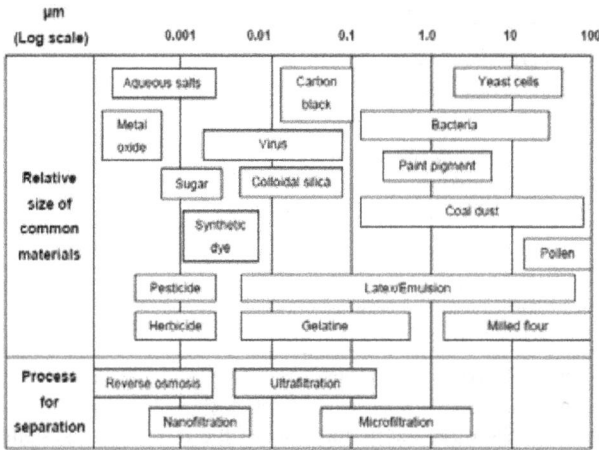

Figure 11. Filtration spectrum [25]

Asymmetric membranes have an excellent mass transfer property which has led to their utilization in numerous industrial applications, especially for MBRs [8, 25]. The most important ultrafiltration (UF) membrane module types are hollow fibre and capillary tube membranes [21, 27]. They are ideal for biofilm growth because of the nutrients permeation gradient along the membrane and provide a shear free environment [8, 15]. The large surface area of these membranes to their small volume allows for high operational capacity which is an advantageous property when used in MBRs [28].

Filamentous microorganisms are commonly used in immobilised MBR systems [3, 21, 27]. These particular microorganisms are able to penetrate the membrane due to their apical growth resulting in effective immobilisation on the surface compared to non-motile bacteria and yeast. Although, comparative studies of ceramic and internally skinned polysulphone (PSu) capillary membranes have shown to provide the best attachment and immobilisation of fungal biofilms than other tubular membrane types [21, 27]. Sheldon and Small [27] showed that the ceramic and internally skinned PSu capillary membrane developed thicker biofilms than tubular membranes. Furthermore, the ceramic capillary membrane can resists mechanical stress caused by the increasing immobilised biofilm build-up on the membrane, as ceramic membranes are mechanically stable and can be chemically and steam sterilized [27].

4.1. Modes of operation and orientation for membrane bioreactors

The membrane modules' mode of operation and orientation are important in determining the overall performance of the process. However, one of the biggest challenges in membrane operations is the effect of fouling and feeding mode, orientation and other mechanisms that would limit their application on a large scale.

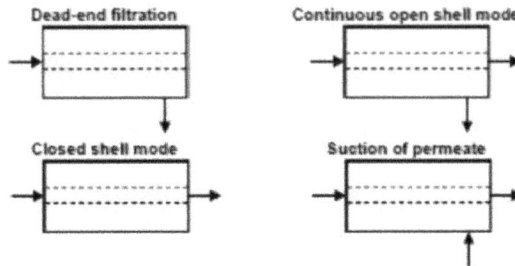

Figure 12. Modes of operation of membrane modules [2]

The effects of membrane modes of operations, as shown in Figure 12, have been studied using both experimental [6, 21, 30] and theoretical approaches [2, 8, 9, 13]. The mode of operation is concerned with the distribution and flow of the fluid in the system, as shown in Figure 12. In a dead-end filtration module the feed stream enters the lumen, permeates through the membrane and exit the shell in a continuous stream, with no retentate stream. In a continuous open shell mode, the feed stream enters the lumen and a portion of the feed permeates through the membrane and leaves the shell in a continuous stream and the other portion exit the lumen in a continuous stream. The amount of feed that will permeate through the membrane is dependent on several factors such as trans-membrane pressure, membrane permeability and feed velocity [25]. The pressure in the lumen is always greater than the shell side pressure in both dead-end filtration and continuous open shell modes [30]. This ensures that the trans-membrane flux, fluid's flow direction between the lumen and shell regions, is directed toward the shell [2].

In a closed shell mode, the feed stream enters the lumen and a portion of the feed permeates through the membrane with no outlet shell stream, while the other portion exits the lumen as a continuous stream. The trans-membrane flux for the initial portion of the membrane is directed towards the shell, but for the end portion it is directed toward the lumen [2]. This is referred to as convective recirculation which results in undesired non-uniform distribution of the biomass in the membrane system [2, 30].

In a suction of permeate, the feed stream enters the shell for an equal distribution along the membrane and permeates through the membrane and exits in a continuous stream [8]. The pressure in the shell is kept at greater than that in the lumen for a trans-membrane flux towards the lumen [30]. There is a negligible pressure gradient in the lumen which results in a relatively uniform trans-membrane flux [30]. The orientation of the MBR is based on its application and is generally operated horizontally or vertically [2, 6, 8, 27]. Studies by Garcin [6] and Ntwampe [21] on Lignin peroxidase and Manganese peroxidase production from *P. chrysosporium* BKMF-1767 (ATCC 24725) in a vertically orientated PSu capillary MBR, had shown the production of these extracellular enzymes were higher in the vertically orientated MBR than in the horizontal orientation. Similarly, the biofilm was denser on the vertically orientated capillary membrane than that in horizontal MBRs [6, 21].

4.2. Membrane bioreactors: Design and application for cyanide remediation

MBRs are integrated biological and separation units for the production of value added products or the bioremediation of toxic components [8]. Immobilised MBRs refers to the fixed microbial biofilm formation on a membrane matrix with the film being fed a nutrient medium passing through the lumen. These MBR's retain the biomass on the membranes in a low shear environment, separate the nutrient supply from the biomass, as well as continuously remove extracellular metabolic products. These immobilised systems ensure that the biomass can be maintained in a state of low or non-proliferation for extended periods, while still producing the products desired [8]. Filamentous microorganisms have had some degree of success on immobilised MBR systems [3, 21]. However, this has not been evaluated for *A. awamori* biomass for cyanide remediation. Four vertically orientated single fibre membrane bioreactors (SFMBRs) (Figure 13) were constructed as described by Edwards *et al.* [4].

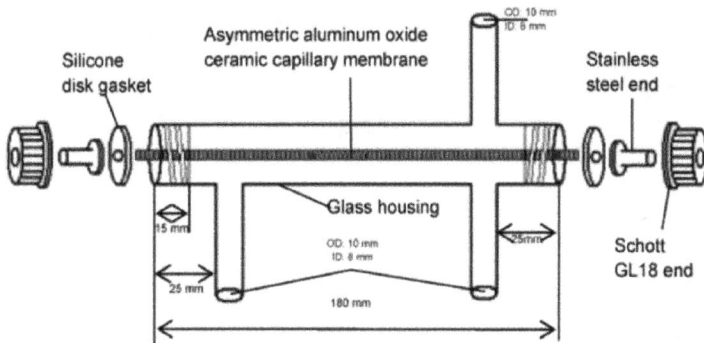

Figure 13. Schematic of SFMBR for filamentous microorganism immobilisation

The glass housing, was produced by Glasschem (South Africa) and the asymmetric aluminium oxide ceramic capillary membranes were produced and supplied by Hyflux CEPAration BV (Netherlands).

Outer diameter (m)	0.0028
Inner diameter (m)	0.0018
Wall thickness (m)	0.0005
Burst pressure (Pa)	5.0×10^6
Maximum temperature (°C)	1000+
Permeability (m/Pa.s)	6.95×10^{-10}

Table 1. Capillary membrane specifications [27]

The two SFMBRs were inoculated with 100 ml of A. *awamori* inoculum (10×10^6 spores) and two other SFMBRs were used as controls. High pressure reverse filtration of the inoculation spore solution on the external surface of the membranes was used to immobilise the spores onto the external surface of the ceramic membranes.

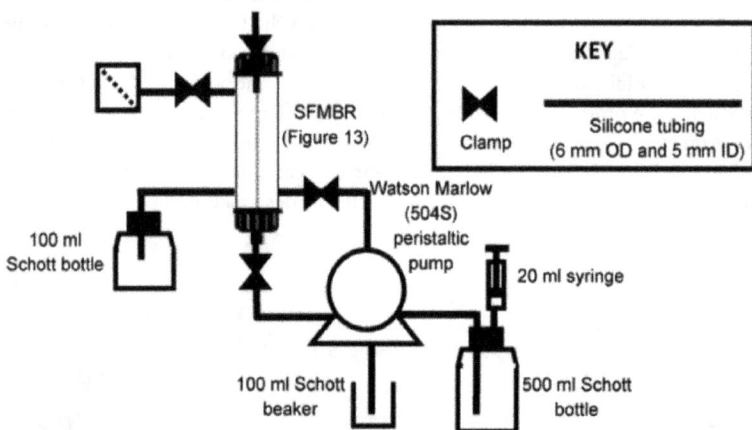

Figure 14. Schematic diagram of the experimental setup

The constructed system was setup in a Scientific Manufacturing (SMC) (160 L unit) low temperature incubator set at 30 °C. A Watson Marlow (504S) peristaltic pump was used to supply the feed solution to the SFMBRs at a flow rate of 3 ml/hr. The SFMBRs were fitted with two way flow 0.2 μm Millipore air filters, to ensure a monoseptic culture in the system and for aeration. The feed solution initially consisted of 1 % (w/v) orange peel solution which was prepared by adding 10 g milled orange peel to a 1 L Schott bottle. The 1 L solution was filtered through a Whatmann No 1 filter paper and the filtrate produced (orange peel extract) was fed to the system for 2 days, in a feed batch mode, to initiate spore germination and biofilm development on the membrane. Thereafter, a feed solution consisting of the orange peel extract and free cyanide solution (280ppm) was fed to the SFMBR.

The total reduced sugars (TRS), such as glucose, fructose, sucrose, were measured using the dinitrosalicylic (DNS) acid colorimetric method [17]. The Merck cyanide (CN⁻) (09701) and Merck ammonium (NH₄⁺) (00683) test kits were used to measure the free cyanide and NH_3/NH_4^+ concentrations, since this is a by-product of the cyanide metabolism. The utilisation of the TRS increased for the pure orange peel extract feed in comparison to the control MBRs (Figure 15). The initial feed was rich with TRS and therefore, the metabolism of the TRS was observed. The change in the TRS utilisation for the orange peel extract containing cyanide concentration of 1.7%, 15.2%, 23.7% and 28.8% at day 2, 4, 6 and 8, respectively and therefore the rate of metabolism was drastically hindered which resulted in reduced metabolism of the TRS.

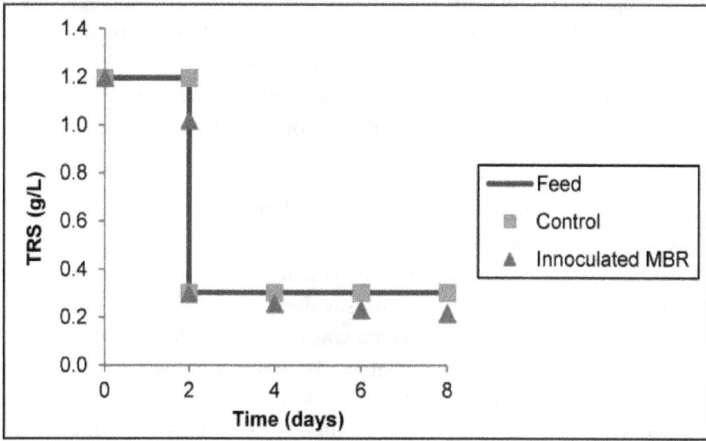

Figure 15. TRS versus time for feed, control and MBR

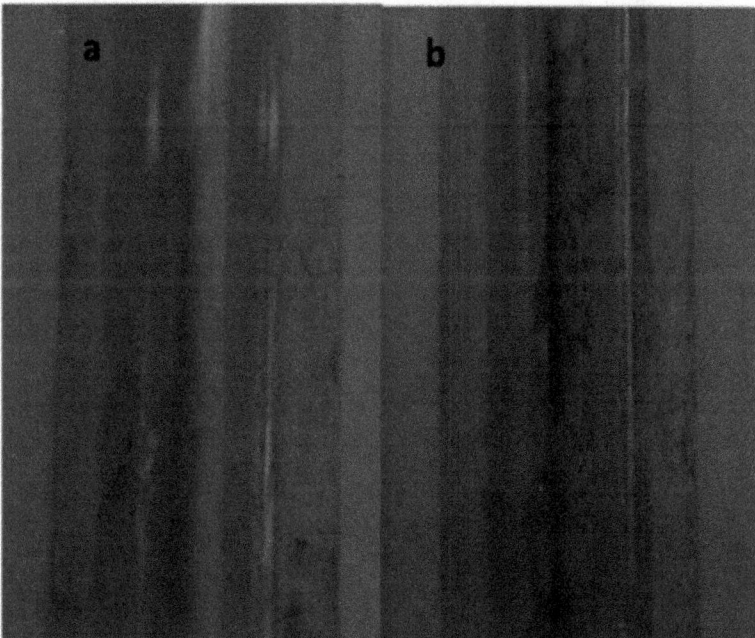

Figure 16. Biofilm development on day (a) 0 and (b) 8

The biofilm development (Figure 16) on the membrane was slow and homogenous along the membrane, as opposed to thick biomass at a particular area of the membrane. The formation of the thin biofilm has its benefits; it allows for better oxygen mass transfer into the biomass, which is vital since *A. awamori*, is an aerobic microorganism. Aerobic microorganisms utilise oxygen for cell maintenance, respiratory oxidation for further growth and for the oxidation of substrates into metabolic products.

The air filter prevented contamination and maintained the oxygen concentration in the MBR due to the gaseous venting, releasing by-products, such as carbon dioxide out of the MBR system and reintroducing oxygen back into the system. The residual cyanide concentration in the permeate decreased with time and a drastic decrease was observed from day 4 as the fungus adapted to the feed (Figure 17). There was also a considerable quantity of cyanide volatilisation compared to actual bioremediation. As shown in the shaken cultures, a majority of the ammonia, produced from cyanide hydrolysis, was consumed by the fungus as a nitrogen source. This was an indication that there was further utilisation of ammonia/ammonium as the fungus adjusts to the cyanide containing feed.

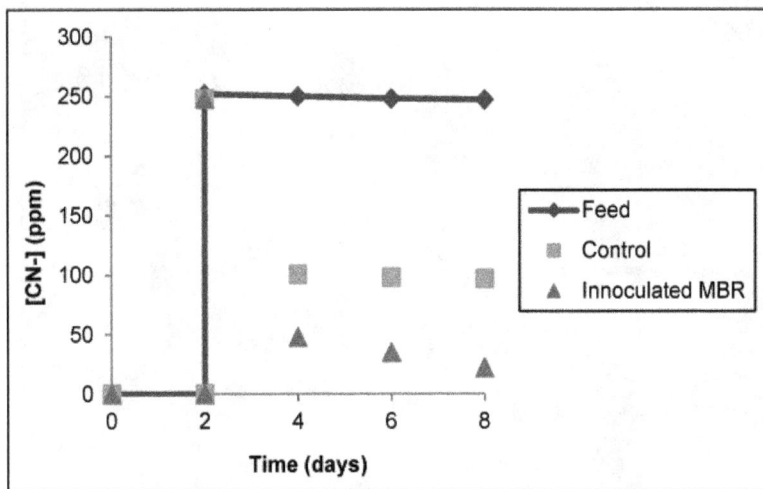

Figure 17. Discharge cyanide concentration versus time for feed, control and MBR

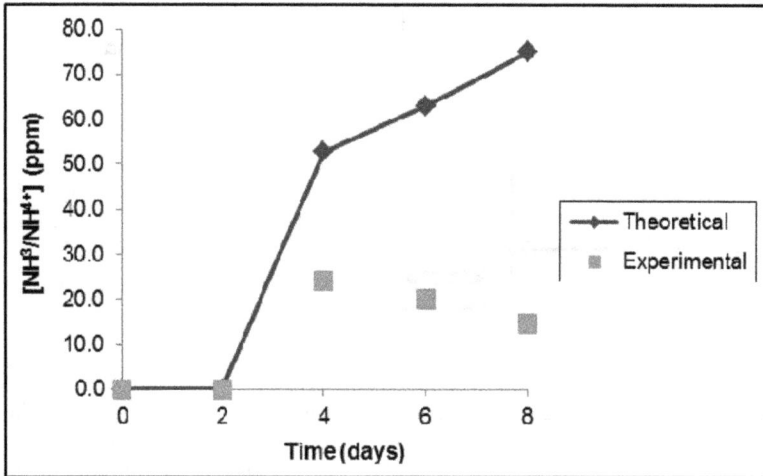

Figure 18. Experimental and theoretical ammonia/ammonium concentration

For the development of a sustainable MBR system for cyanide remediation an alternative MBR system should be used which limits cyanide volatilisation. Immersed MBRs are bioreactors in which the enzyme(s) and/or microorganism(s) or antibiotic(s) are immobilised on membrane(s) and biomass is suspended in the solution and compartmentalised in a reaction vessel [3]. Sidestream MBRs are when the membrane module and bioreactor are separate from each other. Immersed and sidestream MBRs are used for conventional biomass rejection thus allowing for continued biomass utilization [3]. Immersed MBRs require less energy than sidestream MBRs. Membrane modules in a pumped sidestream system utilises more energy due to the high pressures to sustain high volumetric flow rates [11]. However, immersed MBRs may be difficult to operate and remove fouling during operation. The separated system of bioreactor and membrane unit in a sidestream MBRs makes it easier to specifically optimise certain parameters for each unit which cannot be done in immersed MBRs.

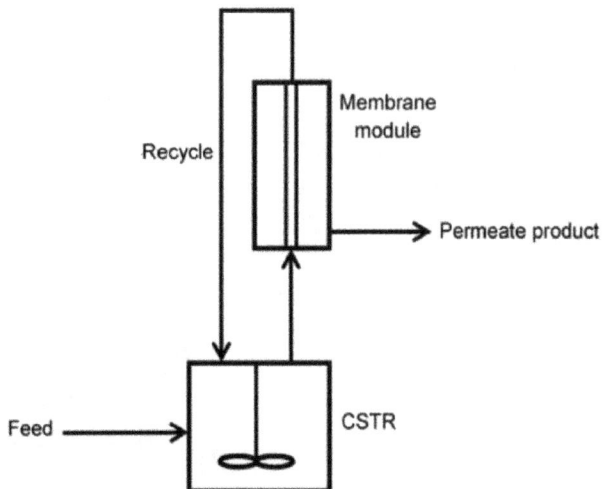

Figure 19. Side stream MBR system

For the continuous bioremediation of cyanide, a sidestream MBR (Figure 19) would be ideal since the solid material, yet to be hydrolysed, can be recycled into the continuous stirred tank reactor (CSTR) and the bioremediated wastewater can be collected as the permeate product. The CSTR can be initially loaded with milled orange peel in a water solution and inoculated with *A. awamori* thus the cyanide containing wastewater can be continuously fed into the system.

5. Conclusion

Large quantities of organic wastes are generated every year, but it has been shown that many of the waste, particularly from the agricultural sector, can be utilised as nutrient source for microbial systems. The use of orange peel has shown to be a rich source for cultivation and supplementation for *A. awamori* (CPUT) isolate for cyanide remediation. The ammonia produced from the cyanide hydrolysis can be used as a nitrogen source by the fungus, although incomplete metabolism of the ammonia was observed. This could be improved by changing operating conditions so that the degradation of cyanide, thus the release and subsequent consumption of the ammonia is improved.

Challenges that are also evident in bioremediation are due to its volatility, especially in open/agitated cultures. The hydrolysis of sugar components from the orange peel by merely boiling it in water is not an effective true solution since this process results in incomplete hydrolysis/liberation of the sugar components and residual solids. This can be improved by hydrolysis using microorganism such as *A. awamori*, as used in this study. The use of an enclosed sidestream MBR would provide a suitable system for complete hydrolysis of the sugars from the orange peel. The unhydrolysed sugar components in the peel can be recycled and further hydrolysed depending on the molecular-weight cut-off size of the membranes used in the reactor. The sidestream MBR can reduce CN⁻ volatilization thus, offer a safe to operate reactor unit which is easier to clean with an added advantage of optimising each unit individually.

Acknowledgements

This research was sponsored by a CPUT University Research Fund (URF RK16).

Author details

Bruno Alexandre Quistorp Santos[1], Seteno Karabo Obed Ntwampe[2] and
James Hamuel Doughari[2]

1 Faculty of Engineering, Department of Chemical Engineering, Cape Peninsula University of Technology, Cape Town, South Africa

2 Faculty of Applied Science, Department of Agriculture and Food Science, Biotechnology programme, Cape Peninsula University of Technology, Cape Town, South Africa

References

[1] Akcil A., Karahan AG., Ciftci H., Sagdic O. Biological Treatment of Cyanide by Natural Isolated Bacteria (*Pseudomonas sp.*). Minerals Engineering 2003; 16(7) 643-649.

[2] Bruining WJ. A General Description of Flows and Pressures in Hollow Fibre Membrane Modules. Chemical Engineering Science 1988; 44(6) 1441-1447.

[3] De Jager D. *Streptomyces coelicolor* Biofilm Growth Kinetics and Oxygen Mass Transfer within a Membrane Gradostat Bioreactor. Masters of Technology thesis. Cape Peninsula University of Technology; 2010.

[4] Edwards W., Leukes WD., Fraser SJ. High Throughput Bioprocess Apparatus, SA patent WO 2007/116266 A1 (2007).

[5] Fogler HS. Elements of Chemical Reaction Engineering, 4ᵗʰ edition. New Jersey: Person Education International; 2006.

[6] Garcin CJ. Design and Manufacturing of Membrane Bioreactor Cultivation of Fungi. Masters of Science thesis. Rhodes University; 2002.

[7] Glass NL., Donaldson GC. 1995. Development of Premier Sets Designed for Use with the PCR to Amplify Conserved Genes from Filamentous *Ascomycetes*. Applied and Environmental Microbiology; 61(4) 1323-1330.

[8] Godongwana B. Momentum Transfer inside a Single Fibre Capillary Membrane Bioreactor. Masters of Technology thesis. Cape Peninsula University of Technology; 2007.

[9] Godongwana B., Solomons D., Sheldon MS. 2010. A Solution of the Convective-Diffusion Equation for Solute Mass Transfer inside a Capillary Membrane Bioreactor. International Journal of Chemical Engineering 2010; 2010(Article ID 738482) 1-12.

[10] Greben HA., Oelofse SHH. Unlocking the Resource Potential of Organic Waste: a South African Perspective. Waste Management & Research 2009; 30(9) 1-9.

[11] Judd S. The MBR Book: Principles and Applications of Membrane Bioreactors in Water and Wastewater Treatment. Amsterdam: Elsevier; 2006.

[12] Kaplan O., Vejvoda V., Plíhal O., Pompach P., Kavan D., Bojarová P., Bezouška K., Macková M., Cantarella M., Jirků V., Křen V., Martínková L. 2006. Purification and Characterization of a Nitriles' from *Aspergillus niger* K10. BMC Biotechnology; 11(2) 1-15.

[13] Kelsey LJ., Pillarella MR., Zydney AL. Theoretical analysis of convective flow profiles in a hollow fibre membrane bioreactor. Chemical Engineering Science 1990, 45(11) 3211-3220.

[14] Liang S., Guo X., Feng N., Tian Q. Effective Removal of Heavy Metals from Aqueous Solutions by Orange Peel Xanthate. Transaction of Nonferrous Metals Society of China 2010; 20(1) 187-191.

[15] Leukes WD. Development and Characterisation of a Membrane Gradostat Bioreactor for the Bioremediation of Aromatic Pollutants using White Rot Fungus. PhD thesis. Rhodes University; 1999.

[16] Locatelli F., Ronco C., Telta C., 2003. Polyethersulfone: Membranes for Multiple Clinical Applications. Contributions to Nephrology 2003; 138(2003) 1-12.

[17] Miller GI. Use of Dinitrosalicylic Acid Reagent for Determination of Reducing Sugars. Analytical Chemistry 1959; 31(3) 426-428.

[18] Mrudula S., Anitharaj R. Pectinase Production in Solid State Fermentation by *Aspergillus niger* using Orange Peel. Global Journal of Biotechnology and Biochemistry 2011; 6(2) 64-71.

[19] Nataraja S., Chetan DM., Krishnappa M. Effect of Temperature on Cellulose Enzyme Activity in Crude Extracts Isolated from Solid Wastes Microbes. International Journal of Microbiology Research 2010; 2(2): 44-47.

[20] Nesbitt AB. Recovery of Metal Cyanides using a Fluidized Bed of Resin. MTech thesis. Cape Technikon; 1996.

[21] Ntwampe SKO. Multicapillary Membrane Design. MTech thesis. Cape Peninsula University of Technology; 2005.

[22] O'Donnell K., Nirenberg HI., Aoki T., Cigelnik E. A Multigene Phylogeny of the *Gibberella fujikuroi* Species Complex: Detection of Additional Phylogenetically Distinct Species. Mycoscience 2000; 41(2000) 61-78.

[23] Patil YB., Paknikar KM. Development of a Process Biodetoxification of Metal Cyanides from Wastewaters. Process Biochemistry 2000; 35(10) 1139-1151.

[24] Patil YB., Paknikar KM. Removal and Recovery of Metal Cyanides using a Combination of Biosorption and Biodegradation Processes. Biotechnology Letters 1999; 21(10) 913-919.

[25] Perry RH., Green DW., Maloney JO. 1988. Perry's Chemical Engineer's Handbook, 7th edition. New York: McGraw Hill; 1988.

[26] Rao MA., Scelza R., Scotti R., Gianfreda L. Role of Enzymes in the Remediation of Polluted Environments. Journal of Soil Science and Plant Nutrition 2010; 10(3) 333-353.

[27] Sheldon MS., Small HJ. Immobilisation and Biofilm Development of *Phanerochaete chrysosporium* on Polysulphone and Ceramic Membranes. Journal of Membrane Science 2005; 263(1-2) 30-37.

[28] Stamatialis DF., Papenburg BJ., Girones M., Saiful S., Bettahalli SNM., Schmitmeier S., Wessling M. 2008. Medical Applications of Membranes: Drug Delivery, Artificial Organs and Tissue Engineering. Journal of Membrane Science 2008; 308(1-2) 1-34.

[29] Sud D., Mahajan G., Kaur MP. Agricultural Waste Material as Potential Adsorbent for Sequestering Heavy Metal Ions from Aqueous Solutions - a Review. BioResource Technology 2007; 99(14) 6017-6027.

[30] Thakaran JP., Chau PC. Operation and Pressure Distribution of Immobilised Hollow Fibre Bioreactors. Biotechnology and Bioengineering 1986; 28(7) 1064-1071.

[31] Torrado AM., Cortés S., Salgado JM., Max B., Rodríguez N., Bibbins BP., Converti A., Domínguez JM. Citric Acid Production from Orange Peel Waste by Solid-State Fermentation. *Brazilian Journal of Microbiology* 2011; 42(1) 394-409.

[32] Umsza-Guez MA., Díaz AB., de Ory I., Blandino A., Gomes E., Caro I. Xylanase Production by *Aspergillus awamori* Under Solid State Fermentation Conditions on Tomato Pomace. Brazilian Journal of Microbiology 2011; 42(4) 1585 - 1597.

[33] Van der Roest HF., Lawrence DP., van Bentem AGN. Membrane Bioreactors for Municipal Wastewater Treatment. London: IWA; 2002.

[34] Varga J., Frisvad JC., Kocsubé S., Brankovics B., Tóth B., Szigeti G., Samson, RA. New and Revisited Species in *Aspergillus* Section *Nigri*. Studies in Mycology 2011; 69(1) 1-17.

[35] Vejvoda V., Kubáč D., Davidová A., Kaplan O., Šulc M., Šveda O., Radka Chaloupko-vá R., Martínková L. Purification and Characterization of Nitrilase from *Fusarium sol-ani* IMI196840. Process Biochemistry 2010; 45(7) 1115-1120.

[36] White TJ., Bruns T., Lee S., Taylor J. Amplification and Direct Identification of Fungal Ribosomal RNA Genes for Phylogenetics. In: Innis MA., Gelfand DH., Sinsky JJ., White TJ. (ed.) PCR Protocols: a Guide to Methods and Applications. San Diego: Academic Press; 1990. p315-322

Bioavailability of High Molecular Weight Polycyclic Aromatic Hydrocarbons Using Renewable Resources

Olusola Solomon Amodu, Tunde Victor Ojumu and
Seteno Karabo Obed Ntwampe

Additional information is available at the end of the chapter

1. Introduction

Polycyclic aromatic hydrocarbons (PAHs) are the world's largest class of carcinogens known to date, not only because of their ability to cause gene mutation and cancer, but due to their persistency in the environment. They are particularly recalcitrant due to their molecular weight, hydrophobic nature and thus, accumulate in various matrices in the environment.

PAHs, also known as polyarenes or polynuclear aromatic hydrocarbons, are formed and released into the environment through natural and anthropogenic sources. Natural sources include volcanoes and forest fires while anthropogenic sources include, majorly, incomplete combustion of fossil fuels, wood burning, municipal and industrial waste incineration. PAHs containing two or three fused benzene rings are classified as low molecular weight (LMW) PAHs and are more water soluble while those with four or more benzene rings are referred to as high molecular weight (HMW) PAHs. They tend to adsorb onto soil and sediment thus, making them recalcitrant in the environment. Sixteen of these organic compounds have been identified as priority pollutants due to their hazardous properties, with HMW PAHs being considered as potential human carcinogens, by the United State Environmental Protection Agency [1].

The cost of biodegradation technology and the low bioavailability including mass transfer limitations of PAHs, especially those with high molecular weight, from several matrices into the aqueous phase for effective enzyme-based microbial biodegradation still constitute major challenges. However, current research efforts have focused on the combined use of biosurfactants and enzymes produced from renewable resources such as agricultural by-products and/or agro-industrial waste, through assisted biostimulation and bioaugmentation, for biodegradation of PAHs. Such methods are relatively inexpensive and less invasive as compared to physico-chemical remediation processes [2-4]. The application of crude biosurfactants produced by *Pseudomonas ssp.*, *Rhodococcus ssp.* and several others, have been observed to achieve a high mobilization rate of PAHs from contaminated environmental matrices [5, 6].

The synergistic effect of biostimulation combined with bioaugmentation using fungal strains of *Rhizopus spp.*, *Penicillium spp.* and *Aspergillus spp.*, isolated from PAH contaminated soil, significantly improve the overall bioavailability for the biodegradation of PAHs in comparison to biostimulation alone [7]. Additionally, several other renewable resources have been used as sources of both carbon and nitrogen by several microorganisms during the expression of biosurfactants thus enhancing PAH bioavailability and enzymatic biodegradation [8-11]. This review describes the environmental behavior of PAHs, and how it affects bioavailability. It also examines the effectiveness of microorganisms used in the production of crude biosurfactants for the biodegradation of HMW PAHs rather than the direct application of refined extracts, with a view to minimize the cost associated with enzymatic biodegradation. The chapter further discusses the effects of bioavailability on the biodegradation of HMW PAHs; bioavailability kinetics to quantitatively estimate PAHs bioavailability using different microorganisms; and enhanced biodegradation of PAHs using crude biosurfactants from renewable resources.

2. Bioavailability of HMW PAHs

2.1. PAHs bioavailability — Definitions and intrinsic factors

Polycyclic aromatic hydrocarbon mobilization and biodegradation are contingent upon their bioavailability from various matrices. Given the legal and regulatory implications of the bioavailability concept as part of a risk assessment framework, the term must be clearly understood in order to establish the minimum level permissible for contaminants like PAHs in the environment. The understanding of bioavailability is also important to be able to assess and evaluate the overall success of PAHs biodegradation. Researchers, however, differ in their opinion as to what the exact definition of bioavailability should be [12].

The following definitions for bioavailability were compiled in Technical Reports published by the European Centre for Ecotoxicology and Toxicology of Chemicals [13] and the United State National Research Council [14]: (i) "The ability of a substance to interact with the biosystem; (ii) "The fraction of the contaminants in the environment that is potentially available for biological action" [15]; (iii) "The amount/percentage of a compound that is actually taken up by an organism as the outcome of a dynamic equilibrium of organism-bound sorption processes, and soil particle-related exchange processes, all in relation to a dynamic set of environmental conditions" [16]; (iv) "The fraction of a chemical accessible to an organism for absorption, the rate at which a substance is absorbed into a living system, or a measure of the potential to cause a toxic effect". In pharmacology and toxicology, the term relates to the systemic availability of a xenobiotic after intravenous or oral dosing [17]. Although these concepts are useful, making direct parallels from the pharmacological usage to contaminants in soil and sediment or biota can be problematic. For example, microorganisms do not have a digestive tract, target organs, or a circulatory system [13]. These concepts may also not be appropriate as contaminants in soil or aquatic environments, being controlled by, for example, the rate of desorption and mobilization from solid matrices, are often continuously supplied to organisms gradually rather than at an acute dose. Environmental scientists often consider bioavailability to represent the accessibility of a soil-bound chemical for assimilation and possible toxicity [18, 19] and thus, have tried to adapt the use of bioavailability concept when considering human exposure to soil-borne contaminants. For example, Ruby [20] including Kramer and Ryan [21] suggested that the bioavailable portion is the amount of compound that is removed from soil through desorption processes under physiological conditions. Another view of bioavailability is represented by the contaminant crossing a cell membrane, entering a cell, and becoming available at a site of biological activity. Others might think of bioavailability more specifically in terms of contaminant binding to or release from a solid phase.

Obviously, the different definitions of the term bioavailability by scientists in various disciplines are capable of causing semantic confusion and thus, garner more attention than proffering solution to its challenges. The authors, in this chapter, have compiled these opinions in order to present a simple and workable definition to the term bioavailability as it is important to estimateF; the extent of contaminants desorption from the sorbed phase, the non-aqueous phase residue as against the minimum level required in the environment for such contaminants, and thus assessing the overall success of biodegradation. Considering these opinions about bioavailability, two words are common to almost all, which are; uptake or absorbed and available. Based on this observation, bioavailability can be defined as the amount of available contaminants in the environment that can be absorbed by microorganisms and /or biological products. Other clauses such as the fraction of contaminants taken up or absorbed, the fraction of contaminant that is potentially available, the mobilization or transportation of contaminants from the sorbed phase, the amount desorbed from the soil matrices, etc., which are often included in the definition of bioavailability and thus causing confusion, are intrinsic factors or features of bioavailability. However, these intrinsic factors are influenced by the physico-chemical properties of the contaminants and those of the sorbent.

Hence, studies on bioavailability are very crucial in order to link the quantity of PAHs taken up by microorganism with the actual amounts that are available to cause adverse effects in the environment. Many factors have been known to affect PAH bioavailability, which are [12, 22, 23];

- Physical and chemical properties of PAHs,

- Soil properties (soil organic matter, dissolved organic matter, moisture content, etc.),

- Aging of PAHs in soil and receptor microorganism.

2.2. Effects of physical and chemical properties of PAHs on bioavailability

Bioavailability is influenced by the molecular structure and size of PAHs. LMW PAHs are removed faster by physico-chemical and biological processes due to their higher solubility, volatility and the ability of many microorganisms to use them as sole carbon sources in comparison to the HMW PAHs [24]. Bioavailability changes with time and weathering [25]. Aging is a central part concerning availability and refers to the process of organic compounds in soil becoming less susceptible to degradation, extractability and other related processes in a time dependent manner [26, 18]. Aging increases sorption propensity of soil contaminants making them more recalcitrant to diffusion and mobility which consequently lead to low bioavailability. Both the physico-chemical properties of the contaminant and the soil characteristics influence aging, which may include several steps and processes such as oxidation thus incorporating the contaminant into natural organic matter [27, 28], slow diffusion into small pores and absorption into organic matter, or entrapment due to the formation of semi-rigid films around non-aqueous-phase liquids (NAPL) with a high resistance toward NAPL-water mass transfer [29].

2.3. Effects of soil or sediment and dredging properties on PAHs bioavailability

Soil properties such as organic matter content, soil texture, soil depth, particle size, pH, porosity, intrinsic permeability, liquid limit, and cation exchange capacity, influence PAH bioavailability. Soil properties can vary greatly from one region to another. They can even vary within the same region spatially and with depth [30]. Microorganisms have different levels of tolerance to these factors, which affect their growth and other metabolic activities. Soil structure such as aggregation has been found to decrease PAH availability through physical sequestration of PAHs in the interior of aggregates [31]. For instance, Nam et al. [32] found that PAH bioavailability to phenanthrene degraders declined with time in soils which had more than two percent organic matter. Hundal et al. [33] reported on the retention of large amounts of phenanthrene by smectite clays. Yang et al. [34] investigated the impact of soil organic matter on PAHs distribution in soils and reported that when the soil organic matter was increased from 0.2 to 7.1% the average non-bioavailable amount of acenaphthene, anthracene, fluoranthene, and pyrene were almost tripled from 436.9 to 1205.8 ng/g.

2.4. Effects of mass transfer on bioavailability of HMW PAHs

The overall PAHs biodegradation can be conceptually divided into the following steps: desorption to the eqeuous phase; mass transfer to biologically accessible regions; and biological uptake and transformation [35, 18], as shown in Figure 1. These steps occur sequentially, such that the overall biodegradation rate can be controlled by any of these steps. Given the sequential nature of the process, the impact of desorption rate on overall biodegradation is expected to be greatest in the case where biodegradation rates are higher relative to desorption rates. This occurs when the active microorganisms are capable of high biodegradation rates and either the media is porous thus having a high capacity for the solute, or media has large diffusion distances [36, 37]. Based on this, it is important to examine how the physical morphology of surface and subsurface soils can impact the biodegradation of sorbed organic chemicals. Naturally, soil contains porous particles of different sizes, many of which are smaller than the size of microorganisms. For example, analysis of one of the coarser sand sizes from the Borden aquifer in Ontario (Canada), indicated that, roughly 50% of the intraparticle pore volume resides in pores that are less than 0.1 μm in diameter [38].

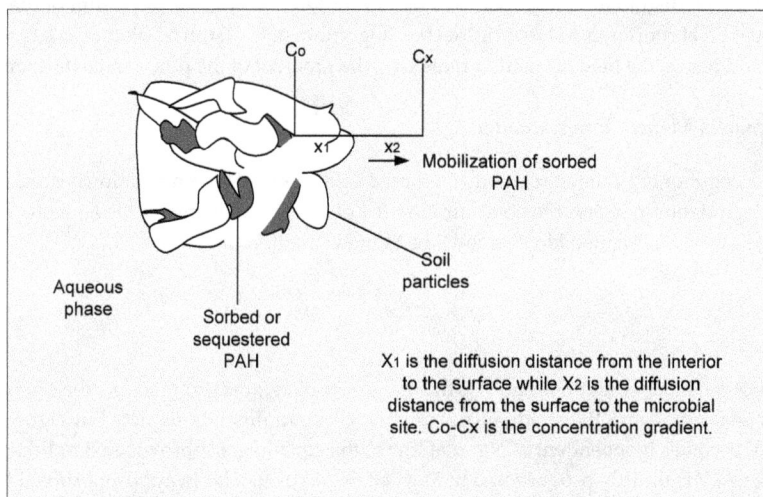

Figure 1. Diffusion of entrapped PAH out of soil micropores into aqueous phase to become available to microorganism.

Pores with diameter larger than 1 μm comprised about 12% of the total pore space, and only about 5% of the pore volume was attributed to pores larger than 2 μm. Considering that most indigenous bacteria found in soil are 0.5 to 1.0 μm in diameter [39], bacteria will be physically excluded from attaching to intraparticle pores of these grains. The mean diameter of intraparticle pores occupied by bacteria has been estimated to be typically larger than 2 μm [40], and this is likely to be larger than intraparticle porespaces of many natural sorbent solids. However, for those pores which are accessible to bacteria, slow mass transfer of contaminants from the pore interior as well as from the pore surface can limit the extent of microbial growth and consequently biodegradation.

3. Kinetic models and assessment of bioavailability of PAHs

3.1. First–order kinetics

Biodegradation rate of PAHs in the aqueous phase and the rate constant (k) can be determined using the reaction rate expression as follows:

$$-\frac{dC}{dt} = kC^n \tag{1}$$

C - is the concentration of PAH (mg/l), t the time (days), k the rate constant for chemical disappearance of PAH (days^{-1}) and n, the reaction order, which is unity for first order kinetics [24, 41]. Based on the assumptions that only dissolved forms are available for biodegradation and that the biodegradation rate follows a first order kinetics [42], the logarithm of the ratio of residual PAH concentration to its initial (i.e., logarithm of C/C_o) can be plotted as a function of time and hence, the biodegradation rate being the gradient of the plot, can be determined.

3.2. Michaelis-Menten kinetic model

The most commonly assumed relationship for the assessment of bioremediation is done using the biodegradation rate and the concentration (C) of the contaminants in the aqueous phase using the Michaelis-Menten kinetic model or Monod equation:

$$V = \frac{(V_{max} S)}{(K_m + S)} \tag{2}$$

where V is the biodegradation rate, V_{max} the maximum biodegradation rate, K_m the Michaelis-Menten constant, and S the residual contaminant concentration. As depicted in Figure 2, at high S, V becomes independent of S and at low S, the equation is approximated to first order kinetics, i.e., V is directly proportional to S. At all values of S, V is always proportional to the biocatalyst concentration available for the biodegradation process.

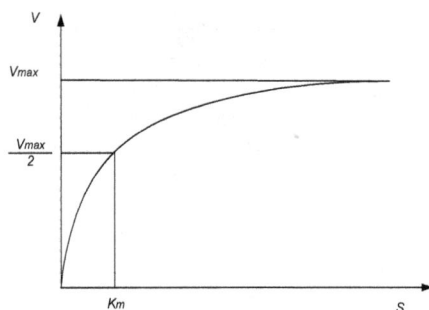

Figure 2. Michaelis-Menten kinetics rate profiles.

At low PAH concentrations, such as when the concentration of the contaminant is in the part per billion ranges usually encountered for underground water contaminations, insufficient energy and carbon source availability can become a limiting factor for microbial growth and maintenance. That is, a threshold may exist below which microbial biomass growth cannot be sustained. Such a minimum concentration for sustainable growth is defined as the concentration at which microbial growth is balanced by decay [36, 43]. Also at such a concentration, the biodegradation rate reverts back to that described by first-order kinetics. However, the application of suitable microorganisms and/or their products can be used to enhance the concentration of PAH in the aqueous phase through increasing the desorption rate and mobilization from the sorbed phase. At such an enhanced concentration, Michaelis–Menten rate kinetics becomes appreciably applicable and from its linearized plot (Figure 3), the maximum PAH degradation rate can easily be determined. The rate equations can also be simplified by making certain assumptions. Particularly, the steady-state approximation which assumes a negligible change in the concentration of the enzyme-substrate complex during the course of the reaction. Furthermore, the Michaelis–Menten reaction mechanism presupposes that catalysis is irreversible and that the enzyme is not subject to any product inhibition. This limits the suitability of using this model to predict biodegradation rates where chemical surfactants have been known to inhibit product formation [44].

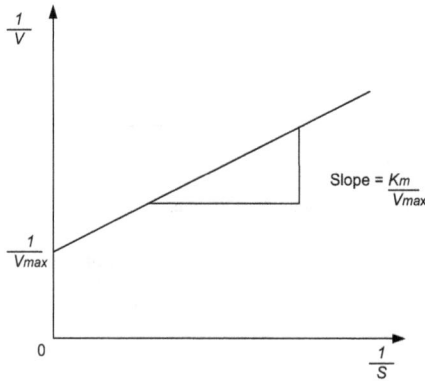

Figure 3. A plot of the linearized form of Michaelis-Menten rate equation.

3.3. Freundlich adsorption isotherm

The Freundlich isotherm is an adsorption isotherm which relates the concentration of a solute on the surface of an adsorbent to the concentration of the solute in the liquid with which it is in contact. The isotherm is mathematically represented as:

$$Q = \frac{x}{m} kC^{\frac{1}{n}} \tag{3}$$

For which the linearized form is:

$$\log Q = \log k + N \log C \quad \left[N = \frac{1}{n} \right] \tag{4}$$

where Q is the sorbed amount of PAH (mg/g soil), x the mass of PAH (mg), m the mass of soil (g), C the equilibrium concentration of PAH in solution (mg/l), k and n are constants for a given PAH and adsorbent, which can be either a solid or a liquid, at a particular temperature and thus define the isotherm's curvature. The amount of the sorbed PAH can be determined by plotting the isotherms' linearized form as a function of the equilibrium concentration of the PAH in solution (Figure 4).

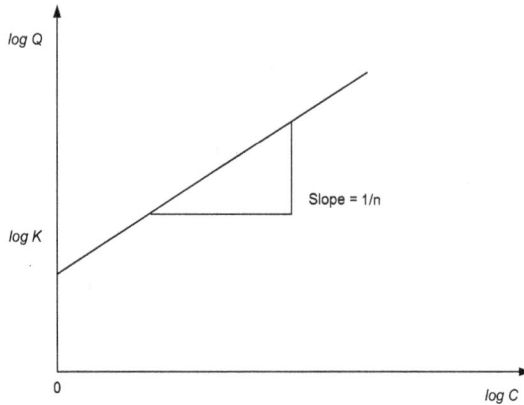

Figure 4. A plot of the linearized form of the Freundlich isotherm.

The isotherm can also be represented in terms of sorbent pore pressure P as:

$$Q = kP^{\frac{1}{n}} \tag{5}$$

At high pressure, the extent of adsorption is independent of pressure (i.e. $1/n = 0$), at low pressure, it is dependent on pressure.

In a recent study of biodegradation of phenanthrene and pyrene in a slurry soil system using *Phanerochaete chrysosporium*, biosorption isotherms were found to fit well with the Freundlich isotherm. The Freundlich n values were approximated to unity, indicating that the biosorption was dominated by partitioning onto fungal biomass and the soils' organic matter [45, 46]. It also showed that the amount of PAH sorption by the microbial biomass was dependent on the concentration of the PAH present in the contaminated medium. The extent of adsorption varied directly with pressure until saturation pressure is reached. Beyond that point, the rate of adsorption reaches a maximum, even after applying higher pressure [47], thus the Freundlich adsorption isotherm will not be suitable at a higher pressure. Although, Langenfeld *et al.* [48] has shown that pressure variation has no effect on the supercritical extraction efficiency of PAHs, the use of Freundlich isotherm in biodegradation studies without considering the effects of pore pressure on sorption of HMW PAHs [45] can be problematic.

3.4. Enhanced bioavailability model

Zhang *et al.* [36] proposed a bioavailability model that can be used to quantitatively estimate the impact of sorption on the biodegradation by taking a mass balance for organic compounds in a batch system containing liquid and solids and substituting the expression into the Michaelis-Menten rate equation. The model assumed that only the organic contaminant in the liquid phase is biodegradable. Park *et al.* [49] extended the model using formulations for reversible and instantaneous sorption–desorption processes with first-order biodegradation reactions in both liquid and solid phases, including the assumption that the liquid-phase degradation rate coefficient is not affected by the presence of solids. Under the sorption–desorption equilibrium assumption, the liquid-phase contaminant disappearance rate can be expressed as:

$$\frac{dC}{dt} = E_f B_f k_1 C \tag{6}$$

Using integration, the concentration of PAHs in the aqueous phase at any particular time during the biodegradation process can be determined as:

$$C = C_0 \exp\left(E_f B_f k_1 t\right) \tag{7}$$

where, E_f is the enhanced transformation factor (dimensionless);

$$E_f = 1 + R_{sl} K_d f_s k_s / k_1 \tag{8}$$

B_f is the bioavailability factor;

$$B_f = 1 \Big/ (1 + R_{sl} K_d) \tag{9}$$

R_{sl} is solid/liquid ratio (mg/l);

$$R_{sl} = m \Big/ V_1 \tag{10}$$

where C is the liquid-phase concentration of contaminant (mg/l), C_0 the initial liquid-phase concentration (mg/l), t time (min), k_1 the first-order liquid-phase degradation rate coefficient (min^{-1}) which can be determined from biodegradation assay, K_d the sorption distribution coefficient (dimensionless), f_s the fraction of attached biomass in the system (dimensionless), k_s the first-order sorbed-phase degradation rate (min^{-1}), m the sorbent mass (g), and V_l is the liquid volume (l).

The bioavailability factor (B_f) range from the above equation is between zero and unity. A B_f approaching unity indicates that the effect of sorption is practically negligible while a B_f tending towards zero shows that biodegradation will be significantly limited by sorption [36]. On the other hand, if there is no sorbed-phase degradation by the attached biomass, E_f is unity and the enhanced bioavailability model reduces to bioavailability model which assumes sorption-desorption and only liquid-phase biodegradation. An $E_f > 1$ indicates that biodegradation is faster than that expected based on the liquid-phase degradation while $E_f < 1$ indicates slower rates [49].

3.5. Effects of temperature on bioavailability

Studies have demonstrated that temperature optimization had a positive effect on biodegradation of hydrophobic organic compounds especially for PAHs in a soil historically contaminated with these hydrocarbons. These findings indicated that in-situ remediation processes are accelerated with elevated temperature in a range between mesophilic and thermophilic temperatures. However, at a high temperature, the activity of fungal and bacterial populations is reduced [50]. Temperatures in the thermophilic range (50 to 60°C) are shown to greatly accelerate decomposition of organic matter, in general [51]. Microbial utilization of hydrocarbons can occur at temperatures ranging from -2 to 70°C. Iqbal *et al.* [52] investigated the ability of indigenous aerobic microorganisms to degrade low and HMW PAHs in sewage sludge fed in continuous bioreactors. It was reported that when the temperature increased from 35 to 45°C and at 55°C, biodegradation of the high molecular weight PAHs was enhanced from 50 to 80%. The results also showed improved kinetic rates with elevated temperature. A reduction in kinetic rates was associated with a decrease in contaminant concentrations over time; however, the kinetic rates were found to be dependent upon the contaminants desorption and environmental factors than concentration; i.e. significant kinetic rates was observed in the ambient and high temperature biodegradation processes.

Temperature elevation coupled with moisture optimization was also determined to enhance the bioremediation of such contaminated soil. Raising the temperature also decreases adsorption, which makes more organic material available for microorganisms to degrade. With the synergistic effects of elevated temperature and sufficient moisture content, White *et al.* [53] determined that moisture and slurrying soil containing organic compounds dramatically enhanced bioavailability and the rate of biodegradation. Reports also show that desorption resistant HMW PAHs were more efficiently metabolized in slurried than in unslurried soil suggesting that temperature and moisture optimization needs to be combined with efficient nutrient delivery systems for soils/sediments.

Furthermore, temperature plays a role when nutrients are added for biodegradation. Previous studies indicate that at 10°C, biodegradation rates were not affected by the addition of phosphorus or nitrogen. However, at 20°C, biodegradation was increased by the addition of phosphorus [54]. This suggests that temperature optimization needs to be combined with sufficient and suitable nutrient amendments. Sartoros *et al.* [55] considered the biodegradation of a mixture of two PAH compounds, pyrene and anthracene, by a developed enrichment culture with the addition of surfactant at a low temperature of 10°C and discovered a negative impact on the biodegradation. It was also observed that the overall extent and the maximum specific rate of PAHs mineralization decreased with the decrease of temperature; the variation of total PAH concentration had negligible effect. However, at 25°C the addition of a surfactant enhanced the mineralization of PAHs. These results have important implications on the use of surfactants for in-situ bioremediation since groundwater temperatures are often at or below 10°C.

In a bioaugmentation system, where allochthonous or genetically engineered microorganisms are added into contaminated soil to aid the operation of indigenous microbes, microbial population can decline rapidly following their introduction into such natural soil, and growth of the engineered populations in microbiologically undisturbed soils is a rare phenomenon due to microbiostasis [56, 57]. Furthermore, temperature variability is one of the principal contributors leading to the decline of microorganisms introduced into a 'hostile' soil microenvironment. In high temperature treatments, the response and survival of the microorganisms will be compromised compared to ambient temperature treatments. The microbial concentration increases under ambient temperature thus significantly increasing remediation activity. Results suggest that the stimulated growth and activity of indigenous microbes with elevated temperature in non-inoculated treatment indicates a positive relationship between optimum temperatures and better biodegradation performance.

4. Enhancing bioavailability of HMW PAHs using biosurfactants from renewable resources

4.1. Effects of biosurfactants on sorbed PAHs

Biosurfactants are surface active agents produced by microorganisms. Table 1 shows some of these microorganisms and the different groups of biosurfactant they produce. All biosurfactants consist of two parts, a polar (hydrophilic) moiety and a non polar (hydrophobic) group. A hydrophilic group consists of mono-, oligo- or polysaccharides, peptides or proteins and a hydrophobic moiety usually contains saturated, unsaturated and hydroxylated fatty acids or fatty alcohols [58]. Due to their amphiphilic structure, biosurfactants show a wide range of properties, including the lowering of surface and interfacial tension of liquids, the ability to form micelles and micro-emulsions between two different phases, the ability to increase the surface area of hydrophobic water-insoluble substances, and thus increase the water bioavailability of such substances. In comparison to their chemically synthesized equivalents they have many advantages; they are environmentally friendly, biodegradable, less toxic and nonhazardous. They have better foaming properties and higher selectivity; they are active at extreme temperatures, pH and salinity, and can be produced from wastes and from various by-products. This feature makes the production of cheap biosurfactants possible and the concomitant effects of utilizing waste substrates and reducing their environmental pollution [59-63]. Biosurfactants increase the bioavailability of PAHs resulting in enhanced growth of the degrading microorganism and the biodegradation of the contaminants.

An important feature of the physico-chemical properties of surfactants is their hydrophilic-lipophilic balance (HLB) [64, 65]. The HLB value indicates whether a surfactant will produce a water-in-oil or oil-in-water emulsion. Emulsifiers with lower HLB values of 3 to 6 are lipophilic and promote water-in-oil emulsification, while emulsifiers with higher HLB values between 10 and 18 are more hydrophilic and promote oil-in-water emulsion formation [5]. A classification based on HLB values has been used to evaluate the suitability of different surfactants for various applications. For example, it has been reported that successful surfactants are those with the ability to promote desorption of contaminants from contaminated soils and are normally those with HLB values above 10 [66].

Biosurfactant		Microorganism	Applications in Environmental Biotechnology	References
Group	Class			
Glycolipids	Rhamnolipids	*Pseudomonas aerugi-nosa, Pseudomonas sp*	Enhancement of the degradation, dispersion, emulsification of different classes of hydrocarbons and vegetable oils; removal of metals from soil.	[69, 71]
	Trehalolipids	*Mycobacterium tuberculosis, Rhodococcus erythropolis, Arthrobacter sp., Nocardia sp., Corynebacterium sp.*	Enhancement of the bioavailability of hydrocarbons.	[72]
	Sophorolipids	*Torulopsis bombicola, Torulopsis petrophilum, Torulopsis apicola*	Recovery of hydrocarbons from dregs and muds; removal of heavy metals from sediments; enhancement of oil recovery.	[73, 74, 69]
Fatty acids, phospholipids and neutral lipids	Corynomycolic acid	*Corynebacterium lepus*	Enhancement of bitumen recovery.	[75]
	Spiculisporic acid	*Penicillium spiculisporum*	Removal of metal ions from aqueous solution; dispersion action for hydrophilic pigments; preparation of emulsion-type organogels, microencapsulation	[76]
	Phosphati-dyle-thanolamine	*Acinetobacter sp., Rhodococcus erythropolis*	Increasing the tolerance of bacteria to heavy metals.	[77]
Lipopeptides	Surfactin	*Bacillus subtilis*	Enhancement of the biodegradation of hydrocarbons and chlorinated pesticides; removal of heavy metals from a contaminated soil, sediment and water; increasing the effectiveness of phytoextraction.	[78]
	Lichenysin	*Bacillus licheniformis*	Enhancement of oil recovery	[79]
Polymeric biosurfactants	Emulsan	*Acinetobacter calcoaceticus RAG-1*	Stabilization of the hydrocarbon-in-water emulsions.	[80]
	Alasan	*Acinetobacter radioresistens KA-53*		[81]
	Biodispersan	*Acinetobacter calcoaceticus A2*	Dispersion of limestone in water.	[82]
	Liposan	*Candida lipolytica*	Stabilization of hydrocarbon-in-water emulsions.	[83]
	Mannoprotein	*Saccharomyces cerevisiae*		[84]

Source: Pacwa-Płociniczak *et al.* [85]

Table 1. Biosurfactants producing organisms: classification and application in environmental biotechnology.

Another feature is the Critical Micelle Concentration (CMC); which is the concentration above which the formation of micelles is thermodynamically favored [67]. The mobilization mechanism occurs at concentrations below the biosurfactant CMC. At such concentrations, biosurfactants reduce the surface and interfacial tension between air-water and soil-water systems. Due to the reduction of the interfacial force, contact of biosurfactants with a soil-oil system increases the contact angle and reduces the capillary force holding the oil and soil together. Above the biosurfactant CMC, the solubilization process takes place. At these concentrations, biosurfactant molecules aggregate to form micelles, which dramatically increase the solubility of the oil. The hydrophobic part of the biosurfactant molecules interconnect inside the micelle while the hydrophilic ends are then exposed to the aqueous phase on the exterior. Consequently, the interior of a micelle creates an environment compatible for hydrophobic organic molecules. The process of incorporation of these molecules into a micelle is known as solubilization [68]. The formation of micelles leads to a significant increase in the apparent solubility of hydrophobic organic compounds, even above their water solubility limit, as these compounds can partition into the central core of a micelle. The effects of such a process are the reduction of surface and interfacial tension, enhancement of mobilization and mass transfer of contaminants from soil particles into the aqueous phase, and consequently the bioavailability of the hydrophobic contaminants for microbial attack [69, 70].

4.2. Biosurfactants production using agricultural and industrial wastes

The bioconversion of waste materials is considered to be of importance for the development of sustainable biotechnology processes in the near future because of its favorable economics, low capital and energy cost, reduction in environmental pollution, and their relative ease of operation [86, 10, 87, 88]. Producing usable products from agricultural and industrial waste is therefore a feasible and favorable option [89, 90]. Modern society produces high quantity of waste materials through activities related to industries such as those in the forestry, agriculture and municipal sectors [86, 91].

The use of the alternative substrates such as agro-based industrial wastes is one of the attractive strategies for the economic production of biosurfactants to enhance biodegradation of environmental hydrophobic contaminants. It has been suggested that successful approaches to more economical production technologies of biosurfactants will be a collaborative approach involving process development and sustainable raw material supplies. According to Marchant and Banat [92], emphasis should be on the cost effective management of downstream processing. These inexpensive agro-industrial wastes substrates include olive oil mill effluent, plant oil extracts and waste, distillery and whey wastes, potato process effluent and cassava wastewater [10]. These waste materials are some examples of food industry by-products or waste that can be used as feedstock for biosurfactants production. Similarly, vegetable oil wastes can be used for biosurfactant production as they are lipidic carbon sources and are mostly comprised of saturated or unsaturated fatty acids with 16 to18 carbon atoms chain.

Studies involving the application of a variety of vegetable oils for biosurfactants production from canola, corn, sunflower, safflower, olive, rapeseed, grape seed, palm, coconut, fish and soybean oil have been reported. The world production of oils and fats is about 2.5 to 3 million tons, 75% of which are derived from plants and oil seeds [93]. The high content of fats, oils and other nutrients in these wastes make them invaluable cheap raw materials for industries involved in useful metabolite production. Furthermore, from an economical point of view, nutrient rich agricultural residues can be employed for producing useful biological products such as biosurfactants. These materials are among the most abundant organic carbon available on earth [94] and they are the major components of different waste streams from various industries.

In recent times, studies have been focused on the application of agro-industrial wastes or by-products for the production of biosurfactants which are used in crude form or the direct use of surfactants producing strains for biodegradation processes. This is due to the high cost of biosurfactant purification and the stability including sustainability provided by these biosur-factants producing strains in biodegradation processes. The production and properties of a biosurfactant, synthesized by *Bacillus subtilis LB5a* strain, using cassava wastewater as a substrate was investigated. The microorganism was able to grow and to produce a surfactant on cassava waste, reducing the surface tension of the medium to 26.6 mN/m and giving a crude surfactant concentration of 3.0 g/L after 48 h [95]. The biosurfactant obtained was capable of forming stable emulsions with various hydrocarbons. Panesar *et al.* [96] investigated the suitability of molasses, the sugar industry by-product, for biosurfactant production using *Pseudomonas aeruginosa* strain *ATCC 2297*. An attempt was also made to replace the costly nitrogen sources with agro-industrial by-products to formulate low cost medium for biosur-factants production. The strain was found to displayed maximum emulsification activity on molasses medium after 120 h of incubation under optimized conditions. Biosurfactants production by a strain of *Pseudomonas aeruginosa* using palm oil as a sole carbon source was also investigated [97]. The *P. aeruginosa* strain gave emulsification index results of 100% when diesel was used as an oil phase and was able to reduce surface tension of three tested inorganic media to approximately 33 mN/m. Wan-Nawawi *et al.* [98] also reported the versatility of a bacterial strain isolated from a hydrocarbon-based source at a palm oil mill. The strain showed a high bacterial growth on sludge palm oil with a surface tension of 36.2 mN/m and was therefore proposed for biosurfactant production by liquid state fermentation.

4.3. Application of biosurfactants for enhancing PAHs bioavailability

Only limited numbers of microorganisms are capable of degrading HMW PAHs. Hence their biodegradation is limited by their low bioavailability to the microorganisms, which is due to their hydrophobicity, low aqueous solubility and strong adsorptive capacity in soil [99, 100]. Berg *et al.* [101] described an emulsifying agent produced by *P. aeruginosa* UG2 that increased the solubility of hexachlorobiphenyl added to soil slurries, resulting in a 31% recovery of the compound in the aqueous phase. Griffin *et al.* [102] demonstrated that rhamnolipids from bacteria, in combination with the oleophilic fertilizer *Inipol EAp*-22, increased the degradation rate of hexadecane, benzene, toluene, o- and p-cresol and naphthalene in aqueous-phase bioreactors and in those containing soil. They also reported increased rates of biodegradation of aliphatic and aromatic hydrocarbons by pure bacterial cultures.

The efficiency of biosurfactants in the remediation of soil contaminated by phenanthrene and polychlorinated biphenyls (PCBs) was also reported [103]. In an investigation of the capacity of PAH-utilizing bacteria to produce biosurfactants using naphthalene and phenanthrene, Deziel et al. [104] quantified biosurfactants production that was responsible for an increase in the aqueous concentration of naphthalene. This indicates a potential role for biosurfactants in increasing the solubility of such compounds. Similarly, Zhang et al. [105] determined the effect of two biosurfactants on the dissolution and bioavailability of phenanthrene and reported increases in both solubility and the degradation rate of phenanthrene. Kanga et al. [106] applied glycolipid biosurfactants produced by Rhodococcus sp. H13A and a synthetic surfactant (Tween 80) for enhanced substrate solubility. Using naphthalene and methyl-substituted derivatives in crude oil as representative of the PAH content, they observed that both surfactants lowered surface tension in solutions from 72 to 30 mN/m. The biosurfactants were efficient in increasing the solubility of the hydrocarbons, particularly the substituted derivative. In a laboratory column study, Noordman et al. [107] applied biosurfactants for the enhanced removal of phenanthrene from phenanthrene-contaminated soil, eluting the contaminant with an electrolyte solution containing rhamnolipid. The enhanced removal of phenanthrene occurred mainly by micellar solubilization.

Microbially produced biosurfactants were studied to enhance crude oil desorption and mobilization in model soil column systems [108]. The results showed that the ability of biosurfactants from Rhodococcus ruber to remove the oil from the soil core was 1.4 to 2.3 times greater than that of a synthetic surfactant (Tween 60), of suitable properties. The biosurfactant was less adsorbed to soil components than synthetic surfactant, thus rapidly penetrating through the soil column and effectively removing 65–82% of the crude oil. Chemical analysis showed that the crude oil removed by the biosurfactant contained a lower proportion of high-molecular-weight paraffins and asphaltenes.

The capability of biosurfactants and biosurfactant-producing microorganisms to enhance organic contaminants' availability and biodegradation rates was reported by several authors [109, 110, 104]. Kang et al. [111] used sophorolipid in studies on biodegradation of aliphatic and aromatic hydrocarbons and Iranian light crude oil under laboratory conditions. The addition of this biosurfactant to soil increased biodegradation of tested hydrocarbons with the rate of degradation ranging from 85% to 97% of the total amount of hydrocarbons. Their results indicated that sophorolipid may have the potential for facilitating the bioremediation of sites contaminated with hydrocarbons having limited water solubility and increasing the viability of microbial consortia for biodegradation. The solubility and utilization of pyrene as a sole carbon source by the biosurfactant-producing bacterial strains; Bacillus subtilis DM-04, Pseudomonas aeruginosa mucoid (M) and nonmucoid (NM), isolated from a petroleum-contaminated soil were studied [112]. It was reported that the biosurfactants produced by the bacteria under the study were capable of enhancing the solubility of pyrene in aqueous media and can influence the cell surface hydrophobicity of the biosurfactant-producing strains that results in a higher uptake of pyrene.

5. Conclusion

Although environmental biotechnology is regarded as an eco-friendly technology for the clean-up of PAH-contaminated ecosystems, the successful application of this technology is still restricted by the enormous costs of its operation and the limited bioavailability of the contaminants to degradative micro-organisms, especially HMW PAHs, due to sorption and sequestration. Many bacteria, fungi and algae and their products have been applied to degrade a range of LMW PAHs, such as naphthalene, fluorene and phenanthrene; however, their activity towards HMW PAHs containing five or more fused benzene rings, such as Benzo(a)pyrene and Benzo(ghi)perylene, is limited. Application of biosurfactants produced using agricultural and industrial wastes may be promising for reducing the costs of biodegradation technology as well as enhancing bioavailability of HMW PAHs. However, there is a dearth of research on this subject. Bioavailability model could be a vital tool to estimate and assess the success of bioremediation; therefore, more research is needed in this area, considering variations in environmental conditions that could limit field simulation of laboratory results. The synergistic effects of increased temperature and moisture with biosurfactants application have been shown to enhance biodegradation efficiency; however, kinetic investigation is important under these conditions to provide an understanding of biodegradation rate in situation where these conditions cannot be controlled such as in a typical field operation.

Author details

Olusola Solomon Amodu[1], Tunde Victor Ojumu[1] and Seteno Karabo Obed Ntwampe[2]

1 Faculty of Engineering, Department of Chemical Engineering, Cape Peninsula University of Technology, Cape Town, South Africa

2 Faculty of Applied Sciences, Department of Agriculture and Food Sciences: Biotechnology Programme, Cape Peninsula University of Technology, Cape Town, South Africa

References

[1] USEPA U. Integrated Risk Information System (IRIS). EPA; 1999.

[2] Acevedo F, Pizzul L, Castillo MP, Cuevas R, Diez MC. Degradation of polycyclic aromatic hydrocarbons by the Chilean white-rot fungus Anthracophyllum discolor. Journal of Hazardous Materials. 2011;185(1):212-9.

[3] Acevedo F, Pizzul L, González M, Cea M, Gianfreda L, Diez M. Degradation of polycyclic aromatic hydrocarbons by free and nanoclay-immobilized manganese peroxidase from Anthracophyllum discolor. Chemosphere. 2010;80(3):271-8.

[4] Fernando Bautista L, Morales G, Sanz R. Immobilization strategies for laccase from Trametes versicolor on mesostructured silica materials and the application to the degradation of naphthalene. Bioresource Technology. 2010;101(22):8541-8.

[5] Nilanjana D, Preethy C. Microbial degradation of petroleum hydrocarbon contaminants: An overview. Biotechnology Research International. 2010;2011.

[6] Cameotra SS, Singh P. Bioremediation of oil sludge using crude biosurfactants. International Biodeterioration and Biodegradation. 2008;62(3):274-80.

[7] Mancera-Lopez M, Esparza-Garcia F, Chavez-Gomez B, Rodriguez-Vazquez R, Saucedo-Castaneda G, Barrera-Cortes J. Bioremediation of an aged hydrocarbon-contaminated soil by a combined system of biostimulation–bioaugmentation with filamentous fungi. International Biodeterioration and Biodegradation. 2008;61(2): 151-60.

[8] Henkel M, Müller MM, Kügler JH, Lovaglio RB, Contiero J, Syldatk C, et al. Rhamnolipids as biosurfactants from renewable resources: Concepts for next-generation rhamnolipid production. Process Biochemistry. 2012.

[9] Rubilar O, Tortella G, Cea M, Acevedo F, Bustamante M, Gianfreda L, et al. Bioremediation of a Chilean Andisol contaminated with pentachlorophenol (PCP) by solid substrate cultures of white-rot fungi. Biodegradation. 2011;22(1):31-41.

[10] Makkar RS, Cameotra SS, Banat IM. Advances in utilization of renewable substrates for biosurfactant production. AMB Express. 2011;1(1):1-19.

[11] Rodríguez-Meza MA, Chávez-Gómez B, Poggi-Varaldo HM, Ríos-Leal E, Barrera-Cortés J. Design of a new rotating drum bioreactor operated at atmospheric pressure on the bioremediation of a polluted soil. Bioprocess and Biosystems Engineering. 2010;33(5):573-82.

[12] Stokes JD, Paton G, Semple KT. Behaviour and assessment of bioavailability of organic contaminants in soil: relevance for risk assessment and remediation. Soil Use and Management. 2005;21:475-86.

[13] ECETOC. European Centre for Ecotoxicology and Toxicology of Chemicals. Scientific principles of soil hazard assessment of substances: a technical report. 2002(84):24–6.

[14] USNRC. Bioavailability of contaminants in soils and sediments: processes, tools and applications. National Academic Press: Washington, DC. 2002.

[15] Van Leeuwen CJH, J. L. M.. Terrestrial toxicity. In Risk Assessment of Chemicals: an introduction. Kluwer Academic, Dordrecht, The Netherlands. . 1995: 211–6.

[16] Herrchen M. Bioavailability as a Key Property in Terrestrial Ecotoxicity Assessment and Evaluation: Major Statements and Abstracts of Presentations of an International European Workshop: Held at the Fraunhofer-Institute for Environmental Chemistry and Ecotoxicology IUCT Schmallenberg, Germany, April 22-23, 1996: Fraunhofer-IRB-Verl.; 1997.

[17] Klaassen CD, Admur MO. Casarett and Doull's toxicology: the basic science of poisons. 2001.

[18] Semple KT, Morriss A, Paton G. Bioavailability of hydrophobic organic contaminants in soils: fundamental concepts and techniques for analysis. European Journal of Soil Science. 2003;54(4):809-18.

[19] Alexander M. Aging, bioavailability, and overestimation of risk from environmental pollutants. Environmental Science and Technology. 2000;34(20):4259-65.

[20] Ruby MV, Davis A, Schoof R, Eberle S, Sellstone CM. Estimation of lead and arsenic bioavailability using a physiologically based extraction test. Environmental Science and Technology. 1996;30(2):422-30.

[21] Kramer BK, Ryan PB. Soxhlet and microwave extraction in determining the bioaccessibility of pesticides from soil and model solids. Proceedings of the 2000 Conference on Hazardous Waste Research 2000:pp 196–210.

[22] Khan MI, Cheema SA, Shen C, Zhang C, Tang X, Shi J, et al. Assessment of phenanthrene bioavailability in aged and unaged soils by mild extraction. Environmental Monitoring and Assessment. 2011.

[23] Harmsen J. Measuring bioavailability: From a scientific approach to standard methods. Journal of Environmental Quality. 2007;36(5):1420-8.

[24] Alexander M. Biodegradation and Bioremediation, Academic Press, Inc., San Diego, CA. 1999.

[25] Uyttebroek M, Spoden A, Ortega-Calvo JJ, Wouters K, Wattiau P, Bastiaens L, et al. Differential Responses of Eubacterial, Mycobacterium, and Sphingomonas Communities in Polycyclic Aromatic Hydrocarbon(PAH)-Contaminated Soil to Artificially Induced Changes in PAH Profile. Journal of Environmental Quality. 2007;36(5):1403-11.

[26] Bergknut M. Characterization of PAH-contaminated soils focusing on availability, chemical composition and biological effects. Disertation at Umeå University. 2006.

[27] Bosma TNP, Middeldorp PJM, Schraa G, Zehnder AJB. Mass transfer limitation of biotransformation: quantifying bioavailability. Environmental Science and Technology. 1996;31(1):248-52.

[28] Burgos WD, Novak JT, Berry DF. Reversible sorption and irreversible binding of naphthalene and α-naphthol to soil: Elucidation of processes. Environmental Science and Technology. 1996;30(4):1205-11.

[29] Luthy RG, Aiken GR, Brusseau ML, Cunningham SD, Gschwend PM, Pignatello JJ, et al. Sequestration of hydrophobic organic contaminants by geosorbents. Environmental Science and Technology. 1997;31(12):3341-7.

[30] McNally DL, Mihelcic JR, Stapleton JM. Bioremediation for soil reclamation. 2007.

[31] Wu SC, Gschwend PM. Sorption kinetics of hydrophobic organic compounds to natural sediments and soils. Environmental Science and Technology. 1986;20(7):717-25.

[32] Nam K, Chung N, Alexander M. Relationship between organic matter content of soil and the sequestration of phenanthrene. Environmental Science and Technology. 1998;32(23):3785-8.

[33] Hundal LS, Thompson ML, Laird DA, Carmo AM. Sorption of phenanthrene by reference smectites. Environmental Science and Technology. 2001;35(17):3456-61.

[34] Yang Y, Zhang N, Xue M, Tao S. Impact of soil organic matter on the distribution of polycyclic aromatic hydrocarbons (PAHs) in soils. Environmental Pollution. 2010;158(6):2170-4.

[35] Reid BJ, Jones KC, Semple KT. Bioavailability of persistent organic pollutants in soils and sediments--a perspective on mechanisms, consequences and assessment. Environmental Pollution. 2000;108(1):103-12.

[36] Zhang W, Bouwer EJ, Ball WP. Bioavailability of Hydrophobic Organic Contaminants: Effects and Implications of Sorption Related Mass Transfer on Bioremediation. Ground Water Monitoring and Remediation. 1998;18(1):126-38.

[37] Johnsen AR, Wick LY, Harms H. Principles of microbial PAH-degradation in soil. Environmental Pollution. 2005;133(1):71-84.

[38] Ball W, Buehler C, Harmon T, Mackay D, Roberts P. Characterization of a sandy aquifer material at the grain scale. Journal of Contaminant Hydrology. 1990;5(3):253-95.

[39] Alexander M. Introduction to soil microbiology. Soil Science. 1978;125(5):331.

[40] Lawrence G, Payne D, Greenland D. Pore size distribution in critical point and freeze dried aggregates from clay subsoils. Journal of Soil Science. 1979;30(3):499-516.

[41] Kwon SH, Kim JH, Cho D. An analysis method for degradation kinetics of lowly concentrated PAH solutions under UV light and ultrasonication. Journal of Industrial and Engineering Chemistry. 2009;15(2):157-62.

[42] Yang Y, Shu L, Wang X, Xing B, Tao S. Effects of composition and domain arrangement of biopolymer components of soil organic matter on the bioavailability of phenanthrene. Environmental Science and Technology. 2010;44(9):3339-44.

[43] Rittmann BE, McCarty PL. Model of steady state biofilm kinetics. Biotechnology and Bioengineering. 1980;22(11):2343-57.

[44] Zhao Z, Wong JWC. Biosurfactants from Acinetobacter calcoaceticus BU03 enhance the solubility and biodegradation of phenanthrene. Environmental Tchnology. 2009;30(3): 291-9.

[45] Chen B, Ding J. Biosorption and biodegradation of phenanthrene and pyrene in sterilized and unsterilized soil slurry systems stimulated by Phanerochaete chrysosporium. Journal of hazardous materials. 2012.

[46] Chen B, Wang Y, Hu D. Biosorption and biodegradation of polycyclic aromatic hydrocarbons in aqueous solutions by a consortium of white-rot fungi. Journal of Hazardous Materials. 2010;179(1):845-51.

[47] Levenspiel O. Chemical reactions engineering. Recherche. 1998;67:02.

[48] Langenfeld JJ, Hawthorne SB, Miller DJ, Pawliszyn J. Effects of temperature and pressure on supercritical fluid extraction efficiencies of polycyclic aromatic hydrocarbons and polychlorinated biphenyls. Analytical Chemistry. 1993;65(4):338-44.

[49] Park JH, Zhao X, Thomas C. Biodegradation of non-desorbable naphthalene in soils. Environmental Science and Technology. 2001;35(13):2734-40.

[50] Wise D, Torantolo D. Remediation engineering of hazardous waste contaminated soil. 1994. Marcel Dekker Inc., NY.

[51] Parr J, Sikora L, Burge W. Factors affecting the degradation and inactivation of waste constituents in soils. Land Treatment of Hazardous Wastes. 1983:20-49.

[52] Iqbal J, Metosh-Dickey C, Portier RJ. Temperature effects on bioremediation of PAHs and PCP contaminated south Louisiana soils: a laboratory mesocosm study. Journal of Soils and Sediments. 2007;7(3):153-8.

[53] White JC, Alexander M, Pignatello JJ. Enhancing the bioavailability of organic compounds sequestered in soil and aquifer solids. Environmental Toxicology and Chemistry. 1999;18(2):182-7.

[54] Walworth J, Reynolds C. Bioremediation of a petroleum-contaminated cryic soil: Effects of phosphorus, nitrogen, and temperature. Soil and Sediment Contamination. 1995;4(3):299-310.

[55] Sartoros C, Yerushalmi L, Béron P, Guiot SR. Effects of surfactant and temperature on biotransformation kinetics of anthracene and pyrene. Chemosphere. 2005;61(7): 1042-50.

[56] Ho W, Ko W. Soil microbiostasis: effects of environmental and edaphic factors. Soil Biology and Biochemistry. 1985;17(2):167-70.

[57] Van Veen J, Van Overbeek L, Van Elsas J. Fate and activity of microorganisms introduced into soil. Microbiology and Molecular Biology Reviews. 1997;61(2):121-35.

[58] Lang S. Biological amphiphiles (microbial biosurfactants). Current Opinion in Colloid and Interface Science. 2002;7(1-2):12-20.

[59] Kosaric N. Biosurfactants in industry. Pure and Applied Chemistry. 1992;64:1731-.

[60] Rahman K, Rahman TJ, Kourkoutas Y, Petsas I, Marchant R, Banat I. Enhanced bioremediation of alkane in petroleum sludge using bacterial consortium amended with rhamnolipid and micronutrients. Bioresource Technology. 2003;90(2):159-68.

[61] Kosaric N. Biosurfactants and their application for soil bioremediation. Food Technology and Biotechnology. 2001;39(4):295-304.

[62] Das K, Mukherjee AK. Comparison of lipopeptide biosurfactants production by Bacillus subtilis strains in submerged and solid state fermentation systems using a cheap carbon source: Some industrial applications of biosurfactants. Process Biochemistry. 2007;42(8):1191-9.

[63] Das P, Mukherjee S, Sen R. Improved bioavailability and biodegradation of a model polyaromatic hydrocarbon by a biosurfactant producing bacterium of marine origin. Chemosphere. 2008;72(9):1229-34.

[64] González N, Simarro R, Molina M, Bautista L, Delgado L, Villa J. Effect of surfactants on PAH biodegradation by a bacterial consortium and on the dynamics of the bacterial community during the process. Bioresource Technology. 2011.

[65] Tiehm A. Degradation of polycyclic aromatic hydrocarbons in the presence of synthetic surfactants. Applied and Environmental Microbiology. 1994;60(1):258-63.

[66] Volkering F, Breure A, Rulkens W. Microbiological aspects of surfactant use for biological soil remediation. Biodegradation. 1997;8(6):401-17.

[67] Haigh SD. A review of the interaction of surfactants with organic contaminants in soil. Science of the Total Environment. 1996;185(1-3):161-70.

[68] Urum K, Pekdemir T. Evaluation of biosurfactants for crude oil contaminated soil washing. Chemosphere. 2004;57(9):1139-50.

[69] Whang LM, Liu PWG, Ma CC, Cheng SS. Application of biosurfactants, rhamnolipid, and surfactin, for enhanced biodegradation of diesel-contaminated water and soil. Journal of Hazardous Materials. 2008;151(1):155-63.

[70] Perfumo AS, T.J.P.; Marchant, R.;, Banat I. Production and Roles of Biosurfactants and Bioemulsifiers in Accessing Hydrophobic Substrates: In Handbook of Hydrocarbon and Lipid Microbiology ed Timmis, K.N. Chap. 47. 2010;Vol. 2(7):1501- 12. Springer-Verlag, Berlin.

[71] Sifour M, Al-Jilawi MH, Aziz GM. Emulsification properties of biosurfactant produced from Pseudomonas aeruginosa RB 28. Pakistan Journal of Biological Sciences: PJBS. 2007;10(8):1331.

[72] Franzetti A, Gandolfi I, Bestetti G, Smyth TJP, Banat IM. Production and applications of trehalose lipid biosurfactants. European Journal of Lipid Science and Technology. 2010;112(6):617-27.

[73] Baviere M, Degouy D, Lecourtier J. Process for washing solid particles comprising a sophoroside solution. Google Patents; 1994.

[74] Pesce L. A biotechnological method for the regeneration of hydrocarbons from dregs and muds, on the base of biosurfactants. WO Patent WO/2002/062,495; 2002.

[75] Gerson D, Zajic J. Surfactant production from hydrocarbons by Corynebacterium lepus, sp. nov. and Pseudomonas asphaltenicus, sp. nov. Development in Industrial Microbiology. 1978;19:577-99.

[76] Hong JJ, Yang SM, Lee CH, Choi YK, Kajiuchi T. Ultrafiltration of divalent metal cations from aqueous solution using polycarboxylic acid type biosurfactant. Journal of Colloid and Interface Science. 1998;202(1):63-73.

[77] Appanna VD, Finn H, Pierre MS. Exocellular phosphatidylethanolamine production and multiple-metal tolerance in Pseudomonas fluorescens. FEMS Microbiology Letters. 1995;131(1):53-6.

[78] Awashti N, Kumar A, Makkar R, Cameotra S. Enhanced biodegradation of endosulfan, a chlorinated pesticide in presence of a biosurfactant. Journal of Environmental Science and Health B. 1999;34:793-803.

[79] Thomas C, Duvall M, Robertson E, Barrett K, Bala G. Surfactant-based EOR mediated by naturally occurring microorganisms. SPE Reservoir Engineering. 1993;8(4):285-91.

[80] Zosim Z, Gutnick D, Rosenberg E. Properties of hydrocarbon in water emulsions stabilized by Acinetobacter RAG1 emulsan. Biotechnology and Bioengineering. 1982;24(2):281-92.

[81] Toren A, Navon-Venezia S, Ron EZ, Rosenberg E. Emulsifying activities of purified Alasan proteins from Acinetobacter radioresistens KA53. Applied and Environmental Microbiology. 2001;67(3):1102-6.

[82] Rosenberg E, Rubinovitz C, Legmann R, Ron E. Purification and chemical properties of Acinetobacter calcoaceticus A2 biodispersan. Applied and Environmental Microbiology. 1988;54(2):323-6.

[83] Cirigliano MC, Carman GM. Purification and characterization of liposan, a bioemulsifier from Candida lipolytica. Applied and Environmental Microbiology. 1985;50(4):846-50.

[84] Cameron DR, Cooper DG, Neufeld R. The mannoprotein of Saccharomyces cerevisiae is an effective bioemulsifier. Applied and Environmental Microbiology. 1988;54(6):1420-5.

[85] Pacwa-Płociniczak M, Płaza GA, Piotrowska-Seget Z, Cameotra SS. Environmental applications of biosurfactants: Recent advances. International Journal of Molecular Sciences. 2011;12(1):633-54.

[86] Montoneri E, Savarino P, Bottigliengo S, Boffa V, Prevot AB, Fabbri D, et al., editors. Biomass wastes as renewable source of energy and chemicals for the industry with friendly environmental impact. 2009: Parlar Scientific Publications.

[87] Lopes FN. Industrial exploitation of renewable resources: from ethanol production to bioproducts development. Journal de la Société de Biologie. 2008;202(3):191.

[88] Deleu M, Paquot M. From renewable vegetables resources to microorganisms: new trends in surfactants. Comptes Rendus Chimie. 2004;7(6):641-6.

[89] Makkar R, Cameotra S. An update on the use of unconventional substrates for biosurfactant production and their new applications. Applied Microbiology and Biotechnology. 2002;58(4):428-34.

[90] Moldes AB, Torrado AM, Barral MT, Domínguez JM. Evaluation of biosurfactant production from various agricultural residues by Lactobacillus pentosus. Journal of Agricultural and Food Chemistry. 2007;55(11):4481-6.

[91] Martins V, Kalil S, Bertolin T, Costa J. Solid state biosurfactant production in a fixed-bed column bioreactor. Zeitschrift fur Naturfoschung. 2006;61(9/10):721.

[92] Marchant TJPSAPR, Banat I. Isolation and Analysis of Low Molecular Weight Microbial Glycolipids. 2010.

[93] Dumont MJ, Narine SS. Soapstock and deodorizer distillates from North American vegetable oils: Review on their characterization, extraction and utilization. Food Research International. 2007;40(8):957-74.

[94] Kukhar V. Biomass–Renewable Feedstock for Organic Chemicals ("White Chemistry"). Kemija u Industriji. 2009;58.

[95] Nitschke M, Pastore GM. Production and properties of a surfactant obtained from Bacillus subtilis grown on cassava wastewater. Bioresource Technology. 2006;97(2):336-41.

[96] Panesar R, Panesar P, Bera M. Development of low cost medium for the production of biosurfactants. Asian Journal of Biotechnology. 2011.

[97] Thaniyavarn J, Chongchin A, Wanitsuksombut N, Thaniyavarn S, Pinphanichakarn P, Leepipatpiboon N, et al. Biosurfactant production by Pseudomonas aeruginosa A41 using palm oil as carbon source. The Journal of General and Applied Microbiology. 2006;52(4):215-22.

[98] Wan-Nawawi WMF, Jamal P, Alam MZ. Utilization of sludge palm oil as a novel substrate for biosurfactant production. Bioresource Technology. 2010;101(23):9241-7.

[99] Harayama S. Polycyclic aromatic hydrocarbon bioremediation design. Current Opinion in Biotechnology. 1997;8(3):268-73.

[100] Volkering F, Breure AM, van Andel JG, Rulkens WH. Influence of nonionic surfactants on bioavailability and biodegradation of polycyclic aromatic hydrocarbons. Applied and Environmental Microbiology. 1995;61(5):1699-705.

[101] Berg G, Seech AG, Lee H, Trevors JT. Identification and characterization of a soil bacterium with extracellular emulsifying activity. Journal of Environmental Science and Health Part A. 1990;25(7):753-64.

[102] Griffin R, Churchill P, Churchill S, Jones L. Biodegradation rate enhancement of hydrocarbons by an oleophilic fertilizer and a rhamnolipid biosurfactant. Journal of Environmental Quality. 1995;24(1):19-28.

[103] Miller R. Surfactant-enhanced bioavailability of slightly soluble organic compounds. 1995.

[104] Deziel E, Paquette G, Villemur R, Lepine F, Bisaillon J. Biosurfactant production by a soil pseudomonas strain growing on polycyclic aromatic hydrocarbons. Applied and Environmental Microbiology. 1996;62(6):1908-12.

[105] Zhang Y, Maier WJ, Miller RM. Effect of rhamnolipids on the dissolution, bioavailability, and biodegradation of phenanthrene. Environmental Science and Technology. 1997;31(8):2211-7.

[106] Kanga SA, Bonner JS, Page CA, Marc A, Autenrieth RL. Solubilization of naphthalene and methyl-substituted naphthalenes from crude oil using biosurfactants. Environmental Science and Technology. 1997;31(2):556-61.

[107] Noordman WH, Ji W, Brusseau ML, Janssen DB. Effects of rhamnolipid biosurfactants on removal of phenanthrene from soil. Environmental Science and Technology. 1998;32(12):1806-12.

[108] Kuyukina MS, Ivshina IB, Makarov SO, Litvinenko LV, Cunningham CJ, Philp JC. Effect of biosurfactants on crude oil desorption and mobilization in a soil system. Environment International. 2005;31(2):155-61.

[109] Rahman K, Rahman T, Lakshmanaperumalsamy P, Marchant R, Banat I. The potential of bacterial isolates for emulsification with a range of hydrocarbons. Acta Biotechnologica. 2003;23(4):335-45.

[110] Inakollu S, Hung HC, Shreve GS. Biosurfactant enhancement of microbial degradation of various structural classes of hydrocarbon in mixed waste systems. Environmental Engineering Science. 2004;21(4):463-9.

[111] Kang SW, Kim YB, Shin JD, Kim EK. Enhanced biodegradation of hydrocarbons in soil by microbial biosurfactant, sophorolipid. Applied Biochemistry and Biotechnology. 2010;160(3):780-90.

[112] Das K, Mukherjee A. Differential utilization of pyrene as the sole source of carbon by Bacillus subtilis and Pseudomonas aeruginosa strains: role of biosurfactants in enhancing bioavailability. Journal of Applied Microbiology. 2007;102(1):195-203.

Biotechnological Procedures for Environmental Protection

Polyhydroxyalkanoate (PHA) Production from Carbon Dioxide by Recombinant Cyanobacteria

Hitoshi Miyasaka, Hiroshi Okuhata, Satoshi Tanaka, Takuo Onizuka and Hideo Akiyama

Additional information is available at the end of the chapter

1. Introduction

Global warming is the urgent issue of our time, and the carbon dioxide is a greenhouse gas of the major concern. There are various research activities for carbon dioxide mitigation, such as CO_2 recovery from the flue gas of industrial sites, underground and undersea CO_2 storage, and also chemical/biological conversion of CO_2 into the industrial materials [1].

On-site CO_2 fixation by bioprocess is based on the activities of photosynthetic organisms. The fixation of CO_2 by photosynthetic microorganisms can be an efficient system for the CO_2 mitigation, but one of the major problems of this system is the effective utilization of the fixed biomass. The biomass produced by photosynthetic microorganisms must be utilized as a resource, or it will be easily degraded by microorganisms into CO_2 again. Cyanobacteria are procaryotic photosynthetic microorganisms and can provide a simple genetic transformation system for the production of useful materials from CO_2. We have established an efficient vector-promoter system for the introduction and expression of foreign genes in the marine cyanobacterium *Synechococcus sp.* PCC7002, and examined the production of biodegradable plastic, polyhydroxyalkanoate (PHA), by genetically engineered cyanobacteria. The PHA is a biopolymer accumulated by various microorganisms as reserves of carbon and reducing equivalents. PHAs are linear head to tail polyesters composed of 3-hydroxy fatty acid mono-mers (Figure 1), have physical properties similar to those of polyethylene, and can replace the chemical plastics in some applications, such as disposable bulk materials in packing films, containers, and paper coatings [2]. PHA applications as implant biomaterials, drug delivery carrier, and biofuel have also been investigated [3]. The commercial mass production of PHA from corn sugar by using the genetically enigineered microorganism was started in 2009 in the United State and China [3].

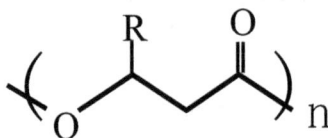

Figure 1. Chemical structure of PHAs

Assimilation and conversion of CO_2 into the biodegradable plastics (biopolymers) by photosynthetic microorganisms is an ideal bioprocess because it converts CO_2 directly into the useful bioplastics with solar energy. It also contributes to the low carbon society by substituting the environmentally unfriendly petroleum-based plastics with the carbon neutral bioplastics, and also by saving the fossil fuel resource required for the petrochemical plastics production. In addition, the biodegradable plastics reduce the burden of plastics waste on landfills and the environment.

2. Shuttle-vector construction

2.1. Construction of a shuttle-vector between *Escherichia coli* and cyanobacteria

Since most of the industrial CO_2 emission sites locate in the seashore area in Japan, we choose a marine cyanobacterial strain for the fixation and utilization of CO_2. The marine cyanobacterial strain *Synechococcus sp.* PCC7002 (*Agmenellum quadruplicatum* PR-6, ATCC 27264) [4] was obtained from the American Type Culture Collection, and cultured at 30 °C in medium A [5] under continuous illumination (50 μmol photons $m^{-2} s^{-1}$) by bubbling with 1% CO_2 in air.

For the construction of a shuttle-vector between *E. coli* and *Synechococcus sp.* PCC7002, we isolated and characterized the smallest endogenous plasmid pAQ1 of this cyanobacterium [6, 7]. The plasmid pAQ1 is 4809 bp long, and has four open reading frames (ORFs): ORF943, ORF64, ORF71, and ORF93 (numbers show putative amino acid number). The construction of the shuttle-vector was done by digesting pAQ1 plasmid and bacterial pUC19 plasmid with restriction enzymes which cleave each plasmid at a unique site, and by ligating the linearized plasmids. The plasmid pUC19 and the plasmid pAQ1 were linearized by *Sma*I and *Stu*I digestions, respectively, and were ligated to generate the shuttle-vector pAQJ6 (Figure 2; both *Sma*I and *Stu*I are blunt-end forming restriction enzymes). The effect of the four ORFs on the transformation efficiency was examined, and ORF943 was found to be important for the maintenance of the shuttle-vector. From this result, the simplified shuttle-vector pAQJ4 with the full ORF943 was designed (Figure 2).

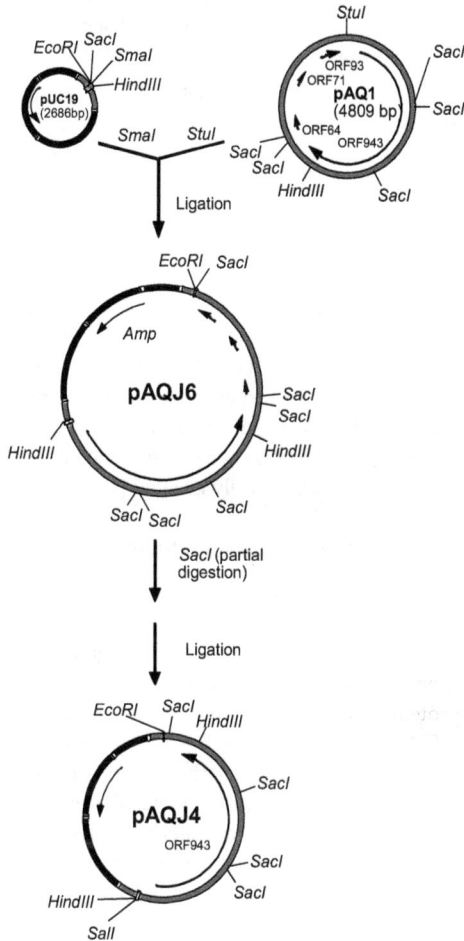

Figure 2. Construction of shuttle-vector between *E. coli* and *Synec hococcus* sp. PCC7002

The stability of the prototype shuttle-vector, pAQJ6, in cyanobacterial cells was relatively low, but that of the simplified shuttle-vector, pAQJ4, was much improved; this vector could be stably retained in cyanobacterial cells after 100 generations of growth without antibiotics selection [8]. This is probably because that there are several hot spots for the homologous recombination between endogenous pAQ1 plasmid and pAQJ6 vector, and these hot spots might have been eliminated in the simplified pAQJ4 vector. The transformation efficiency of the shuttle-vector pAQJ4 was about 4×10^5 (cfu / μg DNA), when we transformed 4×10^7 of cyanobacterial cells with 0.3 μg (0.1 pmol) of pAQJ4 vector in 1 ml solution. This transformation efficiency was 10 to 100 times higher than those of the shuttle-vectors for this cyanobacterium previously reported [4, 9]. With this system, we can obtain several million of cyanobacterial transformant in one experiment, thus this shuttle-vector system can also be applied for the construction of cDNA libraries using cyanobacteria as host cells.

An example of the use of this shuttle-vector for cDNA library is the construction of cDNA library of the halotolerant marine green alga *Chlamydomonas* W80 for the isolation of anti-stress genes [10]. C. W80 shows a surprisingly high oxidative stress tolerance caused by methyl viologen (MV), which is reduced by the photosynthetic apparatus generating highly toxic superoxide (O_2^-). C. W80 tolerates up to 200 µM of MV [11, 12], while other oxygen-evolving photosynthetic organisms such as higher plants, algae and cyanobacteria usually tolerate only less than 5 µM of MV. This alga is a prominent genetic resource of anti-stress genes, and various unique anti-stress genes, such as ascorbate peroxidase [13], glutathione peroxidase [14], and the novel salt and cadmium stress related (*scsr*) gene [15], have been isolated from this alga. Using the cDNA library of C. W80 constructed in pAQJ4 shuttle-vector, we isolated anti-stress genes by a functional expression screening in cyanobacteria. The principle of the screening method was based on the acquisition of stress-tolerance of the cyanobacterial cells carrying the genes of C. W80, and a unique anti-stress gene (a new member of group 3 late embryo-genesis abundant protein genes) was successfully isolated [10].

2.2. Promoter for the expression of the genes on shuttle-vector

An effective promoter is important for the expression of the genes on shuttle-vector. The promoter of the RuBisCO (*rbc*) gene of S. PCC7002 was chosen for the source of strong promoter, and the *rbc* gene was isolated by screening the genomic library of S. PCC7002. RuBisCO is one of the key enzymes of Calvin–Benson cycle (photosynthetic CO_2 fixation cycle), and the most abundant protein in photosynthetic organisms. Our genomic clone of the *rbc* gene (DDBJ Accession No. D13971) is 4234 bp long, and has 962 bp of five prime untranslated region (5' UTR) of *rbc* large subunit (*rbcL*). To determine the location of the promoter activity in the 5'-UTR of *rbc* gene, we constructed CAT (chloramphenicol acetyltransferase) reporter gene construct in pAQJ4 vector, and examined the promoter activities of the 5'-UTR of *rbc* gene by introducing various deletions into this region. The promoter activity was found to be located in the region close to the coding region of *rbcL*, and the 304 bp fragment of the 5' UTR containing the promoter region was used for the promoter for pAQJ4 vector. In addition to the promoter, we also introduced multiple cloning site (MCS) into pAQJ4 vector, and the expression shuttle-vector pAQ-EX1 (DDBJ Accession No. AB071392) was finally developed.

Figure 3 shows the map of the shuttle-vector pAQ-EX1. The transformation efficiency of pAQ-EX1 for S. PCC7002 cells was about 6 x 10^5 cfu / µg DNA [16].

The transformation of the fresh water cyanobacterium S. PCC7942 with pAQ-EX1 vector was also examined [16], and the S. PCC7942 cells were successfully transformed with this vector, although the transformation efficiency (4.0 x 10^2 cfu/ µg DNA) was much lower than that for S. PCC7002. Since S. PCC7942 cells do not have the pAQ1 plasmid, which is the origin of pAQ-EX1 vector, there is no possibility of homologous recombination between pAQ-EX1 and pAQ1 in S. PCC7942 cells. We can, therefore, expect a higher stability of pAQ-EX1 vector in S. PCC7942 cells than in S. PCC7002 cells, and actually the pAQ-EX1 plasmid was quite stably maintained in *S.* PCC7942 cells. The approximate copy numbers of the pAQ-EX1 plasmid in *S.* PCC7002 and *S.* PCC7942 estimated from the yield of plasmid are 15 to 30 copies/cell for *S.* PCC7002, and 5 to 15 copies/cell for *S.* PCC7942, depending on the growth phase of the culture. The *rbc* promoter on pAQ-EX1 vector worked well also in *E. coli* cells, thus the inserted gene on the pAQ-EX1 vector can be efficiently expressed in *E. coli,* in marine cyanobacterium *S.* PCC7002, and in fresh water cyanobacterium *S.* PCC7942.

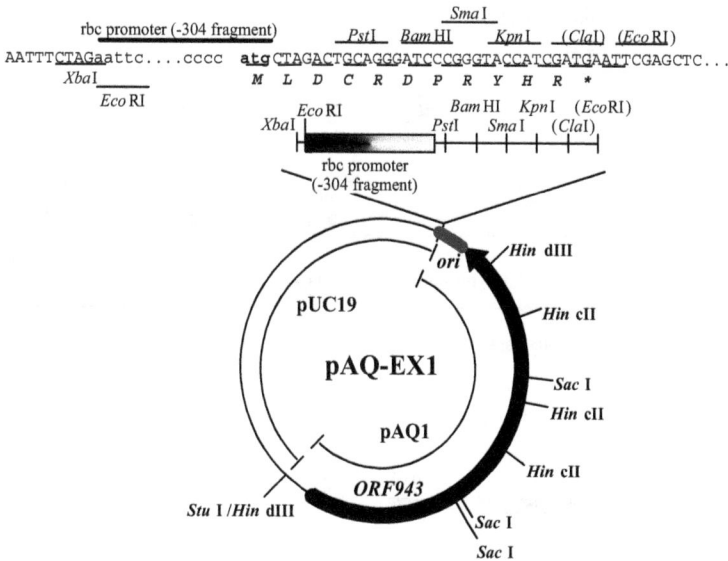

Figure 3. Map of the shuttle-vector pAQ-EX 1

2.3. CO₂ response element in *rbc* promoter

The changes in *rbc* promoter activity in response to CO_2 condition were examined because the down-regulation of *rbc* gene expression by elevated CO_2 concentration has been reported in several photosynthetic organisms. Table 1 shows the comparison of mRNA levels in *S.* PCC7002 cells cultured under various CO_2 conditions (0.03%, 1%, and 15%). The mRNA levels of *rbcL* gene were determined by RT-PCR, and compared to those of the reference gene (*ATPaseA* gene). The mRNA levels of the *rbcL* gene significantly decreased under the higher CO_2 conditions, suggesting the presence of some elements, which down regulates the transcription in response to CO_2 concentration. To examine the CO_2-regulatory element in the *rbc* promoter region, various deletions were introduced into this region, and cyanobacterial CAT assay was done by using pAQJ4-CAT vector [17]. Figure 4 shows the tested promoter fragments and the results of CAT assay. The core promoter region was shown to be located in the

-228 through -132 region, because the promoter activities were drastically decreased in the R3, R4, and R5 fragments, which lack this region. The promoter fragment, designated as R7, which contains whole -304 through -1 region, showed down regulation in promoter activity by elevated CO_2 condition (1% CO_2), while this down regulation was not observed in the R6 fragment lacking the -304 through -250 region. These results indicate that a CO_2-regulatory *cis* element exists in the -304 through -250 region, and a high expression level can be retained with R6 promoter fragment even under high CO_2 condition. The -304 through -250 region is quite A/T rich, and we also identified, by a DNA affinity precipitation assay, the 16-kDa protein which acts as a *trans*-element in CO_2 regulation [17].

Culture condition (CO_2 %)	rbcL	ATPase A	rbcL / ATPase A
	Relative mRNA level [a]		
0.03	1.0	1.0	1.0
1	0.80	1.0	0.80
15	0.57	2.2	0.26

[a]mRNA level is shown as the relative value to the mRNA level at 0.03% CO_2

Table 1. Transcript levels of *rbc*L and reference (ARPase A) genes of S. PCC7002 under various CO_2 conditions

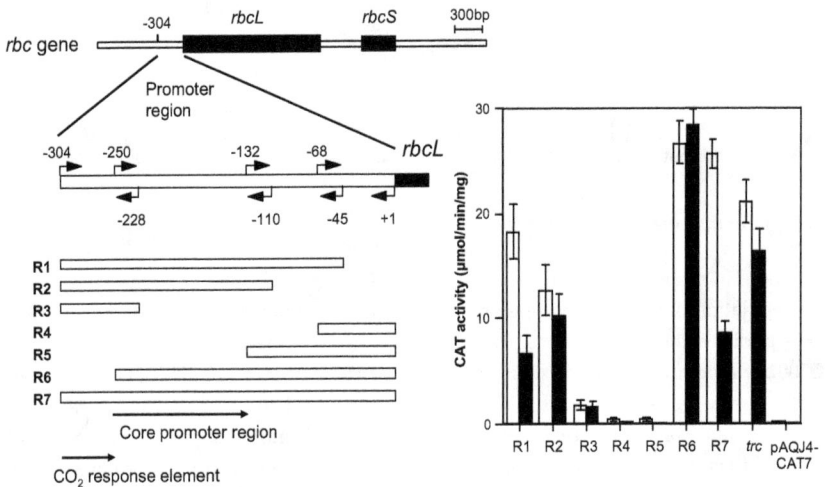

Figure 4. Promoter activities of various fragments of *rbc* promoter of *Synechococcus sp.* PCC7002

3. *recA* complementation as a selection pressure for plasmid stability

There are two foreign gene expression systems for cyanobacteria [18, 19]; one is the plasmid vector system, as we describe in this chapter, and another one is the integration of the foreign DNA into the cyanobacterial genome through homologous recombination. The advantages of the plasmid system are i) the higher copy numbers of the foreign genes in cyanobacterial cells compared to the genome integration method, ii) the well established procedure for the modification of the genes on plasmid, such as point mutation, insertion and deletion, and iii) the wide range of expression host with a shuttle vector system. On the other hand the limitation of plasmid system is the necessity for antibiotics for the maintenance of plasmid. Especially when the genes on plasmid cause a heavy metabolic load, such as PHA production, to the host cells, the plasmids are easily excluded from the cells in the absence of antibiotics pressure. The use of antibiotics is, however, not realistic for the large scale cyanobacterial culture for CO_2 mitigation with respect to its cost. In *E. coli* cells, the *parB* (*hok/sok*) locus of plasmid R which mediates stabilization *via* post-segregational killing of plasmid-free cells is effective for the antibiotics-independent stable maintenance of the plasmid [20], but in cyanobacteria there has been no such a practical plasmid stabilization system reported. We developed a practical plasmid stabilization system by utilizing the *recA* complementation mechanism. RecA is a multifunctional protein that plays key roles in various cellular processes, such as recombination and DNA repair in bacteria [21, 22]. The amino acid sequences of RecA proteins from the different microorganisms are well conserved, and there are several reports on the complementation of *recA* mutation in some bacteria by *E. coli recA* gene. Murphy et al. [23] reported that a *recA* null mutation is lethal in the cyanobacterium, but the *E. coli recA* gene in trans on a plasmid can complement the function of *recA* resulting in segregation of cyanobacterial *recA* null mutant. We expected that this complementation mechanism can be used as a selection pressure to prevent the loss of the plasmid which causes a significant metabolic burden to the host cells.

Figure 5 shows the principles of the selective pressure for the maintainance of plasmid in conventional antibiotics selection (Figure 5A) and our *recA* complementation (Figure 5B) systems. In the conventional antibiotics selection system, the antibiotics-resistant (Ab-R) gene casette is introduced in plasmid, and the loss of the plasmid makes the host cells sensitive to antibiotics. In the *recA* complementation system, the *recA* gene in the genome of the host cells is inactivated by homologous recombination, and the function of genomic *recA* gene is complemented by the *recA* gene on the plasmid. In cyanobacteria, *recA* null mutation is lethal, and the host cell without plasmid can not survive. Generally, cyanobacteria have several copies of genome [24] and *recA* null mutant cells carrying a plasmid with *E. coli recA* gene were generated by three steps as follows (Figure 6); i) generation of *recA* partial mutant of *S.* PCC7002 by homologous recombination with kanamycin resistance gene (*km*) cassette, ii) introduction of shuttle-vector with *E. coli recA* gene into the *recA* partial mutant cells, and iii) conversion of the partial *recA* mutant into the *recA* null mutant by further homologous recombination. The use of *recA* complementation as a selection pressure is a simple and versatile method; only the *E. coli recA* gene on the plasmid, and the partial *recA* mutant host are required, and the *recA* null mutant can easily be obtained by subculturing the cells in the medium with kanamycin (Km). Unlike cyanobacteria, *recA* null mutation is not lethal in *E. coli*, thus this complementation system is not applicable for *E. coli*.

(A) Selection by antibiotics pressure (conventional method)

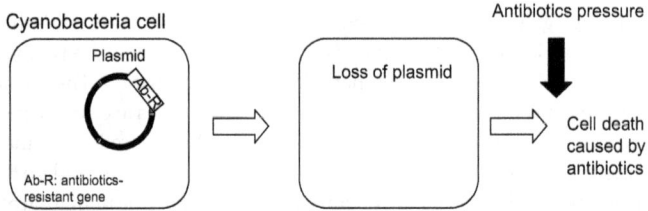

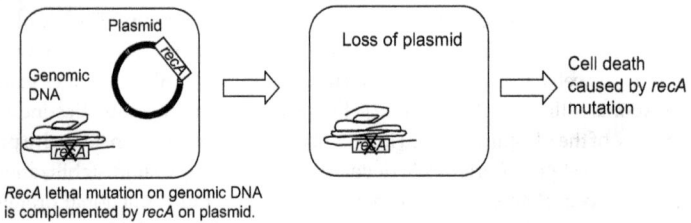

Figure 5. recA complementation as a selection pressure for plasmid stability

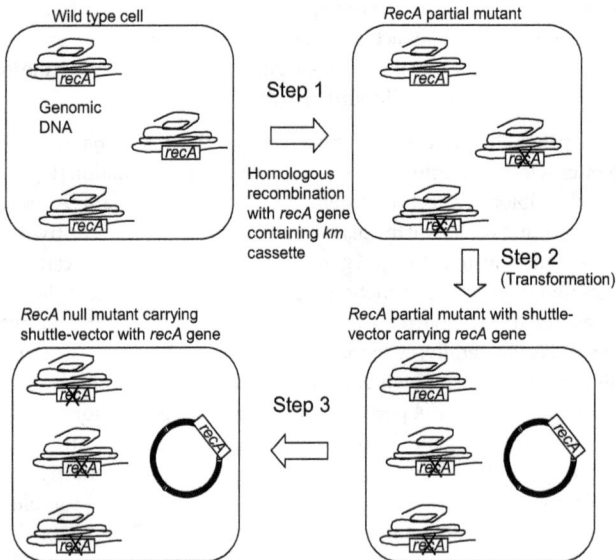

Figure 6. Procedure for generation of *rec*A null mutant cells carrying plasmids with *E. coli rec*A gene

4. PHA production by recombinant cyanobacteria

4.1. Vector construct with *recA* complementation system for PHA production

PHAs are linear head to tail polyesters composed of 3-hydroxy fatty acid monomers (Figure 1), and there are at least 100 different 3-hydroxy alkanoic acids among the PHA constituents [25]. The first PHA discovered was poly(3-hydroxy-butyrate) (PHB). It is a highly crystalline thermoplastic sharing many properties with polypropylene, and the most abundant of the PHAs in nature. The PHB biosynthetic pathway consists of three enzymatic reactions catalyzed by three distinct enzymes (Figure 7A), 3-ketothiolase (PhaA), acetoacetyl-CoA reductase (PhaB), and PHA synthase (PhaC). These three enzymes are encoded by the genes of the *phbCAB* operon (Figure 7B). There are several well established PHA production systems using natural microorganisms such as *Wautersia eutropha*, *Methylobacterium*, and *Pseudmonas*, and also using recombinant bacteria such as *E. coli* [2, 26], and intracellular accumulation of PHA of over 90% of the cell dry weight has been reported. The use of agroindustrial by-products [27], forest biomass [28], and glycerol (by-product of bio-diesel production) [29] for the substrates of microbial PHA production has also been reported. The production of PHAs in transgenic plants carrying bacterial *phb* genes has also been investigated in *Arabidopsis thaliana*, *Gossypium hirsutum* (cotton), and *Zea mays* (corn) [30].

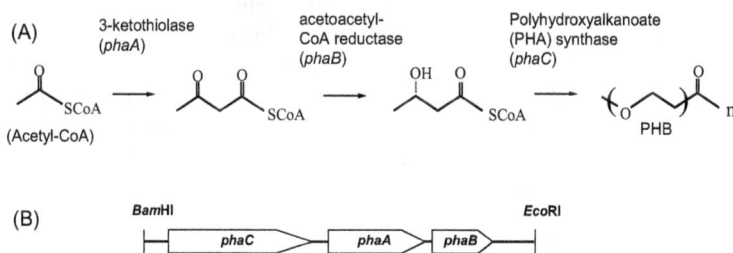

Figure 7. Biosynthetic pathway for PHA (A) and structure of pha genes (B)

There are several cyanobacterial strains which can naturally accumulate PHAs, but generally the PHA productivity in these strains are low [31, 32]. Several attempts have also been made to introduce PHA genes into non-PHA-producing cyanobacterial strains [33, 34].

We investigated the production of PHA by the recombinant cyanobacteria with the *recA* complementation antibiotics-free cyanobacterial expression system [35].

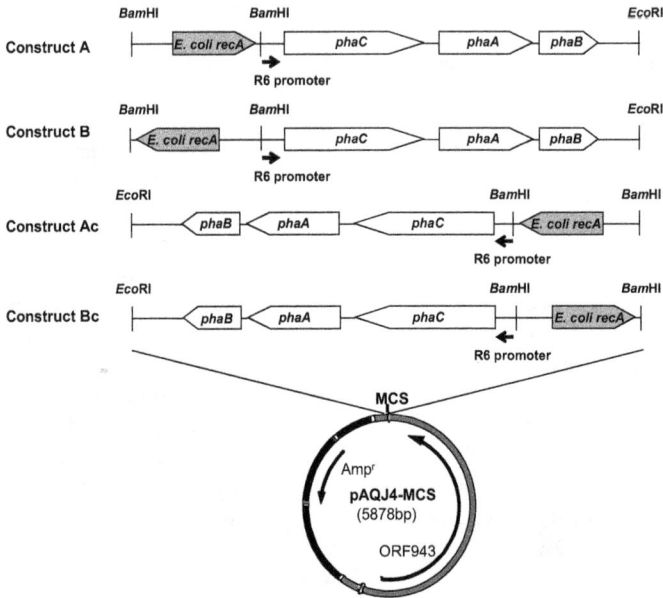

Figure 8. DNA constructs on pAQJ4 vector for PHA production

The *pha* genes and the *E. coli recA* gene were integrated on the shuttle vector pAQJ4, and the genomic *recA* genes of the transformant cyanobacteria were inactivated by an homologous recombination with the cyanobacterial *recA* gene containing a kanamycin resistance (*km*) cassette insertion. Figure 8 shows the *pha* and *E. coli recA* genes constructs on the pAQJ4 vector. For the *pha* genes, the *phaCAB* operon of *Wautersia eutropha* was used, and the *E. coli recA* gene (1.66 kb) with its upstream promoter region was connected to the *pha* genes in the same or opposite directions. For the expression of *pha* genes, the R6 promoter fragment (Figure 4), which lacks the CO_2-down-regulating element, was used. These two constructs were introduced into the pAQJ4-MCS (MCS: multiple cloning site) and pAQJ4-MCS(c) vectors (GenBank accession numbers AB480231 and AB480232, each contains the MCS in a different orientation), generating the four kinds of vector constructs A, B, A-complementary (Ac) and B-complementary (Bc) as shown in Figure 8. The partial *recA* mutant cells of *S.* PCC7002 was used for the transformation. The *recA* genes of the host cyanobacteria were partially inactivated by a homologous recombination with the *recA* gene containing a *km* cassette, and the partial *recA* mutant cells were transformed with the four kinds of *pha-recA* constructs (A, B, Ac, and Bc). The transformant cyanobacteria were obtained only for constructs B, and Ac, and these transformants were designated as Syn-pha/B and Syn-pha/Ac, respectively. The reason for the failure in the isolation of the transformants in the other constructs is not clear, but we speculate that the expression efficiency of *pha* genes might be too high in these constructs, and as a result the transformants could not gain enough energy and/or cellular metabolites for growth.

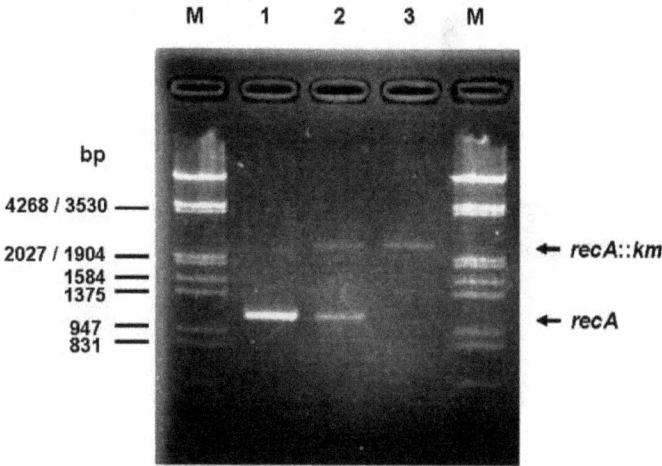

Figure 9. Integrity of the genomic *recA* gene in transformant cyanobacteria. The genomic DNA was isolated from the wild type *S.* PCC7002, Syn-pha/B transformant, and Syn-pha/Ac transformant, and the *recA* gene and *recA* with a *km* cassette (*recA::km*) where amplified by PCR. The PCR products were analyzed with 1% agarose gel electrophoresis. M: Molecular weight marker, lane 1: wild type *S.* PCC7002, lane 2: transformant Syn-pha/B (*recA* partial mutant), and lane 3: transformant Syn-pha/ac (*recA* null mutant)

To obtain the *recA* null mutant of Syn-pha/B and Syn-pha/Ac transformants, the transformant cells were subcultured in the liquid medium with carbenicillin (4 µg / ml) and Km (200 µg / ml). Each liquid culture was allowed to grow into the late stationary phase prior to subculturing to enhance the efficiency of homologous recombination. After five times subculturing, the integrity of the genomic *recA* gene was examined with PCR (Figure 9). In the Syn-pha/Ac transformant cells, only the DNA fragment corresponding to the *recA* with *km* cassette was amplified, and no DNA fragment of wild type *recA* gene was detected on the agarose gel, indicating that the Syn-pha/Ac transformant was changed to a *recA* null mutant. On the other hand, in the Syn-pha/B transformant cells, both *recA* with *km* cassette and wild type *recA* fragments were amplified, thus the *recA* gene in cyanobacterial genome was not completely inactivated. The reason for the failure of *recA* null mutant segregation in the Syn-pha/B transformant is not clear, but a possible explanation is the insufficient complementation of RecA protein by the *E. coli recA* gene on the plasmid. The Syn-pha/Ac transformant was used for the following experiments for the *pha* gene stability and PHA production.

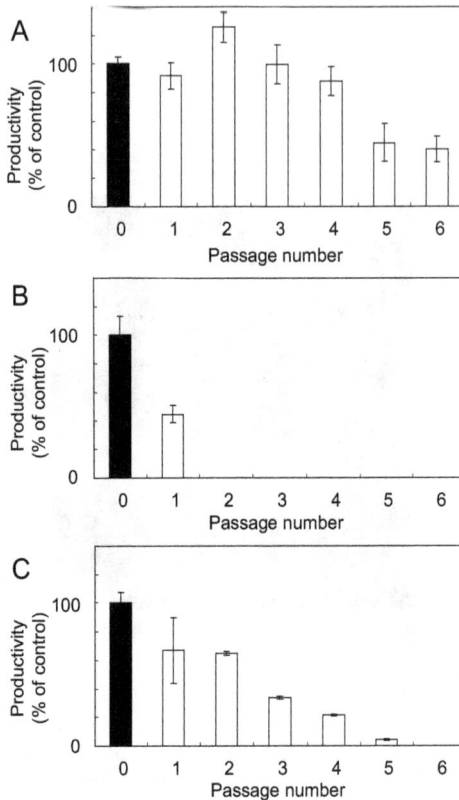

Figure 10. The stability of PHA productivity in the *recA* null mutant transformant (A), wild type (non-*recA*-mutant) transformant (B), and *recA* partial mutant transformant (C). All the transformants (*recA* null mutant, wild type, and *recA* partial mutant) carry the construct Ac plasmid of figure 8. The cells were subcultured in the antibiotics free medium for five times at one week intervals, and the PHA contents in the cells were determined at the end of each culture. The PHA productivities were expressed as the percentage to the PHA content in the cells cultured with antibiotics (passage number 0, shown as black bars).

The stability of PHA productivity in the *recA* null mutant of Syn-pha/Ac transformant was examined in comparison to the wild type (non- *recA* -mutant) transformant cells carrying the construct Ac plasmid of Figure 8, and *recA* partial mutant of Syn-pha/Ac transformant. The cells were subcultured in the antibiotics free medium for five times at one week intervals, and the PHA contents in the cells were determined at the end of each culture. The cell densities at the end of cultures were 7.5×10^8 to 1×10^9 /ml, and the passage of culture was done by diluting the culture into a fresh medium at a dilution ratio of 1:1,000. Figure 10 shows the changes in the PHA productivities in the transformant cells of the *recA* null mutant (Figure 10A), wild type (non- *recA* -mutant) (Figure 10B), and *recA* partial mutant (Figure 10C). The PHA productivities were expressed as the percentage to the PHA content in the cells cultured with antibiotics (passage number 0, shown as black bars in Figure 10). The PHA productivities in the *recA* null mutant (Figure 10A) were kept at the approximately same level with that of passage number 0 at the passage numbers 1 through 4 in the antibiotics free medium, but suddenly decreased to 45% of the passage number 0 at the passage number 5. On the other hand, the PHA productivities in the wild type (non- *recA* -mutant) transformant significantly decreased at the passage number 1 (45% of the passage number 0), and no PHA production was detected at the end of passage number 2 (Figure 10B). Interestingly a partial positive effect for the PHA productivity was observed in the *recA* partial mutant (Figure 10C); the PHA productivity decreased gradually during the consecutive culture passages to a trace level of PHA production at the passage number 6. These results indicate that the *recA* complementation effectively acted as a selection pressure in the *recA* null mutant for the maintenance of the plasmid carrying the *pha* genes, at least for 3 to 4 passages at a dilution rate of 1:1,000. The cell number (and also culture scale) can be increased 10^9 times with three culture passages at a dilution rate of 1:1,000, and therefore this antibiotics-free PHA production system is applicable to the large scale PHA production. The reason for the sudden decrease in PHA productivity in the *recA* null mutant at the passage 5 is not clear, but this might not be caused by the loss of plasmid in the cyanobacterial cells because the cells of passage numbers 4 (high PHA productivity) and 5 (low productivity) did not show any difference in colony forming ability on the antibiotics (carbenicillin) plates. The decrease in the PHA productivity at the passage numbers 4 and 5 might, therefore, be attributed to the other reasons, such as the mutation in *pha* genes and/or the inhibition of the expression of *pha* genes on the plasmid.

4.2. PHA production by transformant cyanobacteria cells

Figure 11 shows the electron micrograph of the control wild type *S.* PCC7002 (A), and the PHA accumulating *recA* null mutant transformant (Syn-pha/Ac) (B) cells. The small PHA granules aligning along the thilacoid membrane were observed in the Syn-phaAc transformant cell. The molecular mass distribution of the PHA was estimated with the gel permeation chromatography (GPC). The molecular mass distribution of the PHA from the Syn-pha/Ac transformant was a little shifted to the higher side compared to that of the standard PHA from *W. eutropha* (Figure 12), but in the range previously reported for various microbial PHAs. The main component of the hydroxyalkanoic acid of the PHA from the Syn-phaAc transformant was hydroxybutyric acid (more than 98%), and a small amount of lactic acid (0.5 to 1.5 %), and a trace amount of hydroxyvaleric acid were also contained.

To obtain a higher PHA accumulation in the cyanobacterial cells, the nutrient condition of the culture was also examined. Since it is reported that the nitrogen and phosphorus supplies, much affect the PHA accumulation in microorganisms [36, 37], the Syn-phaAc transformant cells were cultured in the medium containing various concentrations of nitrogen ($NaNO_3$) and phosphorus (KH_2PO_4) sources, and the cell growth and PHA accumulation were compared (Figure 13). There was a clear negative relationship between cell growth and PHA accumulation, and nitrogen limitation seemed to be effective for the accumulation of PHA although the cell growth was significantly suppressed in the nitrogen limited medium. The maximum PHA accumulation was 52% of cell dry weight, the highest among the ever reported PHA accumulation in cyanobacteria. Accordingly the two-staged culture system consisting of cell growth and PHA production phases should be applied to increase the total productivity (g per liter culture) of PHA. Asada et al. reported that acetyl-CoA flux is the limiting factor in PHA production by genetically engineered cyanobacterium [32], and the high PHA productivity in Syn-phaAc transformant cells suggests the abundant intracellular supply of acetyl-CoA in S. PCC7002.

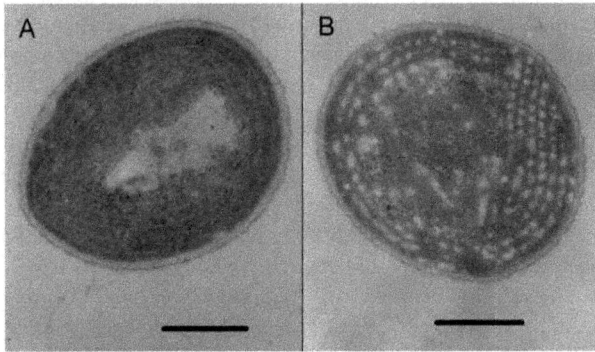

Figure 11. Electron micrograph of wild type S. PCC7002 (A) and PHA accumulating Syn-pha/Ac transformant cells (B) in the early exponential growth phase (OD550=2). The PHA content in the Syn-pha/Ac transformant cell is approximetly 10% of the dry weight. Scale bars represent 0.5 μm.

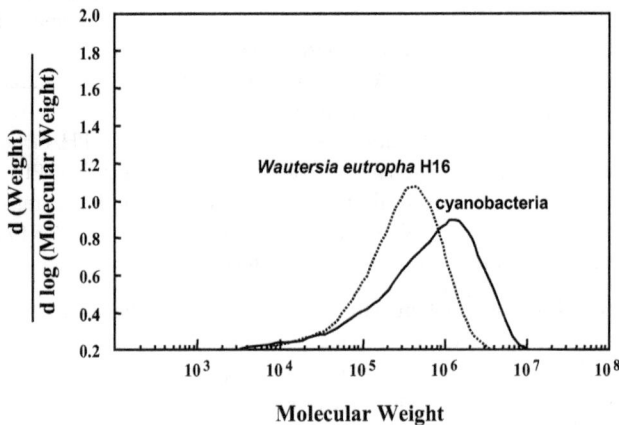

Figure 12. Molecular distribution of the PHA from *recA* null mutant Syn-pha/Ac transformant cells (solid line), and *Wautersia eutropha* H16 (broken line). Molecular weight of PHA samples was determined by gel permeation chromatography (GPC)

Our study is the first practical approach for the antibiotics-free maintenance of plasmid in cyanobacteria, and with this system the fixation and direct conversion of CO_2 into the useful bioplastics can be realized under low maintenance and low cost conditions.

The future research subjects to realize the on-site CO2 fixation and utilization system with recombinant cyanobacteria are the followings.

1. Although the promoter derived from the *rbc* gene was found to be quite effective both in cyanobacterial and bacterial cells, and the PHA production by the transformed cyano-bacterial cells was also quite successful with this promoter, the use of a switchable promoter (ON/OFF type promoter) might further enhance the PHA production.

2. Method and system for the efficient harvesting of cyanobacteria at low energy and low cost should be developed.

3. Photosynthetic CO_2 assimilation only occurs during the day, and the productivity of biomaterials is much influenced by light condition. The use of cyanobacteria capable of growing photoautotrophic and also heterotrophic (with waste water) is one possible solution to this limitation.

4. Generally biomaterials, such as fuel and plastic, produced from CO_2 are low price. Simultaneous prodcuction of higher value products, such as fine chemicals, can lower the cost for the production of biomaterials from CO_2.

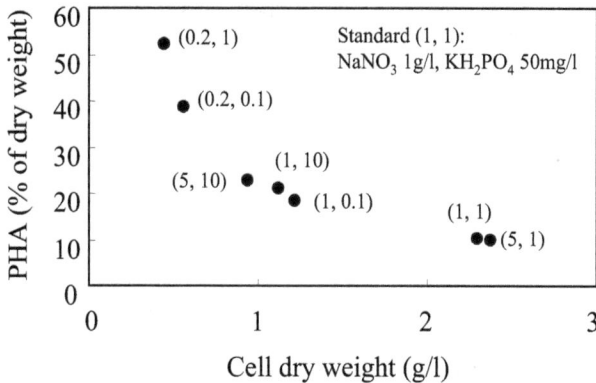

Figure 13. Effect of nitrogen and phosphorus concentrations in medium on cell growth and PHA accumulation in Transformant cyanovacteria. The cells (initial cell density; 5×10^6 cells/ml) were cultured in the 50 ml medium containing various concentration of $NaNO_3$ and KH_2PO_4. The standard concentration of $NaNO_3$ and KH_2PO_4 are 1g/l and 50 mg/l, respectively. The standard concentrations of $NaNO_3$ and KH2PO4 are shown as (1,1), and the ratio of each nutrient to the standard concentration is shown in the parenthesis.

5. Conclusion

The fixation and direct conversion of CO_2 into the useful biomaterials by the transgenic cyanobacteria are two processes of a promising technology for the coming low carbon economy. We have developed an efficient shuttle-vector between the marine cyanobacterium *Synechococcus sp.* PCC7002 and *E. coli*, and also a practical antibiotics-free cyanobacterial plasmid expression system by using the complementation of the cyanobacterial recA null mutation with the E. coli recA gene on the plasmid. Although considerable researches are still required to realize the practical on-site applications of the present system to the industrial emission sites, such as thermal power plants, this technology can be a promising option for the biological conversion of CO_2 into useful industrial materials.

Author details

Hitoshi Miyasaka[1], Hiroshi Okuhata[1], Satoshi Tanaka[1], Takuo Onizuka[2] and Hideo Akiyama[2]

*Address all correspondence to: miyasaka.hitoshi@a4.kepco.co.jp

1 The Kansai Electric Power Co., Environmental Research Center, Seikacho, Japan

2 Toray Research Center, Inc., Kamakura, Japan

Current affiliation for Hideo Akiyama: New Projects Development Division, Toray Industries, Inc., Kamakura, Japan

References

[1] Lal R. Carbon sequestration. Philos. Trans. R. Soc. Lond. B. Biol. Sci. 2008; 363 815-830.

[2] Madison, L. L., Huisman, G. W. Metabolic engineering of poly(3-hydroxyalkanoates): from DNA to plastic. Microbiol. Mol. Biol. Rev. 1999; 63 21-53.

[3] Chen G. Q. A microbial polyhydroxyalkanoates (PHA) based bio- and materials industry. Chem. Soc. Rev. 2009; 38 2434-2446.

[4] Buzby J. S., Porter R. D., Stevens S. E. Jr. Plasmid transformation in *Agmenellum quadruplicatum* PR-6: construction of biphasic plasmids and characterization of their transformation properties. J. Bacteriol. 1983; 154 1446-1450.

[5] Tabita R. F., Stevens S. E. Jr., Quijano R. D-ribulose 1, 5-diphosphate carboxylase from blue-green algae. Biochem. Biophys. Res. Commun. 1974; 61 45-52.

[6] Akiyama H., Kanai S., Hirano M., Miyasaka H. Nucleotide sequences of plasmid pAQ1 of marine cyanobacterium *Synechococcus* sp. PCC7002. DNA Res. 1998; 5 127-129.

[7] Miyasaka H., Nakano H., Akiyama H., Kanai S., Hirano M. Production of PHA (poly hydroxyalkanoate) by the genetically engineered marine cyanobacterium. Stud. Surf. Sci. Catal. 1998; 114 237-242.

[8] Akiyama H., Kanai S., Hirano M., Miyasaka H. A novel plasmid recombination mechanism of the marine cyanobacterium *Synechococcus* sp. PCC7002. DNA Res. 1998; 5 327-334.

[9] Lorimier R., Guglielmi G., Bryant D. A., Stevens S. E. Jr. Functional expression of plastid allophycocyanin genes in a cyanobacterium. J. Bacteriol. 1987; 169 1830-1835.

[10] Tanaka S., Ikeda K. Miyasaka H. Isolation of a new member of group 3 late embryo-genesis abundant (LEA) protein gene from a halotorelant green alga by a functional expression screening with cyanobacterial cells. FEMS Microbiol. Lett. 2004; 236 41-45.

[11] Miyasaka H., Kanaboshi H., Ikeda K. Isolation of several anti-stress genes from the halotolerant green alga *Chlamydomonas* by simple functional expression screening with *E. coli*. World J. Microbiol. Biotechnol. 2000; 16 23-29.

[12] Tanaka S., Ikeda K., Miyasaka H., Shioi Y., Suzuki Y., Tamoi M., Takeda T., Shigeoka S., Harada K., Hirata K. Comparison of three Chlamydomonas strains which show distinctive oxidative stress tolerance. J. Biosci. Bioeng. 2011; 112 462-468.

[13] Takeda T., Yoshimura K., Yoshii M., Kanaboshi H., Miyasaka H., Shigeoka S. Molecular characterization and physiological role of ascorbate peroxidase from halotolerant *Chlamydomonas* sp. W80 strain. Arch. Biochem. Biophys. 2000; 376 82-90.

[14] Takeda T., Miyao K., Tamoi M., Kanaboshi H., Miyasaka H., Shigeoka S. Molecular characterization of glutathione peroxidase-like protein in halotolerant *Chlamydomonas* sp. W80. Physiol. Plant 2003; 117 467-475.

[15] Tanaka S., Suda Y., Ikeda K., Ono M., Miyasaka H., Watanabe M., Sasaki K., Miyamoto K., Hirata K. A novel gene with anti-salt and anti-cadmium stress activites from a halotolerant marine green alga *Chlamydomonas* sp. W80. FEMS Microbiol. Lett. 2007; 271 48-52.

[16] Ikeda K., Ono M., Akiyama H., Onizuka M., Tanaka S., Miyasaka H. Transformation of the fresh water cyanobacterium *Synechococcus* PCC7942 with the shuttle-vector pAQ-EX1 developed for the marine cyanobacterium *Synechococcus* PCC7002. World J. Microbiol. Biotechnol. 2002; 18 55-56.

[17] Onizuka T., Akiyama H., Endo S., Kanai S., Hirano M., Tanaka S., Miyasaka H. CO_2 response element and corresponding trans-acting factor of the promoter for ribu-lose-1,5-bisphosphate carboxylase/oxygenase genes in *Synechococcus* sp. PCC7002 found by an improved electrophoretic mobility shift assay. Plant Cell Physiol. 2002; 43 660-667.

[18] Koksharova O. A., Wolk C. P. Genetic tools for cyanobacteria. Appl. Microbiol. Biotechnol. 2002; 58 123-137.

[19] Vioque A. Transformation of cyanobacteria. Adv. Exp. Med. Biol. 2007; 616 12-22.

[20] Gerdes K., Rasmussen P. B., Molin S. Unique type of plasmid maintenance function: postsegregational killing of plasmid-free cells. Proc. Natl. Acad. Sci. USA 1986; 83 3116-3120.

[21] Cox M. M. Recombinational DNA repair in bacteria and the RecA protein. Prog. Nucleic Acid Res. Mol. Biol. 1999; 63 311-366.

[22] Lusetti S. L., Cox M. M. The bacterial RecA protein and the recombinational DNA repair of stalled replication forks. Annu. Rev. Biochem. 2002; 71 71-100.

[23] Murphy R. C., Gasparich G. E., Bryant D. A., Porter R. D. Nucleotide sequence and further characterization of the *Synechococcus* sp. strain PCC 7002 *recA* gene: complementation of a cyanobacterial *recA* mutation by the *Escherichia coli recA* gene. J. Bacteriol. 1990; 172 967-976.

[24] Herdman, M., Evolution and genetic properties of cyanobacterial genomes. In: Carr, N.G. and Whitton, B.A (eds.), The biology of cyanobacteria. Blackwell Scientific Publications; Oxford; 1982. p263-305.

[25] Steinbuechel A., Valentin H. E. Diversity of bacterial polyhydroxyalkanoic acids. FEMS Microbiol. Lett. 1995; 128 219-228.

[26] Reddy, C.S.K., Ghai, R., Rashmi, Kalia, V.C., Polyhydroxyalkanoates: an overview. Bioresour. Technol. 2003; 87 137-146

[27] Gomez J.G.C., Méndez B.S., Nikel P.I., Pettinari M.J., Prieto M.A., Silva L.F. Making Green polymers even greener:towards sustainable production of polyhydroxyalkanoates from agroindustrial by-products In: Petre M. (ed.) Advances in Applied Biotechnology . InTech; 2012. p41-62

[28] Keenan T. M., Nakas J. P., Tanenbaum S. W. Polyhydroxyalkanoate copolymers from forest biomass. J. Ind. Microbiol. Biotechnol. 2006; 33 616-626.

[29] Ashby R. D., Solaiman D. K., Foglia T. A. Synthesis of short-/medium-chain-length poly(hydroxyalkanoate) blends by mixed culture fermentation of glycerol. Biomacromol. 2005; 6 2106-2112.

[30] Salehizadeh H., Van Loosdrecht M.C.M. Production of polyhydroxyalkanoates by mixed culture: recent trends and biotechnological importance. Biotechnol. Adv. 2004; 22 261-279.

[31] Miyake M., Takase K., Narato M., Khatipov E., Schnackenberg J., Shirai M., Kurane R., Asada Y. Polyhydroxybutyrate production from carbon dioxide by cyanobacteria. Appl. Biochem. Biotechnol. 2000; 84-86 991-1002.

[32] Asada Y., Miyake M., Miyake J., Kurane R., Tokiwa Y. Photosynthetic accumulation of poly-(hydroxybutyrate) by cyanobacteria--the metabolism and potential for CO2 recycling. Int. J. Biol. Macromol. 1999; 25 37-42.

[33] Sudesh K., Taguchi K., Doi Y. Effect of increased PHA synthase activity on polyhydroxyalkanoates biosynthesis in Synechocystis sp. PCC6803. Int. J. Biol. Macromol. 2002; 30 97-104.

[34] Takahashi H., Miyake M., Tokiwa Y., Asada Y. Improved accumulation of poly-3-hydroxybutyrate by a recombinant cyanobacterium. Biotechnol. Lett. 1998; 20 183-186.

[35] Akiyama H., Okuhata H., Onizuka T., Kanai S., Hirano M., Tanaka S., Sasaki K., Miyasaka H. Antibiotics-free stable polyhydroxyalkanoate (PHA) production from carbon dioxide by recombinant cyanobacteria. Bioresour. Technol. 2011; 102 11039–11042

[36] Hankermeyer C. R., Tjeerdema R. S. Polyhydroxybutyrate: plastic made and degraded by microorganisms. Rev. Environ. Contam. Toxicol. 1999; 159 1-24.

[37] Panda B., Jain P., Sharma L., Mallick N. Optimization of cultural and nutritional conditions for accumulation of poly-beta-hydroxybutyrate in *Synechocystis* sp. PCC 6803. Bioresour. Technol. 2006; 97 1296-1301.

Chapter 10

Role of Biotechnology for Protection of Endangered Medicinal Plants

Krasimira Tasheva and Georgina Kosturkova

Additional information is available at the end of the chapter

1. Introduction

The last two centuries of industrialization, urbanization and changes in land use converting agricultural and natural areas to artificial surface have led to European plants being considered amongst the most threatened in the world. In some countries, more than two-thirds of the existing habitat types are considered endangered. Human activity is the primary cause of risk for 83% of endangered plant species. Habitat destruction and loss are also a problem because they lead to the fragmentation of the remaining habitat resulting in futher isolation of plant population [1]. From another side during the last 10 years an intense interest has emerged in "nutraceuticals" (or "functional foods") in which phytochemical constituents can have long-term health promoting or medicinal qualities. Although the distinction between medicinal plants and nutraceuticals can sometimes be vague, a primary characteristic of the latter is that nutraceuticals have a nutritional role in the diet and the benefits to health may arise from long-term use as foods (i.e. chemoprevention) [2]. In contrast, many medicinal plants possess specific medicinal benefits without serving a nutritional role in the human diet and may be used in response to specific health problems over short- or long-term intervals [3].

There is indisputable interest towards traditional and alternative medicine world-wide [4] and at the same time an increasing application of herbs in medical practices, reported by World Health Organization (WHO) [5]. Nowadays the centuries-old tradition of medicinal plants application has turned into a highly profitable business on the world market. Numerous herbal products have been released like patented medical goods, food additives, herbal teas, extracts, essential oils, etc [6 - 9].

There is an expansion of the market of herbs and herbs based medical preparations all over the world. The income a decade ago in the North American market for sales of medicinal plants has climbed to about $3 billion/year [10]. In South America, Brazil is outstanding with 160 millions USD for 2007 while in Asia, China is at the leading trade position with 14 billions USD for 2005, etc. [11]. Similar increase was observed in Western Europe with 6 billion USD income for a period of two years from 2003 to 2004. The sales increased in Czech Republic by 22 % from 1999 to 2001 and jumped twice in Bulgaria [12].

Medicinal plants are precious part of the world flora. More than 80 000 species out of the 2 500 000 higher plants on Earth are reported to have at least some medicinal value and around 5 000 species have specific therapeutic value. The contemporary phytotherapy and the modern allopathic medicine use raw materials from more than 50 000 plant species [13]. About two thirds of these fifty thousand plants utilized in the pharmacological industry are harvested from nature [14]. Small portions like 10%-20% of the plants used for remedies preparations are cultivated in fields or under controlled conditions [15]. Ages-old exploitation of the natural resources and the dramatically increased interest are a real thread for the biological diversity. Bad harvesting management and insufficient cultivation practices may lead to extinction of endangered species or to destruction of natural resources. Science has already recorded diminishing natural populations, lost in the genetic diversity, local extinction of many species and/or degeneration of their natural habitats [16]. This alarming situation is raising the questions about special efforts which should be paid both to protection of the plant populations and to up-to-date knowledge concerning more reasonable and effective utilization of these plants [12].

Bulgaria as a country with a rich and diverse flora (comprising of 7 835 species) and with old traditions in herbs' use faces the same global problems. One of the most serious challenges is the control and the limitation of the expanding gathering of endangered medicinal plants [17]. The Biodiversitry Act covers *Sideritis scardica* (mursala tea), *Alchemilla vulgaris, Acorus calamus, Rhodiola rosea, Leucojum aestivum, Gentiana sp., Glycyrrhiza glabra, Ruta graveolens*, and some medicinal plants under special rules of protection and use e.g. *Inula helenium, Carlina acanthifolia, Berberis vulgaris, Rhammus frangula, Rubia tinctorum, Atropa belladona, Origanum heracleoticum* etc. More than 750 herbs are used in the folk medicine while 150 - 250 are used in the official medicine and can be found at the market [18 - 20]. A considerable number of the wild species are rare, endangered or under protection [21, 22], and 12.8% are endemics [23]. Recently 120 herbs have been traditionally collected from their natural populations, 47 are under protection, 38 are included in the Red Data Book of Bulgaria, 60 have been cultivated, 35 are main industrial crops [24]. Bulgarian medicinal plants are famous for their high content of biologically active substances. Their high value qualities are due to the unique combinations of specific soil and climatic conditions in the different sites of the country [25]. Bulgaria is the European leader in herbs export and occupies the 8th world position with trade in 40 countries all over the world. The greatest export of 50% is to Germany being 3 600 tones in 1991 and doubling to 6 000 tones in 2 000 [8]. Spain, Italy, France, Austria and USA are also major trade partners. The export is increasing steadily from 6 - 7 t in 1992 to 12 tones in 2000 – 2003, to 15 - 17 000 tones in 2007 [22, 26 - 28,]. These amounts represented about 70% - 80% out of all harvested and processed medicinal plants in Bulgaria [27]. The bigger number of these species is wild growing [29] but recently cultivation in fields has been applied as a measure to protect medicinal plants included in the list of protected species. At present about 20% of the medicinal plants are cultivated, but this share comprises about 40% of the export [24].

Worldwide the constant expansion of herbs' trade, the insufficient cultivation fields, and the bad management of harvesting and overharvesting have led to exhaustion of the natural resources and reduction of the biodiversity. According to the data of the Food and Agricultural Organization (FAO) at the United Nations annually the flora bares irretrievable losses which destroy the natural resources and the ecological equilibrium [30]. Four thousand to 10 000 medicinal species were endangered of disappearing at the beginning of this century [14]. To stop the violence against nature, efforts should be directed both to preservation of the plant populations and to elevating the level of knowledge for sustainable utilization of these plants in traditional, alternative, and allopathic medicine [12].

This great issue is in the focus of science which offers different decisions to solve the global problem. Cultivation of the valuable species in experimental conditions is one of the approaches. The latter refers to application of classical methods for multiplication by cuttings, bulbs, and so forth, as well as by biotechnological methods of *in vitro* cultures and clonal propagation for production of enormous number of identical plants. The micropropagation is considered to have the greatest commercial and iconomical importance for the rapid propagation and *ex situ* conservation of rare, endemic, and endangered medicinal plants [31 - 34]. Except for clonal multiplication and maintaining the genetic structure biotechnology is powerful for modifying genetic information and gene expression to obtain new valuable compounds with new properties or with increased amounts [35 - 37]. Micropropagation, cell and callus cultures, metabolic engineering and genetic manipulations are especially appropriate for species which are difficult to propagate *in vivo* [36].

In Bulgaria quite successful investigations have been performed for *in vitro* clonal multiplication of valuable, endemic, rare and endangered medicinal species: *Rhodiola rosea, Gentiana lutea, Sideritis scardica, Pancratium maritimum, Scabiosa argentea, Cionura erecta, Jurinea albicaulis* subsp. *kilaea, Peucedanum arenarium, Linum tauricum* subsp. *bulgaricum, Aurinia uechtritziana, Silene thymifolia, Glaucium flavum, Stachys maritima, Astrodaucus littoralis, Otanthus maritimus, Plantago arenaria, Verbascum purpureum, Alchemilla sps*, etc. [38 - 44]

More than 2 000 different species are used in Europe for production of medicinal and other herbal preparations. Seventy percents of these species are growing in wild nature [17, 29] with already limiting resources which demands search for alternative methods friendly to nature. Biotechnological methods seem appropriate ones with their potential for multiplication, selection and protection of medicinal plants. In this respect biotechnological approaches are convenient for use of cells, tissue, organs or entire organism which grow and develop in *in vitro* controlled conditions, and can be subjected to *in vitro* and genetic manipulations [33] to obtain desired substances [45]. These methods are especially appropriate and reasonable to apply when the targeted species have high economical or trade value, or plant resources are limited concerning the availability of wild area or good healthy plants, or when the plants are difficult to grow [46, 47].

In vitro cultivation may be directed to development of different systems depending on the practical needs. At present production of a large number of identical plants by clonal micropropagation is the most prominent one. Complex and integrated approaches for cultivation of plant systems may be the basis for future development of new, effective, safe and high quality products. These scientific achievements might be used for the establishment of *ex situ* and *in vitro* collections, multiplication of desired species and to obtain raw material for the pharmaceutical and cosmetic industries [48]. *In vitro* technologies offer some or most of the following advantages: easier extractions and purification of valuable substances from temporary sources; new products which may not be found in nature; absence of various environmental and seasonal effects, automation, better control of the biosynthetic pathways and flexibility in obtaining desired product; shorter production cycles and cheaper less costly products. Here should also be mentioned the potential of the sophisticated techniques of genetic engineering, which might be applied respecting the rules of contained use [33; 47]. At present the methods of plant cell and tissue cultures have found many proper sites for application in the medicinal plants utilization. The achieved results and the confidence for further success drive the efforts for wider application of plant biotechnologies in more spheres concerning medicinal plants [37].

2. Essence of *in vitro* culture

Plant cell methods and techniques were initially used in fundamental scientific investigations at the beginning of their development in the early 60-ties of the last century. Plant biotechnology is based on the totypotence of the plant cell [35; 49]. This process of *de novo* reconstruction of an organism from a cell in differentiated stage is highly linked to the process of dedifferentiation when the cell is returning back to its early embryogenic/ meristematic stage. In this stage cells undergo division and may form nondifferentiated callus tissue or may redifferentiate to form new tissue, organs and an entire organism. Morphogenesis *in vitro* is realized via two major pathways: (i) organogenesis when a group of cells is involved for *de novo* formation of organs and (ii) somatic embryogenesis when the new organism is initiated from a single cell.

3. Micropropagation

Micropropagation is a vegetative propagation of the plants *in vitro* conditions (in glass vessels under controlled conditions) leading to development of numerous plants from the excised tissue and reproducing the genetic potential of the initial donor plant.

Usually tissues containing meristematic cells are used for induction of axilary or adventitious shoots but induction of somatic embryos can be achieved from differentiated cells as well.

Micropropagation is used routinely for many species to obtain a large number of plants with high quality. It is widely applied to agricultural plants, vegetable and ornamental species, and in some less extent to plantation crops. One of the substantial advantages of micropropagation over traditional clonal propagation is the potential of combining rapid large-scale propagation of new genotypes, the use of small amounts of original germplasm (particularly at the early breeding and/or transformation stage, when only a few plants are available), and the generation of pathogen-free propagules. [50]. Compared to the other spheres of *in vitro* technologies clonal propagation has proved the greatest economical and market importance in industry including pharmaceutical industry which needs for raw material from the medicinal plants is increasing constantly. It offers faster and alternative way for production of raw material and from another side overcoming the problems arising from the limited natural resources.

At present, there is a long list of research groups worldwide investigating hundreds of medicinal species. Various success procedures and recipes for many of these species have been developed. However, there is not a universal protocol applicable to each species, ecotype, and explant tissue. From another side all these continuous tedious studies on the standardization of explant sources, media composition and physical state, environmental conditions and acclimatization of *in vitro* plants have accumulated information, continuously enriched, which is a good basis for elaboration of successful protocols for more species. Wider practical application of micropropagation depends on reduction of costs so that it can become compatative with seed production or traditional vegetative propagation methods (e.g., cuttings, tubers and bulbs, grafting) [50].

4. Metabolic engineering and biotransformation

The plant cell culture systems have potential for commercial exploitation of secondary metabolites. Similar to the fermentation industry using microorganisms and their enzymes [35, 51, 52] to obtain a desired product plant cells are able to biotransform a suitable substrate compound to the desired product. The latter can be obtained as well by addition of a precursor (a particular compound) into the culture medium of plant cells. In the process of biotransformation, the physicochemical and biological properties of some natural products can be modified [53]. Thus, biotransformation and its ability to release products into the cells or out of them provide an alternative method of supplying valuable natural products that occur in nature at low levels. Generally, the plant products of commercial interest are secondary metabolites, which in turn belong to three main categories: essential oils, glycosides and alkaloids [51]. Plant cell cultures as biotransformation systems have been highlighted for production of pharmaceuticals but other uses have also been suggested as new route for synthesis, for products from plants difficult to grow, or in short supply, as a source of novel chemicals. It is expected that the use, production of market price and structure would bring some of the other compounds to a commercial scale more rapidly and *in vitro* culture products may see further commercialization [54]. The application of molecular biology techniques to produce transgenic cells and to effect the expression and regulation of biosynthetic pathways is also a significant step towards making *in vitro* cultures more generally applicable to the commercial production of secondary metabolites [54]. However, because of the complex and incompletely understood nature of plant cells growing in *in vitro* cultures, case-by-case studies have been used to explain the problems occurring in the production of secondary metabolites from cultured plant cells.

Genetic manipulations (direct and indirect genetic transformation) are other different approaches to increase the content biological active substances in plants. Genetic engineering covers a complex of methods and techniques applied to the genome in order to modify it to obtain cells and organisms with improved qualities or possessing desired traits. These might refer to better yield or resistance, as well as, to higher metabolite production or synthesis of valuable biologically active substances [55]. Gene transfer may be direct when isolated desired DNA fragments are inserted into the cell most often by electrical field or adhesion. This method is less used in medicinal plants. Indirect genetic transformation of plants uses DNA vectors naturally presenting in plant pathogens to transfer the isolated genes of interest and to trigger special metabolic pathways [56]. *Agrobacterium rhizogenes* induces formation of "hairs" at the roots of dicotyledonous plants. Genetically modified "hairy" roots produce new substances, which very often are in low content. Hairy roots are characterized with genetic stability and are potential highly productive source for valuable secondary metabolites necessary for the pharmaceutical industry [57, 58]. Manipulations and optimization of the productivity of the transformed hairy roots are usually the same as for the other systems for *in vitro* cultivation [59]. They also depend on the species, the ecotype, the explant, the nutrient media, cultivation conditions, etc [60].

All these application of the principles of plant cell division and regeneration to practical plant propagation and further manipulations could be possible if there are reliable *in vitro* cultures, which efficiency depends on many various factors.

5. Factors influencing cell growth *in vitro*

The ability of the plant cell to realize its totypotence is influenced in greatest extend by the genotype, mother/donor plant, explant, and growth regulators what was confirmed by the tedious empirical work of *in vitro* investigations [61, 62]. Here, some of the specific and most important requirements will be mentioned in order of understanding the efforts and originality of some ideas when establishing *in vitro* cultures of medicinal plants.

Genotypes. Morphogenetic potential of excised tissue subjected to cultivation *in vitro* is in strong dependence of the genotype [63]. Genetically plants demonstrate different organogenic abilities, which were observed for all plants groups including medicinal plants [64 - 72]. Some of the species (like tobacco and carrot) are easy to initiate in *in vitro* cultures while others are more difficult - reculcitrant (cereals, grain legumes, bulbous plants). Many of the wild species like most of the medicinal plants and especially those producing phenols are more difficult or extremely difficult to handle.

Donor plant. The donor plant should be healthy, in the first stages of its intensive growth, not in dormancy. Rhyzomes and bulbs usually need pretreatment with low or high temperatures for different periods of time [35, 73].

Explant. The explant type might determine the organogenesis potential and the genetical stability of the clonal material. Physiological age of the explant is also crucial. Immature organs and differentiated cells excised from stem tips, axilary buds, embryos and other meristematic tissues are the most appropriate [35, 62, 73]. However, despite the development of cell and molecular biology the limits still exist in receiving easy information about the genetic, epigenetic and physiological status of the explant. Empirical approach is the most common to specify the chemical and physical stimuli triggering cell totypotence.

Nutrient media. Although more than 50 different media formulations have been used for the *in vitro* culture of tissues of various plant species the formulation described by Murashige and Skoog (MS medium) [74] is the most commonly used, often with relatively minor changes. Other famous media are those of Gamborg [75; 76], Huang and Murashige [77] Nischt and Nischt etc. The nutrient medium usually consists of all the essential macro- and micro salts, vitamins, plant growth regulators, a carbohydrate, and some other organic substances if necessary [62].

Plant growth regulators. Plant growth regulators, including the phytochormones, are essential for cell dedifferentiation, division and redifferention leading to callus tissue and organ formation. The auxins and cytokinins are the most important for *in vitro* development and morphogenesis. However, the most appropriate plant regulators and their concentrations in the nutrient media depend on the genotype, explants type and the donor plant physiological status. Hence, numerous combinations could be designed and the optimal ones are validated empirically. All that creates the difficulties of the experimental work, which is dedicated to find the balance between the factors determining reliable *in vitro* development.

Cytokinins. Different groups of cytokinins might be used but the most efficient ones for induction of organogenesis and a large number of buds are the natural cytokinins (zeatin and kinetin) or the synthetic ones - 6-benzylaminopurine (benzyl adenine (BA, BAP), 6-γ(-dimethylallyl-amino)-purine (2iP) and thidiazuron (TDZ).

Auxins. The auxins also are obtained from natural plant materials like indolyl-3-acetic acid (IAA), indole 3-butyric acid (IBA), α- naphthyl acetic acid (NAA) or are chemically produced like 2,4-dichlorophenoxyacetic acid (2,4-D), 2,4,5-trichlorophenoxyacetic acid (2,4,5-T), picloram, etc. The auxins have a wide spectrum of effects on different processes of plant development and morphogenesis. Depending on their chemical structure and concentration, they induce or inhibit cell division, stimulate callus or root formation.

Gibberellins. The group of gibberellins includes more than 80 compounds, which stimulate cell division and elongation. The most commonly used one is gibberellic acid (GA_3).

Vitamins and supplements. Growth regulatory functions are attributed to some of the vitamins B group – thiamine (B1), niacin (vit B_3, nicotinic acid, vitamin PP), piridoxin (vit B_6), which in fact are the most popular for *in vitro* recipes. Supplements like yeast extract, coconut milk, maize extract and some other might effect tissue growth and bud development.

The best morphogenesis could be achieved when the optimal balance between the effect of genotype, explant and growth regulators is identified.

6. Rooting, aclimatization and adaptation

The processes of root formation and adaptation have their specific requirements and not all of the quoted cases of organogenesis, embryogenesis, regeneration are followed by rhizogenesis and adaptation. These processes depend on the genotype and in most of the cases on the ecotype of the species [62], whereas the necessary culture conditions are chosen in an empirical way. The reduction of the sucrose from 2 % - 3% to 1% - 0.5% stimulates root induction and formation. Aclimatization of the obtained *in vitro* plants is a critical moment for establishment of good protocol for micropropagtion. Adaptation of plants in greenhouse, field or in the nature is another delicate and difficult stage. Usually in *in vitro* conditions, regenerants formed well-developed root system. However, they quickly loose their turgor after transfer to soil. Their leaves withered and dried. These plants underwent stress due to the changes in humidity and culture medium.

7. *In vitro* cultures conditions

The light, temperature and air humidity are important parameters for *in vitro* cultivation of the plant cells and tissues. The light is one of the important factors for morphogenetic process like bud and shoots formation, root induction and somatic embryogenesis. Light spectrum and intensity as well as the photoperiod are very important for successful cultivation [78]. The recommended temperature in the cultivation rooms or phytothrone chamber is about 23-25 °C but the cultures of tropical species require higher temperature (27-30°C), while arctic plants cultures – lower (18-21° C).

Efficient protocolos for *in vitro* propagation (plant cloning) were established for a long list of medicinal plants like *Panax ginseng* [79, 80], *Aloe vera* [81], *Angelica sinensis, Gentiana davidii* [82], *Chlorophytum borivilianum* [83, - 86), *Tylophora indica* [87, 88], *Catharanthus roseus* [89], *Holostemma ada-kodien and Ipomoea mauritiana* [90], *Saussurea involucrata* [91], *Kniphofia leucocephala* [92], *Podophyllum hexandrum* [93], *Saussurea obvallata* [94], *Ceropegia candela-brum* [95], *Syzygium alternifolium* [96], *Chlorophytum arundinaceum* [97], *Rotula aquatica* [98, 99], etc.

Establishment of micropropagation system is a base for conservation of the species and for protection of the genefund, as well as for studies of valuable substances in important medicinal plants. Different strategies are developed as well for establishment of cell cultures aiming at production of biologically active compounds. These systems could be used for large scale cultivation of plant cells for obtaining of secondary metabolites. These methods are reliable and give possibility for continuous supply of raw materials for production of natural products [45, 82].

8. *In vitro* cultures and application of biotechnology in *Gentiana, Leucojum* and *Rhodiola*

In this chapter a small part of the successful *in vitro* research in medicinal plants and the application of "green biotechnology" methods for protection of endangered species will be illustrated by examples from the investigations in the genera of *Gentiana, Leucojum* and *Rhodiola*. These groups of medicinal plants were chosen because the three of them are with outstanding importance for the pharmaceutical and nutraceutical industries. The species belonging to them are worshiped for their multiple beneficial health effects and have been used for thousands of years in folk medicine all over the world. However, their distribution is at different parts of the Earth – *Gentians* are the most widely spread in various climatic zones, *Rhodiola* covers less territory, predominantly in the cold regions in north and high mountains, while *Leucojum* can be found in limited warm and south regions in Europe. Many species from *Gentiana, Rhodiola* and *Leucojum* genera can be found in Bulgaria but most of them are endangered and included in the Red Book like *Gentiana lutea, Rhodiola rosea* and *Leucojum aestivum*. In world scale level of protection – *Leucojum spp* are in the list of the most threatend with heavy measures of restriction. Nearly all *Gentiana* species are endangered while many *Rhodiola* species are under special regime of use. However, one and the same *Rhodiola* species may be close to extinction in one country but widely spread (even as a weed) in another country. Another consideration of ours is the ability of the plants from these genera to be cultivated in field what was possible for *Gentiana*, partially for *Rhodiola* and not possible for *Leucojum*. Described here examples illustrate different levels of development of *in vitro* cultures and application of biotechnology to the three chosen groups of herbs. Development of *in vitro* cultures started about 40 years ago in *Gentiana*, 25 years ago in *Rhodiola* and 20 years ago in *Leucojum*.

The most intensive *in vitro* research was carried in *Gentiana* obtaining all kinds of *in vitro* cultures, including somatic embryo cultures, with success in cryopreservation, in biotransformation and genetic metabolic engineering. *Rhodiola* occupies a middle position with considerable success in callus, suspension and micropropagation systems. Cultivation in bioreactors, biotransformation and genetic transformation were successful.

Leucojum seems to be the most difficult, though protocols for clonal propagation have been established and gene bank *in vitro* has been reported (in Bulgaria). Callus, suspension and organogenic cultures could be obtained and growth in bioreactors with possibilities for biotransformation and even genetic transformation (though at this stage without synthesis of galanthamine).

Bulgaria is a pioneer in *Leucojum aestivum* biotechnology. It is in the frontiers of micropropagation of *Rhodiola rosea* and has less investigation in *Gentiana in vitro* cultures.

Genus Gentiana belongs to family Gentianaceae and is a group of medicinal plants of special interest. It is a large genus comprising of about 400 species widely distributed in the mountain areas of temperate zones [100], including Central and South Europe. Most of the species are interesting to horticulture for their beautiful and attractive flowers but they have more important medicinal value, which is due to the production of secondary metabolites in their roots (Radix Gentianae). The most efficient ones are the bitter secoiridoid glucosides (gentiopicroside, amarogentin), xanthones, di- and trisacharides, pyridine alkaloids [66, 101]. Traditionally, the pharmaceutical industry largely depends on wild sources exploiting intensively the natural areals. The annual drug demands have been much higher than the production from wild sources [66]. At the same time many gentians are either difficult to grow outside their wild habitat or their cultivation (if possible) proved to be not economic. Continuous collection of plant material from natural habitats has led to the depletion of *Gentiana* population and many representatives of the genus are protected by law. Some of the gentians having the status of endangered species, for example, are: *Gentiana lutea* L. - included in the Red Book of Bulgaria and of other European and world countries [66]; *Gentiana kurroo* Royle - close to extinction and legally protected by law [102]; *Gentiana dinarica* Beck - a rare and endangered species of the Balkan Dinaric Alps; *Gentiana asclepiadea* L. - distributed in South and Central Europe, *Gentiana triflora*, *Gentiana punctata*, *Gentiana pneumonanthea* - under protection of its progressively decreasing habitats; *Gentiana dahurica* Fisch – with exhausted natural resources though this species could be cultivated in some areas of the northwest of China; *Gentiana straminea* Maxim an endangered medicinal plant in the Qinghai-Tibet Plateau [103]. Due to problems with germination of seeds in *in vivo* conditions as well as the high variability of generatively propagated plants these species have attracted the attention of scientists being aware of the potential of biotechnology. The genetic variability of endemic or endangered species is usually very low and methods (like *in vitro* micropropagation) of conservation and restoration of natural resources have been given much attention in the last years. Despite the remarkable success of the tedious and wide investigations worldwide *in vitro* cultures of *Gentiana* species proved to be very difficult to achieve because of their low natural capacity of regeneration which was manifested in the multiplication *in vitro*, too [104].

First investigations on establishment of *in vitro* cultures of *Gentiana* were reported a quarter of century ago. Wesolowska et al. [105] succeeded in induction of callogenesis in *G. punctata* and *G. panonica* and of organogenesis and rhizogenesis in *G. cruciata* and *G. purpurea*. Authors observed that regenerated plants synthesize secoiridoids, which could not be found in the wild plants. This raised the hopes for the application of biotechnology techniques to other species of the *Gentiana* genus. The next decade the scientists explored the basic factors and plant requirements for establishment of *in vitro* cultures and micropropagation in various gentians. Different explants were tested for development of efficient regeneration schemes.

Using shoots and node fragments as explants, regeneration systems of *Gentiana scabra var buergeri* [106], *Gentiana kurroo* [107], *Gentiana cerina*, *Gentiana corymbifera* [65], *Gentiana punctata* [108], *Gentiana triflora* [109, 110] and *Gentiana ligularia* [111] were established. Stem segments with meristem tissue were appropriate explants to initiate tissue cultures and to induce formation of shoots *de novo* in four other species of *Gentiana*: *G. lutea* G. *cruciata*, *G. acaulus* and *G. purpurea* [66]. Different explants (shoot tips, lateral green buds, and root segments) were tested in *Gentiana lutea* [112]. Leaf explants were used as well to induce shoots of *Gentiana macrophylla* [113] and *G. kurroo* Royle [107, 114]. Vinterhalter et al. [115] micropropagated *Gentiana dinarica* Beck using axillary buds as explants.

Seeds in different stage of maturity were object of quite strong interest as an initial plant material for *in vitro* cultures. Considerably high germination of 54 % was achieved when seeds of *G. corymbifera* were cultured on a Murashige and Skoog (MS) medium containing 100 mg/l gibberellic acid (GA₃) for 70 days. In the absence of GA₃ germination did not exceed 5% [65].

Immature seeds in different stages of ripening were tested in order to find out the most suitable initial material to obtain *in vitro* cultures and multiplication of *Gentiana lutea*. Despite the addition of 0.5 mg/l of gibberellic acid to the MS medium, the average germination was quite low 21% [112]. Seedlings from immature and mature seeds of *Gentiana pneumonanthe* and *Gentiana punctata* were also chosen as initial material to excise shoot tips and one-nodal cuttings for induction of organogenesis and further clonal propagation [104, 116]. Petrova et al. [40] studied the possibility for micropropagation of Bulgarian ecotype of *Gentiana lutea* using stem segments with two leaves and apical or axillary buds excised from mature seeds germinated *in vitro* (Figure 1). To increase the germination seeds were treated with 0.03% GA₃ for 24 hours. Some of the seeds were mechanically scarified in the micropile region. Germination was initiated on three variants of nutrient media based on MS and different concentration from 25 to 100 mg/l of GA₃. In these investigations, GA₃ and scarification stimulate *G. lutea* seed germination. Only 20 % of the non-scarified and 33.33 % of the scarified seeds germinated on the control medium. Giberrellic acid in concentration of 50 mg/l proved to have optimum effect resulting in 42.5 % germination for the nonscarified seeds and 60 % for the scarified ones. Lower and higher levels of GA3 stimulated in a less extend the seed germination but the response to GA3 of the scarified seeds was stronger than that of the non-scarified ones.

In vitro response is determined, as mentioned before, not only by the explant type but by the media composition as well and by the effect of the plant growth regulators on the dedifferentiation and redifferentiation processes undergoing in the explants cultured *in vitro*. Many reports, especially at the beginning of the *in vitro* investigations of gentians, pointed out that the cytokinine benzyl aminopurine BAP (or benzyl adenine BA) and the auxines indolacetic acid (IAA) or naphtilacetic acid (NAA) were the best plant growth regulators for induction of organogenesis and regeneration of plants which allowed establishment of a micropropagation schemes. Among the numerous examples, some of them were mentioned below as an illustration.

The initial results of Sharma et al. [107] were very promising reporting fifteen-fold shoot multiplication of *Gentiana kurroo*, which was obtained every 6 weeks on Murashige and Skoog's medium (MS) containing 8.9 μM benzyladenine and 1.1 μM 1-naphthaleneacetic acid. The efficiency of these plant growth regulators were confirmed in the experiments with other species. Optimal shoot multiplication of *G. dinarica* was achieved on MS medium enriched with 1.0 mg/l BA and 0.1 mg/l NAA [115]. The ideal medium for adventitious buds formation and for differentiation of calli contained 0.6 mg/l BA and 0.1 mg/l NAA [117] while the ideal medium for induction of calli from tender stems of *Gentiana scabra* was by substitution of NAA with 2,4 D in concentrations of 1.0-1.5 mg/l at the background of the same cytokinin –BA at lower concentration of 0.2 mg/l.

Momcilovic et al. [66] observed that the optimal concentrations of the two plant hormones BAP and IAA were slightly different in the four investigated species *Gentiana acaulis* L., *G. crucia-*

ta L., *G. lutea* L. and *G. purpurea* after different combinations of concentrations were tested (1.14 µM IAA with BA in various concentrations of 1.11-17.75 µM, or 8.88 µM BA with various IAA concentrations 0.57-9.13 µM). Excised nodal segments of axenically germinated seedlings were initially transferred to MS, supplemented with 8.88 µM BA and 1.14 µM IAA. Axillary buds started to grow on all node segments within a few days. Their stems remained short (5 to 15 mm for *G. acaulis and G. cruciata,* respectively) though the leaves reached a length between 25 mm (*G. acaulis*) and 120 mm (*G. cruciata*). Since only the shoots of 5-10 mm were chosen for subculturing, a four to six-fold multiplication was achieved every 4 weeks. Production of well-developed shoots was stimulated by increasing BA concentrations in the presence of 1.14 µM IAA. Indoleacetic acid concentrations higher than 2.28 µM suppressed shoot size in all investigated species. Similar observations were made by Zeleznik et al. [112] who induced shoots proliferation from *Gentiana lutea* shoot tips on MS medium supplemented with 1 mg/l of indoleacetic acid (IAA) and 0.1 mg/l benzyladenin (BA) which caused proliferation in one third of the cultured shoots in a period of 21 days.

The experiments went further in investigating the effect of more plant growth regulators. Based on the well known Murashige and Skoog nutrient medium and commonly used BAP and IAA a comparison was made with other cytokinins and auxins.

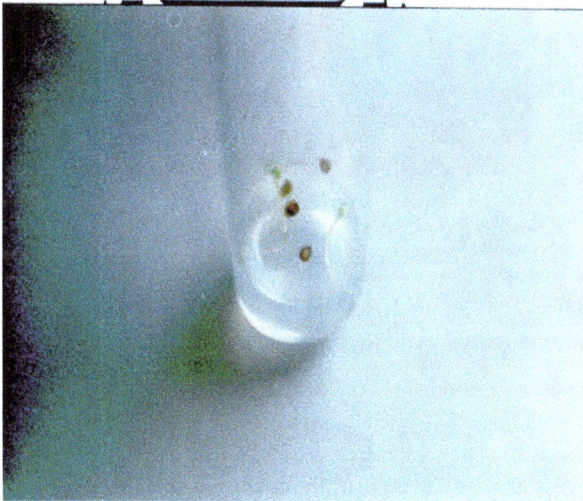

Figure 1. Seed germination of *G. lutea* on MG2 medium (MS basal medium enriched with 50 mg/l GA$_3$) [40].

Bach and Pawlowska [116] studied the efficacy of four cytokinins (BA, kinetin, thidiazuron, 2-iP) and gibberellin at the concentration of 1.5 µM for propagation of *Gentiana pneumonanthe.* The highest multiplication rate was achieved in the culture of the one-nodal cuttings on medium supplemented with 10.0 µM BA. In other experiments [104] media supplemented

Figure 2. *In vitro* micropropagation of *Gentiana lutea* L. on MPl medium (MS basal medium enriched with 2 mg/l zea-   7
tin and 0.2 mg/l IAA) (The mean number of shoots per explant reaching 9 in Vth passage.) [40].

with 2-iP or zeatin and IBA ensured a low multiplication of *Gentiana punctata*. Clonal propagation was slightly improved by addition of maize extract to the culture media.

Different concentrations and combinations of BAP (1 – 2 mg/l), zeatin (1 – 2 mg/l), IAA (0.1 – 0.2 mg/l), 2-iP (0.5 mg/l), and 2,4-D (0.5 mg/l) were used for bud induction and shoot multiplication of *Gentiana lutea* [40]. Best results were recorded on MP1 nutrient medium supplemented with 2 mg/l zeatin and 0.2 mg/ IAA. The mean shoot number per explant was relatively high reaching 4.57 and the average shoot height - 3.90 cm. Second in efficiency was MP3 nutrient medium supplemented with 2 mg/l BA and 0.2 mg/l IAA inducing 4.00 shoots on average per explant (Figure 2).

In vitro response may be influenced by other characteristics of the culture media like medium consistence. Sadiye Hayta et al [118] observed that efficient production of multiple shoots of *G. cruciata* L. directly from nodal segments, inducing 3.9 shoots per explants on average was stimulated on a semi-solidified Murashige and Skoog (MS) basic medium enriched with 2.22 μM 6-benzyladenine (BA), 2.46 μM indole-3-butyric acid (IBA) [118].

In gentiana's experiments plant growth regulators were investigated not only as a factor for establishment of *in vitro* cultures but as a factor which may effect biosynthesis of the biologically active substances in the regenerated plantlets or shoots induced *in vitro*.

Similar observation about the influence of the plant growth regulators on the synthesis of secoiridoids, flavonoids and xantones was studied by Mencovic et al. [119]. There was tendency for a negative correlation between the levels of biologically active substances produced by the regenerants and the concentration of BAP and IAA added into the culture media.

Dević et al. [120] were interested in the effect of applied phytohormones on content of mangiferin in *Gentiana asclepiadea* L. *in vitro* cultures. The content of mangiferin in different plant material was determined by High Performance Liquid Chromatography (HPLC) analysis revealed that the content of mangiferin in the shoots obtained *in vitro* varied with different concentration of applied cytokinine and different auxins. There was no detectable content of mangiferin in roots obtained *in vitro* [120].

Rooting is the next crucial step after successful regeneration and multiplication of plants. Rooting was accomplished successfully in excised *Gentiana kurroo* shoots grown on MS basal medium containing 6% sucrose [107]. Pawlowska and Bach [116] observed too that in vitro multiplied shoots of *Gentiana pneumonanthe* formed roots on a medium without growth regulators. However, the auxins IAA, NAA, and IBA at a concentration of 0.5 μM and 1 μM stimulated rhizogenesis in excised axillary shoots with IAA demonstrating the best effect. Relatively high percentage of 52 % formation of roots from multiplied shoots of *Gentiana lutea* was achieved on MS medium supplemented with 2 mg/l of naphtalenacetic acid (NAA) [112]. Better results were reported by Petrova et al. [40] for *Gentiana lutea* when shoots were transferred to half strength MS medium enriched with either IAA (1 or 2 mg/l), IBA (2 or 3 mg/l) or NAA (0.5 or 3 mg/l). The best results of 92% and 91% rooting were obtained on half strength MS nutrient media containing 3 mg/l IBA or 3 mg/l NAA, respectively. Mean root length was almost equal in the both cases varying from 1.48 cm to 1.95 cm. Spontaneous rooting on plant *Gentiana dinarica* growth regulator-free medium occurred in some 30 % of shoot explants. Rooting was stimulated mostly by decreased mineral salt nutrition and a medium with half strength MS salts, 2% sucrose and 0.5–1.0 mg/l IBA was considered to be optimal for rooting. Wen Wei and Yang Ji [117] confirmed that the ideal medium for the rooting culture and rooting sub-culture of G. scabra tube seedling was 1/2 MS with 0.1 mg/l IAA and 0.3 mg/l NAA. The highest rooting of 81.7% of G. cruciata regenerants was also observed on half-strength MS medium supplemented with 2.46 μM IBA [118]. Beside the successful combinations of plant growth regulators inducing rooting there were reports on less favorable culture media. Butiuc-Keul et al. [104] report about failure in rhizogenesis induction in *Gentiana punctata* shoots transferred on medium supplemented with 1.0 mg/l each NAA and 2iP [104].

Acclimatization and adaptation efficiency varied with the species. In the early experiments, Pawlowska and Bach [116] achieved 65 % survival of rooted plantlets of *Gentiana pneumo-*

nanthe after being potted in soil in a greenhouse. Further, the plants were successfully planted outdoors in field conditions. These *in vitro* regenerants had a greater number of flowers and stems than plants grown in a natural habitat. *In vitro* plantlets of *Gentiana punctata* have been transferred to soil after six weeks of culture and acclimatization was successfully obtained, too [104]. Peat-based substrate for rooting plantlets of *Gentiana dinarica* was successfully used, too [115].

Turf/vermiculite mixtures were very appropriate for acclimatization of plants with well-developed roots transferred to pots in growth chambers. All the acclimatized plants (100%) survived, remained healthy and analysis of the content of secondary metabolites in the clones was determined by HPLC. The presence of gentiopicroside, loganic acid, swertiamarin, and sweroside in the samples was confirmed. Gentiopicroside was found to be the major compound [118].

For the purposes of conservation of the endangered species and for restoration of their habitats it is of a great importance to maintain the genetic stability of the regenerated plants *in vitro*. In this aspect the investigations of Kaur R et al [121] are very interesting. Genetic stability of *Gentiana kurroo* micropropagated plants maintained *in vitro* for more than 10 years was studied using randomly amplified polymorphic DNA (RAPD) and karyotype analysis. A large number of micropropagated plantlets developed from nodal segment explants were assessed for genetic variations and compared with donor mother plant maintained in the arboretum. Out of 20 RAPD primers, 5 displayed the same banding profile within all the micropropagated plants and donor mother plant. No chromosomal variations were observed by the karyotype analysis. High multiplication rate of healthy plant material associated with molecular and karyotypic stability ensures the efficacy of the protocol to be used across a long period for *in vitro* propagation of this important medicinal plant species. These results are extremely important for the application of biotechnological methods and especially of micropropagation for the multiplication of the species for their conservation when *in vitro* clones should be identical to the donor mother plants from the natural habitats.

Somatic embryogenesis is another morphogenetic pathway for regeneration of plants, which is considered the most efficient way to regenerate plants [122]. In contrast to organogenesis when the buds and shoots are not formed obligatory from one cell, a somatic embryo derives from a single cell. This way of development assures greater genetic stability and identity with the initial plant. It opened new possibilities for large-scale multiplication of valuable plants with many expectations for mass production of artificial seeds.

However, somatic embryogenesis is more difficult to obtain. Nevertheless, it was successfully induced in a number of *Gentiana* species: *Gentiana lutea* [122, 123], *Gentiana crassicaulis*, *Gentiana cruciata* [123, 124], *Gentiana pannonica* [123], *Gentiana tibetica* [123], *Gentiana pneumonanthe* and *G. kurroo* Royle [116, 123, 125, 126, 127], *Gentiana davidii* var. *formosana* (Hayata) [128], *Gentiana straminea* [103, 129].

Like in the previously described experiments for micropropagation, one of the requirements leading to success is the appropriate choice of explants. The most commonly used explants were: leaves from the first and second whorls, the apical dome, and axenic shoot culture used for *Gentiana kurroo* (Royle), *Gentiana cruciata* (L.), *Gentiana tibetica* (King. ex Hook. f.), *Gentiana lutea* (L.), and *Gentiana pannonica* (Scop.) [123]; stem explants for initiation of callus and cell suspension cultures of G. *davidii* var. *formosana* [128]; hypocotyl (adjacent to cotyledons) of 10 days old seedlings of *Gentiana cruciata* [124]; seedling explants (root, hypocotyl and cotyledons) for *Gentiana kurroo* (Royle) embryogenic callus [125]; immature seeds (claimed to be superior initial material) of *Gentiana straminea* Maxim [129].

Plant growth regulators are the other very important factor for triggering the totipotence of the plant cell to develop somatic embryo. Unlike organogenesis and shoot formation in gentians where among the numerous tested plant growth regulators several cytokinines and auxins could be distinguished as more prominent, in the case with somatic embryogenesis it was difficult to point out the best ones. In a large number of combinations a wide spectrum of natural phytohormones and synthetic phytoregulators were examined: auxins like α-naphthaleneacetic acid (NAA), 2,4-dichlorophenoxyacetic acid (2,4-D), 3,6-dichloro-o-anisic acid (dicamba), and cytokinins: zeatin, 6-furfurylamonopurine (kinetin), N-phenyl-N′-1,2,3-thiadiazol-5-ylurea (TDZ), N-(2-chloro-4-pyridyl)N′-phenylurea, 6-benzylaminopurine (BAP) or benzyladenine (BA), and adenine sulfate. However, the natural auxin indoleacetic acid is not seen in this list. It makes impression that more auxins of synthetic origin are involved in the studies.

The role of the plant growth regulators will be illustrated by several examples of establishment of cell suspension cultures and somatic embryogenesis. One of the pioneer investigations was performed by Fu-Shin Chuen et al. [128]. Fast-growing suspension cell cultures of G. *davidii* var. *formosana* were established by subculturing callus, which was initiated from stem explants on MS basal medium supplemented with 0.2 mg/l kinetin and 1.0 mg/l NAA. Cell suspension growth was maintained in liquid MS basal medium supplemented with 0.2 mg/l kinetin and 3% sucrose. The cultures were incubated on an orbital shaker (80-100 rev/min) at $25 \pm 1°C$ and low light intensity (2.33 $\mu Em^{-2}s^{-1}$). The low pH of 4.2-5.2 was crucial for the successful cell division and growth.

Quite interesting work from the early period of somatic embryogenesis was that one of Mikula et al., [124]. Authors investigated the effect of phytoregulators on *Gentiana cruciata* structure and ultrastructure changes occurring during tissue culture. MS induction medium containing 1.0 mg/l dicamba, 0.1 mg/l NAA, 2.0 mg/l BAP and 80 mg/l adenine sulfate was used for culturing of hypocotyl (adjacent to cotyledons) explants from 10 days old seedlings. During the first 2 days of culture cell division of epidermis and primary cortex was the first response. Numerous disturbances of karyo- and cytokinesis were observed, leading to formation of multinuclear cells. With time, the divisions ceased, and cortex cells underwent strong expansion, vacuolization and degradation. About the 6th day of culture, callus tissue was formed and the initial normal divisions of vascular cylinder cells were observed. Cells originating from that tissue were small, weakly vacuolated, with dense cytoplasm containing active-looking cell organelles and actively dividing leading to formation of embryogenic callus tissue. During the 6–8th week of culture, in the proximal end of the explant, masses of somatic embryos were formed from outer parts of intensively proliferating tissue. Production of somatic embryos was more effective from suspension culture than from agar medium. Liquid culture made it possible to maintain the cell suspension's embryogenic competence for 5 years.

Quite vast and extensive studies on the establishment of gentians embryogenic cultures and their biotechnological potentials were carried by a research group with impressive publishing activity [114, 126, 130, 131]. Culture initiation and intensive callus proliferation of *Gentiana cruciata* were stimulated by 2,4-D and kinetin using various explants [130]. However, only some of the tissues of initial explant were able to form embryogenic callus. Cytological, ultrastructural and scanning analysis brought evidences that almost each of the cotyledon cells responded by callus formation and somatic embryo differentiation. Central cylinder of the hypocotyls gave the best response for embryogenic proliferation compared to other tissues of hypocotyls. Another medium containing 1.0 mg/l dicamba, 0.1 mg/l NAA, 2.0 mg/l BAP and 80 mg/l SA proved to be very efficient to maintain very long-term cell suspension cultures of proembryogenic masses. Long-term culture provided opportunities for numerous analysis to have evidences of the single cell origin of somatic embryos which originated from freely suspend single cells or single cells from the embryogenic clusters. Medium supplemented with GA_3 helped to complete development and stimulated the somatic embryo conversion in germlings. Embryogenic potential was genotype dependent with *G. tibetica* and *G. kurroo* being outstanding generating more than hundreds somatic embryos from 100 mg of tissue for more than two years. Interestingly the regeneration ability was maintained not only in the long-term suspension cultures but it was demonstrated in the protoplast cultures, too [126].

Protoplasts with very high viability ranging from 88 to 96 % were isolated from cell suspensions derived from cotyledon and hypocotyl of *Gentiana kurroo* [126]. Three techniques of culture and six media were evaluated in terms of their efficiency in producing viable cultures and regenerating entire plants. The best results of plating efficiency (68.7% and 58.1% for cotyledon and hypocotyl derived suspensions, respectively) were obtained with agarose bead cultures in medium containing 0.5 mg/l 2,4-D and 1.0 mg/l kinetin. Regeneration of plants was also possible when embryos were transferred to half-strength MS medium. However, flow cytometry analysis revealed increased amounts of DNA in about one third of the regenerants which limits the application of isolated protoplasts in the programs for conservation and reproduction of an endangered species. Hence, the efforts were directed again to cell and tissue cultures examining the factors for efficient and reliable plant regeneration, even to examining photosynthetic activity in dependence of the sucrose content in the emryogenic culture media [126].

Fiuk and Rybczyn'ski [123, 125] expanded their studies using leaves derived from axenic shoot culture of five *Gentiana* species (*Gentiana kurroo, Gentiana cruciata, Gentiana tibetica, Gentiana lutea,* and *Gentiana pannonica*) and cultured on MS basal medium supplemented with three different auxins: 2,4-D, NAA, or dicamba in three concentrations of 0.5, 1.0, or 2.0 mg/l; and five different cytokinins: zeatin, kinetin, BAP, TDZ, and N-(2-chloro-4-pyridyl)N'-phenylurea in concentrations between 0.25 and 3.0 mg/l depending on the cytokinin activity. After two months the percentage of embryogenesis was the highest for *G. kurroo* reaching 54.7% and depending on plant growth regulators. This gentian was the only species responding to the all tested combinations of auxins and cytokinins, while none of the 189 induction media stimulated somatic embryogenesis from *G. lutea* explants. Efficiency of embryogenesis was genotype dependent *G. tibetica* and *G. cruciata* both produced an average of 6.6 somatic embryos per explant, while *G. pannonica* and *G. kurroo* regenerated at 15.7 and 14.2 somatic embryos per explant, respectively. Optimum regeneration was achieved in the presence of NAA combined with BAP or TDZ. NAA also stimulated abundant rhizogenesis. Somatic embryos were also regenerated from adventitious roots of *G. kurroo, G. cruciata,* and *G. pannonica.* Somatic embryos developed easily into plantlets on half strength MS medium.

The same research group extended its investigations on the factors influencing efficiency of somatic embryogenesis in cell suspension of *Gentiana kurroo* (Royle) - the species revealing the best morphogenic potential in their previous studies [125]. Suspension cultures were initiated in liquid MS medium supplemented with 0.5 mg/l 2,4-D and 1.0 mg/l kinetin from embryogenic callus derived from seedling roots, hypocotyls and cotyledons. Unexpectedly the highest growth rate was observed for root derived cell suspensions. Further more differences in aggregate structure depending on their size were detected by microscopic analysis. In order to assess the embryogenic capability of the particular culture, 100 mg of cell aggregates were implanted on MS agar medium supplemented with 0–2 mg/l kinetin, 0–2 mg l/l GA$_3$ and 80 mg/l adenine sulfate. The highest number of somatic embryos was obtained for cotyledon-derived cell suspension on GA$_3$-free medium, but the presence of the other plant growth regulators (0.5–1.0 mg/l kinetin, 0.5 mg/l GA$_3$ and 80 mg/l adenine sulfate) determined the best morphological quality of embryos. The morphogenic competence of cultures also depended on the size of the aggregate fraction and was lower when size of aggregates decreased. Flow cytometry analysis revealed 100% uniformity for regenerants derived from cotyledon suspension but lack of uniformity of plantlets obtained from hypocotyls suspension. These observations were of great significance for the choice of appropriate explants and culture media conditions for the multiplication of a particular gentian species via somatic embryogenesis.

Cai YunFei et al. [129] confirmed the role of the explant and its interaction with the plant growth regulators added into the media of *Gentiana straminea* Maxim. They observed that calli induced from immature seeds were superior to those from hypocotyls or young leaves in regeneration via somatic embryogenesis and demonstrated that 2,4-D was efficient for both callus induction and embryogenesis, IAA is suitable for embryogenic callus proliferation, and BAP promotes both embryo development and the accumulation of gentiopicroside in the cultures. Experiments went further in exploring *Gentiana in vitro* cultures potentials by selecting regenerated plants for high gentiopicroside content. A highly productive clone was selected. Its cells contained 5.82 % of gentiopicroside, which levels were two folds higher than the control plants (1.20-3.73 %). Genetic stability of the regenerated plants was also proved both by cytological and random amplified polymorphic DNA analyses.

Similar experiments were performed with *Gentiana straminea* Maxim. MS medium supplemented with 2 mg/l 2,4-D and 0.5 mg/l BA was the best medium for embryogenic callus induction from leaf explants [103]. Genetic stability of the regenerants was assessed by 25 inter simple sequence repeat (ISSR) markers. Out of 25 ISSR markers, 14 produced clear, reproducible bands with a mean of 6.9 bands per marker confirming that the regenerants maintained high genetic fidelity.

One of the recent reports [122] presented interesting results for the possibility to use recurrent somatic embryogenesis in long-term cultures of *Gentiana lutea* for production of synthetic seeds. After induction of somatic embryogenesis in the presence of auxins in the first cycle of *in vitro* cultures, recurrent somatic embryogenesis was performed in long-term cultures in the absence of phytohormones but in the presence of the sugar alcohols mannitol and sorbitol. Adventive somatic embryos were generated continuously at a high rate along with maturation, germination and development into plants.

One of the possibilities of biotechnology for conservation of rare species is the establishment of *in vitro* germplasm banks, which may include cryopreservation of *in vitro* multiplied valuable plant material. There are several interesting publications of one research group dedicated to this problem [124].

For preservation of proembryogenic masses of G. *cruciata*, four protocols of cryopreservation were studied: direct cooling, sorbitol/DMSO treatment, vitrification, and encapsulation. Direct cooling and sorbitol/DMSO treatment was unsuccessful. Vitrified tissue required a minimum 3 weeks culture on solid medium for cell proliferation to reach the proper fresh weight for manipulation. Alginate beads with PEMs were transferred directly to liquid medium for post-freezing culture. Vitrification and encapsulation maintained high viability of post-freezing proembryogenic masses, but encapsulation ensured faster restoration of G. *cruciata* cell suspension [124]. A reliable technique for cryopreservation by encapsulation was developed for two suspension cultures of *Gentiana* species (*Gentiana tibetica* and G. *cruciata*) of different ages and embryogenic potential. A water content of 24-30% (fresh weight basis) after 5-6 h dehydration of encapsulated cells of gentians yielded the highest survival (68% for G. *tibetica* and 83% for G. *cruciata*) after cryopreservation. Flow cytometry showed that cryopreservation did not change the genome size neither of the somatic embryos nor of the regenerants [132]. The embryogenic cell suspension culture of *Gentiana cruciata*, cryopreserved by the encapsulation/dehydration method, survived both short- (48 h) and long-term (1.5 years) cryostorage with more than 80% viability. The (epi)genetic stability of 288 regenerants derived from: non-cryotreated, short-term, and long-term cryo-stored tissue was studied using metAFLP markers and ten primer combinations. AFLP alterations were observed but they were not associated with the use of cryopreservation, but were probably related to the *in vitro* culture processes [133]. These results gave great hopes for the use of cryo-techniques in preservation of valuable medicinal species.

Genetic transformation was also applied to gentiana species aiming at obtaining higher production of biologically active substances or biosynthesis of new valuable compounds. *Agrobacterium rhizogenes* mediated transformation was achieved in shoots of micropropagated *Gentiana acaulis*, G. *cruciata*, G. *lutea*, and G. *purpurea* inoculated with suspensions of *Agrobacterium rhizogenes* cells [134, 135]. Few years later Menkovic et al [119] after infection with *Agrobacterium rhizogenes* managed to obtain nine hairy root clones which differed in the amount of secondary metabolites. *Agrobacterium tumefaciens* was also used for inoculation of *Gentiana punctata* [136] and *Gentiana dahurica* Fisch by A. *tumefaciens* [137]. However, due to the great opposition in many countries against the genetically modified organisms, especially these ones with potential use in food and nutraceutical industries genetic transformation experiments remained more in the laboratory mainly to study the metabolic pathways.

Genus Leucojum. Leucojum aestivum (summer snowflake) is one of the most worshiped medicinal plants on the Balkan region and in the world. Leucojum aestivum L. (Amaryllidaceae family) is a polycarpic geophyte distributed in the wetlands of Central and South Europe (Mediterranean and the Balkans) and in West Asia. L. aestivum grows on alluvial soils with high nitrogen levels. The mean size of the plants increased with the water content of the soil. Seed reproduction is whimsical. Seed set of the plants was not influenced by the size of a population, but strongly increased with the density of flowering plants. Optimal temperature for seed germination is 20-25°C [138]. Overharvesting of its bulbs for medical purposes has brought to a destruction or alteration of its habitats across Europe [138]. Therefore, summer snowflake has turned into an endangered species and is protected in several European countries (e.g. Bulgaria, Hungary and Ukraine).

Leucojum aestivum L. is used as a source of galanthamine - an isoquinoline alkaloid produced exclusively by plants of the family Amaryllidaceae (mainly belonging to the genus Galanthus, Leucojum and Narcissus). Due to its acetylcholinesterase inhibitory activity, galanthamine is used for various medical preparations for the treatment of neurological disorders and especially for senile dementia (Alzheimer's disease) and infantile paralysis (poliomyelitis). A very effective Bulgarian remedy to cure poliomyelitis was produced from L. aestuvum in the middle of the XXth century. This marked tremendous interest and respect of the plant and enormous demands for raw material. Despite the possibility for organic synthesis, galantha-mine is still extracted from natural sources. For industrial purposes L. aestivum plants are harvested from wild populations in their natural habitats which causes increasing problems regarding quality of the plant material as well as natural populations depletion. The limited availability of the plants and the increasing demands for this valuable metabolite has imposed urgent search for alternative approaches both for protection of the species and for production of galanthamine. In this respect biotechnology methods could be used for in vitro storage of genotype accessions from different natural populations with proven alkaloid profiles, for rapid propagation of this threatened medicinal plant for both industry and natural resource protection, and for production of its valuable biologically active substances under controlled conditions. However, not so much data are available in the literature concerning in vitro cultures of this plant.

The Bulgarian scientists Stanilova et al. [38] and Zagorska et al [139] are pioneer in establish-ment of in vitro cultures and micropropagataion of Leucojum aestivum. One of the prerequisites for their success was the elaboration of a successful procedure for decontamination of the plant material gathered from nature. Plant material should be used 42 days after collecting. The bulbs were rinsed for 16 hours with stream water followed by immersion in 70% ethanol for 30 s and sterilized with 0.1% $HgCl_2$ for 3 min. Relatively good decontamination was achieved for leaf explants applying hypochlorites – 47.46% using 5% Ca $(ClO)_2$ for 6 min and 54.76% - 15% NaClO for 20 min. During their initial studies of the morphogenetic potential of the basal and apical parts of bulbs, stems, leaves and ovaries it was observed that the scales of L. aestivum possessed the highest regenerative ability producing 4.08 - 4.19 regenerants per explant. Whereas leaf explants had lower regeneration potential – 1.67 regenerants per explant. Murashige and Skoog (MS) medium supplemented with 1 mg/l benzyladenine (BA) and 1 mg/

l kinetin as well as Linsmaier and Skoog (LS) medium enriched with 0.5 mg/l NAA and 0.1 mg/l kinetin proved to be the most suitable for direct organogenesis [38]. Rhyzogenesis was induced on MS basal medium with reduced sugar content of 15 g/l and enriched with 0.1 mg/l NAA, 0.1 mg/l kinetin and 0.1 mg/l BAP. Further investigations focused on *in vitro* clonal propagation of *L. aestivum*. Twenty four clones were obtained and most of them demonstrated high regeneration rates and stable alkaloid profiles. Galanthamine levels of some of the *in vitro* obtained clones was as high as galantamine levels of commercially important representative of Bulgarian *L. aestivum* populations. Five clones: four galanthamine-type and one lycorine-type were selected as promising for further investigations [140].

In Turkey *Karaog'lu* [141] confirmed the effectiveness of bulb-scales explants for micropropagation of *Leucojum aestivum* and tested immature embryos for initiation of *in vitro* cultures. Using 2 and 4 bulb-scales explants the highest number of bulblets (6.67 and 5.83) were achieved on MS medium containing 1 mg/l BA and 1 mg/l NAA or 2 mg/l BAP and 0.5 mg/l NAA, respectively. Regeneration capacity of immature embryos was twice lower reaching 2.27 bulblets on MS medium containing 0.5 mg/l BA and 4 mg/l NAA. The best rooting of bulblets regenerated from bulb scales was obtained on MS medium containing 1 mg/l NAA. Rooted bulbs were finally transferred to compost and acclimatized to ambient conditions [141].

Later *in vitro* cultures of *Leucojum aestivum* were reported in Hungary. Kohut et al. [142] succeeded to obtain from 81 % to 92 % contamination free material. Prior to surface sterilization the old leaves and roots were dissected from the bulbs and they were stored at low temperature of 2–3°C for 1 and 5 week periods. The bulbs, bulb scales and leaves of the bulbs were placed on *MS* medium containing 1 mg/l BA and 0.1 mg/l NAA.

Shoot *in vitro* cultures were initiated also from bulb explants in others' experiments [143]. However, Gamborg's B5 medium was used for the initiation and maintaining of the cultures, which were kept in darkness. This medium contained 30 g/l of sucrose, 1 mg/l 2,4-D, 0.5 g/l casein hydrolysate, 2 mg/l adenine, and 10 mg/l glutathione. The *in vitro* cultures were subcultured at 2.5 month intervals in MS medium supplemented with 1 g/l Ca(NO3)2, 0.5 mg/l BAP, 0.01 mg/l IBA, and 2.93 mg/l paclobutrazol. During the subcultures, shoot-clumps which were formed were cut to increase the number of explants, and the newly formed shoot clumps were separated. The *in vitro* cultures were maintained at 23-25° C with a 16/8 h light /dark photoperiod. Later the same research group [144] offered a three step protocol for *in vitro* long-term conservation of *L. aestivum* which was used to create a genebank with accessions from 31 Bulgarian populations. For *in vitro* cultures dormant bulbs were used, which were cut into 8, 16 or more segments. For sterilization, these segments called "twin-scale" were treated with 70% ethanol for 30 s and sterilized with 1% HgCl2 for 3 min. The development of the shoot-clumps started from the basal parts of the scales at the end of the first week. The development of *in vitro* shoot-clump cultures was tested on three nutrient media: MS, B5, and QL with or without plant growth regulators, BAP (0.5 - 3.0 mg/l), IBA (0.01 - 1 mg/l) NAA (0.2 - 2 mg/l) and TDZ (1 - 2 mg/l), sucrose (0 - 120 g/l), and charcoal (2g/l). Shoot-clumps were obtained, from explants cultivated on B5 medium (6), supplemented with 0.5 g casein hydrolyzate, 1 mg/l 2,4-D, 10 mg/l adenine, 10 mg/l glutathione, 30 g/l sucrose, 6 g/l agar. The fastest multiplication however was observed on MS medium with 30 g/l sucrose, 2 mg/l BAP, 1.15 mg/l

NAA. Increasing sucrose concentration up to 90 g/l resulted in higher mass of the obtained bulbs. About 1000 regenerated bulbs with well-developed roots were successfully adapted at *ex vitro* conditions. Authors observed that plant *ex vitro* adaptation depended on the bulb size. The biggest bulbs (over 1.5 cm in size) were the most adapted (99 %) whereas about 60% of the medium size bulbs (0.5-1.5 cm) and 20% of the small bulbs (less than 0.5 cm) survived. Mainly easily rooting bulbs were formed on hormone free nutrient medium (MS with vitamins, sucrose-30 g/l, charcoal - 2 g/l, and pH-5.6) [144].

Callus cultures from young fruits of *Leucojum aestivum* L. were obtained, too [145]. Non-differentiated cell growth was stimulated by high concentrations of the auxin 2,4-D (4 mg/l) and the cytokinin BA 2 mg/l. Callus tissue formed regenerants when 1.15 mg/l NAA and 2 mg/l BA were added to the MS medium.

Somatic embryos were formed from callus tissues cultivated on MS medium containing 2 μM or 5 μM picloram (4-amino- 3,5,6-trichloropicolinic acid) and 0.5 μM BAP [146]. Regeneration of plants was possible on medium enriched with zeatin (0.5 μM). Authors observed that the processes of differentiated or non-differentiated growth leading to somatic embryogenesis or callus growth, respectively, were influenced by ethylene or its precursor ACC (1-aminocyclopropane-1-carboxylic acid). At higher concentrations (25 μM) of picloram callus cultures produced ethylene (9.5 nL/g fresh weight: F.W.) whereas no ethylene was detected in cultures of somatic embryos cultivated on medium supplemented with 0.5 μM NAA and 5 μM zeatin. Application of ACC increased ethylene production thus suppressing callus growth and enhancing somatic embryos induction and globular embryos development. Another effect of ACC was to induce galanthamine production in somatic embryo cultures (2% dry weight). However, galanthamine production in callus cultures was induced by silver thiosulphate (STS) though in low levels (0.1% dry weight). These results are promising for use of somatic embryos cultures in bioreactors for production of galanthamine [146].

Alkaloid content in *Leucojum aestivum* wild plants and their *in vitro* cultures was studied in a series of experiments carried out by a Bulgarian research group [143, 145, 147 - 150]. Callus cultures were obtained from young fruits of *Leucojum aestivum* on MS nutrient medium supplemented with 4 mg/l 2,4 D and 2 mg/l BAP. Further, shoot cultures were established by subculturing the obtained calli on the same nutrient medium supplemented with 1.15 mg/l NAA and 2.0 mg/l BAP. These *in vitro* systems were used to study the growth and galanthamine accumulation. The authors observed that the amount of accumulated galanthamine strongly depended on the level of tissues differentiation. The maximum yield of biomass (17.8 g/l) and the maximum amount of accumulated galanthamine (2.5 mg/l) were achieved under illumination after the 35th day of submerged cultivation of one of the lines *L. aestivum* -80 shoot culture.

The alkaloids of intact plants, calli and shoot-clump cultures of *L. aestivum* were analyzed by capillary gas chromatography – mass spectrometry (CGC-MS). In one series of experiments fourteen alkaloids of galanthamine, lycorine and crinane types were identified (11 in the intact plants and eight in the *in vitro* cultures) in alkaloid mixtures extracted from intact plants and *in vitro* cultures. Excellent peak resolution for the alkaloids was exhibited and isomers of galanthamine and N-formylnorgalanthamine were well separated [147]. Applying the same methods of CGC-MS, extracts from bulbs collected from 18 Bulgarian populations and from shoot-clumps obtained *in vitro* from eight different populations were subjected to analysis and nineteen alkaloids were detected. Typically, galanthamine type compounds dominated in the alkaloid fractions of *L. aestivum* bulbs but lycorine, haemanthamine and homolycorine type alkaloids were also found as dominant compounds in some of the samples. Galanthamine or lycorine as main alkaloids presented in the extracts from the shoot-clumps obtained in vitro. The galanthamine content ranged from traces to 454 µg/g dry weights in the shoot-clumps while it was from 28 to 2104 µg/ g dry weight in the bulbs [143]. In other investigations twenty-four alkaloids were detected analizing intact plants, calli and shoot-clump cultures. Shoot-clumps had similar profiles to those of the intact plant while calli were characterized with sparse alkaloid profiles. Seven shoot-clump clones produced galanthamine predominantly whereas another three were dominated by lycorine. It was also observed that illumination stimulated accumulation of galanthamine (an average of 74 µg/g of dry weight) in shoot-clump strains while in darkness galanthamine levels were two folds less (an average of 39 µg/g of dry weight). The shoot-clumps, compared to intact plants, accumulated 5-folds less galanthamine. The high variability of both the galanthamine content (67% and 75% of coefficient of variation under light and darkness conditions, respectively) and alkaloid patterns indicated that the shoot-clump cultures initiated from callus could be used as a tool for improvement of the *in vitro* cultures production of the valuable substances [148]. The investigations extended on the alkaloid patterns in *L. aestivum* shoot culture cultivated at temporary immersion conditions where 18 alkaloids were identified, too. The temperature of cultivation influenced enzyme activities, catalyzing phenol oxidative coupling of 4'-O-methylnorbelladine and formation of the different groups *Amaryllidaceae* alkaloids. Decreasing the temperature of cultivation of *L. aestivum* 80 shoot culture led to activation of para-ortho' phenol oxidative coupling (formation of galanthamine type alkaloids) and inhibited ortho-para' and para-para' phenol oxidative coupling (formation of lycorine and haemanthamine types alkaloids). The *L. aestivum* 80 shoot culture, cultivated at temporary immersion conditions, was considered a prospective biological matrix for obtaining wide range of *Amaryllidaceae* alkaloids, showing valuable biological and pharmacological activities [150]. The most recent report was about successful cultivation of shoot culture of summer snowflake in an advanced modified glass-column bioreactor with internal sections for production of Amaryllidaceae alkaloids. The highest amounts of dry biomass (20.8 g/l) and galanthamine (1.7 mg/l) were obtained when shoots were cultured at temperature of 22°C and 18 l/(l h) flow rate of inlet air. At these conditions, the *L. aestivum* shoot culture possessed mixotrophic-type nutrition, synthesizing the highest amounts of chlorophyll (0.24 mg/g DW (dry weight) chlorophyll A and 0.13 mg/g DW chlorophyll B). The alkaloids extract of shoot biomass showed high acetylcholinesterase inhibitory activity (IC_{50} = 4.6 mg). The gas chromatography–mass spectrometry (GC/MS) profiling of biosynthesized alkaloids revealed that galanthamine and related compounds were presented in higher extracellular proportions while lycorine and hemanthamine-type compounds had higher intracellular proportions. The developed modified bubble-column bioreactor with internal sections provided conditions ensuring the growth and galanthamine production by *L. aestivum* shoot culture

[149]. The influence of the nutrient medium, weight of inoculum, and size of bioreactor on both growth and galanthamine production was studied in different bioreactor systems (shaking and nonshaking batch culture, temporary immersion system, bubble bioreactor, continuous and discontinuous gassing bioreactor) under different culture conditions. The maximal yield of galanthamine (19.416 mg) was achieved by cultivating the L. aestivum shoots (10 g of fresh inoculum) in a temporary immersion system in a 1l bioreactor vessel which was used as an airlift culture vessel, gassing 12 times per day (5 min) [151].

Completely different types of experiments were the attempts of genetic transformation with Agrobacterium. Agrobacterium rhizogenes strain LBA 9402 has been tested [152] for its capacity to induce hairy roots of this monocotyledonae plant. Diop et al. [152] have developed an efficient transformation system for L. aestivum, which could be used to introduce genes encoding enzymes of isoquinoline alkaloid biosynthesis into L. aestivum to enhance the production of target molecules in this medicinal plant. However, the transformed roots obtained did not synthesize galanthamine.

At this stage of in vitro research establishment of organogenic cultures and optimization of galanthamine production by differentiated cells using the methods of biotransformation are more promising and reliable.

Genus Rhodiola is highly varied among others in family Crassulaceae (comprising of 1500 species in 35 genus). The genus Rhodiola includes over 200 quite polymorphic species, out of which 20 species (Rh. alterna, Rh. brevipetiolata, Rh. crenulata, Rh. kirilowii, Rh. quadrifida, Rh. sachalinensis, Rh. Sacra etc.) have pharmacological properties and are used for production of medical preparations [153].

Intensive and unscrupulous exploitation of the natural habitats in many countries has led to extinction of these species in these regions [154]. This provoked nature-protecting measures to be undertaken like (1) cultivation under appropriate conditions, (2) protection of the populations in the protected areas, (3) including the species in Red Books of rare and endangered plants species. Rhodiola species contain various quantities of salidrosid – one of the most important ingredients in the biological active complex [155 - 159]. Salidroside content in plants varies depending on the genetical structure, the developmental stages, the plant age, the ecological and agrobiological conditions [160] what is one of the reasons for the scientists to look for conditions minimizing these effects by biotechnological way of more controlled production of this biologically active compound. From another hand extracts from medicinal plants are rich in other metabolites bringing to the multiple health benefits [159] what stimulates search and identification of more biologically active substances which can be produced in cultures in vitro.

In Bulgaria Rhodiola rosea (Golden root, Roose root) (Sedum roseum (L.) Scop., S. rhodiola DC.) is under protection of the Act for biological diversity [161]. Rhodiola rosea is included in the Red Books of Republic of Buryatia AR, of Yakut ASSR, of Mongolia; "Rare and Extinct Plant Species in Tyva Republic," "Rare and Extinct Plant Species in Siberia," in Great Britain—Cheffings & Farrell, in Finland—category "last concerned."(according to IUCN Red List Categories and Criteria: Version 3.1 (IUCN, 2001).

Rhodiola rosea species are worshiped for their roots and rhizomes therapethical role in many diseaases. *Rhadix et Rhizoma Rhodiolae* of *Rh. rosea* are used in medicine for optimization of own-body biochemical and functional reserves of the organism, for stimulation of body's nonspecific resistance for regulation of the metabolism, central nervous system, cardiovascular system and the hormonal system [162 – 164] for rehabilitation after heavy diseases, for prophylactics of onco disease [153, 159, 165], etc. *Rhodiola quadrifida* (Pall.) Fisch. et May is a perennial grassy plant growing predominantly in some highland regions of the former USSR (Altai, Sayan), in East Siberia, in some mountainous regions of China (Sichuan) and in high mountain regions of Mongolia. It is used in traditional medicine of Mongolia and Tibet, against fatigue, stress, infections, inflammatory diseases and protection of people against cardiopulmonary function problems when moving to high altitude [166; 167]. The phytochemical composition of the ingredients (without cinnamic alcohol and rosiridin) is similar to that of *Rh. rosea* [168]. *Rhodiola kirilowii is a* Chinese medicinal herb. Roots and rhizomes extracts are used in Asiatic medicine independently of their adaptogenic properties also as antimicrobial and anti-inflammatory drugs [169, 170]. *Rhodiola sacra* grow in the Changbai Mountain area, Tibet and Xinjiang autonomous regions in China. In Tibetan folk medicine, *Rhodiola Radix* is used as a hemostatic, tonic and contusion releaf factor. Positive effects on learning and memory have been reported, too [171, 172]. *Rhodiola crenulata* is distributed in the high cold region of the Northern Hemisphere in the high plateau region of southwestern China, especially the Hengduan Mountains region including eastern Tibet, northern Yunnan and western Sichuan. It has strong activities of anti-anoxia, antifatigue, anti-toxic, anti-radiation, anti-tumour, anti-aging, and active-oxygen scavenging [173, 174]. *Rhodiola sachalinesis* A. Bor. is used as a drug of "source of adaptation to environment" in Chinese traditional medicine. Salidroside can effectively enhance the body's ability to resist anoxia, microwave radiation, and fatigue. Furthermore, its effect on extending human life was also found [175]. *Rhodiola imbricata* Edgew commonly known as rose root, is found in the high altitude regions (more than 4000 m altitude) of India. The radioprotective effect, along with its relevant superoxide ion scavenging, metal chelation, antioxidant, anti-lipid peroxidation and anti-hemolytic activities were evaluated under both *in vitro* and *in vivo* conditions [176]. *Rhodiola iremelica* Boriss. – is an endemic plant of Middle and South Ural mountain. It is included in the Red Book of Republic of Bashkortostan (Bashkiria) in the category of rare and endangered species. *Rh. iremelica* is located in places with different climatic conditions making them unique [177].

Despite the incontestable/undisputed interest to *Rh. rosea* and the wide intensive research in phytochemistry, the potential area of the plant biotechnologies, remains less studied and exploited in comparison to other medicinal species. Some of the researchers studied the possibility for induction of calli cultures, biotransformation and organogenesis. Other authors focus their research on *Rhodiola* potential for regeneration and investigation of biologically active substances. Experiments are focused mainly in two directions: 1) looking for possibilities for *in vitro* synthesis of valuable metabolites and/or 2) establishment of effective systems for micropropagation, for reintroduction of the plant in nature or in the field for protection of the species.

Pioneer experiments on golden root *in vitro* cultures were initiated 20 years ago [178] from a Russian scientist who described rooting of assimilating of sprouts *R. rosea*. Later a few other reports have appeared concerning the effect of culture media composition and of explant type on the ability for callogenesis, organogenesis and regeneration of *R. rosea*, as well as other factors influencing growth and morphogenesis. Using leaf segments Kirichenko et al. [179] studied callus and regeneration ability for propagation *in vitro* of rose root while Bazuk et al. [180] focused on the rooting potential of shoots obtained from stem segments with two adjacent leaves. Investigations that are more detailed were carried using seeds and rhizomes from three ecotypes from the High Altai and South Ural region, which served as the explant source to study induction of callogenesis and organogenesis [181]. Explant development was observed on MS media containing various phytoregulators (BAP, IAA, NAA, IBA, 2,4-D). Very high percentage of 86% of the explants formed abundant calli on MS medium supplemented with 0.1 - 0.2 mg/l IAA. BAP and IAA in concentrations of 0.2mg/l and 0.1mg/l, respectively, was the optimal combination for multiple bud formation in *Rhodiola rosea* from stem segments, while for *Rhodiola iremelica* the efficient concentrations were lower—0.1 mg/l BAP and 0.05 mg/l IAA. The processes of efficient callogenesis and organogenesis were influenced by ecotype differences. Adaptation of regenerants in vermiculite for two weeks in conditions of high humidity (85–90%) and later in mixture of soil, peat, and vermiculite in proportion of 1 : 1 : 1 was successful, but with considerable differences in the survival rate (from 10% to 95%). In the later experiments [182], the effect of 5% or 10% v/v liquid extracts of *Rh. rosea* extracts on the morphogenic abilities of *Rh. rosea* and *Rh. iremelica* were studied. Different *in vitro* responses were provoked. Bud induction was stimulated by the lower concentration and inhibited by the higher ones leading to formation of 8.5 shoots per explant in the first case and 1.1 in the second case.

Unlike the previously described investigations with the Altai ecotype of *Rhodiola rosea* the optimal concentrations of the cytokinin BAP were 10–15-fold higher for induction of *in vitro* cultures from immature leaves explants from a Tibetan ecotype of golden root [183]. The authors noted interaction between the growth regulators and the illumination of the cultures. Two mg/l BA and 0.2 mg/l NAA added to the MS medium stimulated formation of incompact callus tissue. However, when explants were cultivated under dark conditions, higher concentrations of the same phytoregulators BA (3 mg/l) and NAA (0.25 mg/l) were more efficient. MS medium containing 2 mg/l BA and 0.25 mg/l NAA induced shoot multiplication while rooting was induced on MS medium containing 0.5 mg/l or 1 mg/l IAA.

In Bulgaria the first investigations on *Rhodiola rosea* were on the content of polyfenols and salidrosid in the local populations in Rila, Pirin and Balkan Mountain [184]. The highest salidrosid levels in the rhizome and root were found in the plants from Rila Mountain while the lowest ones in the plants from Pirin Mountain 1.55 % and 0.72 %, respectively. From another side polyphenols were in highest concentration in rose root from Pirin population. Salidrosid is accumulated in roots and rhizomes while polyphenol content is equal in all parts of the plant [185, 186].

Seeds of *Rhodiola rosea* lose their germination potential for a relatively short perod of time compared to other species. Stratification is one of the approaches for overcoming this problem. Revina et al. [187] reported about higher germination up to 75 % after treatment of seeds for one month at temperatures of 2-4 °C. Other authors confirm the role of stratification [188] and report

about stimulation of germination up to 50-75 % after subjecting seeds to lower temperatures of
-5°C for a period of 3 months [189]. Dimitrov et al. [190] applied a new approach for *in vitro*
cultivation of seeds. Golden root germination of seeds started on the 7th day of cultivation and
lasted until the 40th day reaching from 37.5% to 97.0% depending on media composition.
Germination was stimulated when seeds were cultured on MS basal medium enriched with
50-100 mg/l giberrellic acid. These high concentrations of GA_3 enhanced germination while lower
concentrations of 5-25 mg/l GA_3 favored obtaining of seedlings with bigger size [43].

The initial investigations for establishment of golden root *in viro* cultures in Bulgaria were
dedicated to find out a suitable ecotype for *in vitro* experiments [190 – 192]. Tasheva et al. [193,
194] optimized seed germination *in vitro* and later report the first successful results for *in
vitro* propagation of *Rhodiola rosea*. A large number of explants isolated from *in vitro* seedlings
(stem segment with leaf node, apical bud, explants excised from the seedling root basal area)
and *in vivo* plant (apical bud, adventitious shoots, stem expalnt, rhizome buds, rhizome
segments) were used to study *in vitro* response [195] on Murashige and Skoog (MS) basal
medium containing various hormonal combinations including benzyladenine, kinetin, zeatin,
2-ip etc. *In vitro* development led to formation of plantlets, leaf rosette, various type of callus
(compact green, pale, soft liquidy) and callus degeneration without bud formation. The
authors observed that the explants of seedling and apical bud are more suitable for mass clonal
propagation. Multiple shoots proliferation from leaf node explants was most effective on
nutrient medium containing 1.0 or 2.0 mg/l zeatin, 0.1 mg/l IAA and 0.4 mg/l GA_3 (Figure 3
a, Figure 3 b). Rooting *in vitro* proved to be the most efficient on nutrient medium containing
IBA (2.0 mg/l), IAA (0.2 mg/l) and GA_3 (0.4 mg/l) [195]. Interestingly, it was observed that the
coefficient of propagation varied during the different seasons. Highest level of proliferation
was recorded in May-June, when the mean number of shoots per explant was 6.78, while
during the cold seasons multiplication was relatively lower with 2.11 shoots per explant [196].
Adaptation of obtained plants was done under controlled conditions in a cultivation room for
2-3 months and later grown plants were transferred to green house, where survival rate
reached high levels of 85% (Figure 3 c). After 6 months, these regenerants were rooted in natural
conditions in the Rhodopes Mountains experimental field where the survival rate was 68%,
after winter has passed. In April, vigorous vegetation was observed with formation of sprouts,
floweres, seeds and rhizomes like plants in their natural environment.

Genetic stability of *in vitro* regenerated plants is very important for micropropagation aiming
production of elite plant material or conservation of the species. Chromosome number in the
root tip cells of *in vitro* regenerats of *Rhodiola rosea* was examined. All the plantlets though
obtained on different media had 22 chromosomes which number was identical with the diploid
chromosome number of 2n = 22 of the wild plant. These results indicate that the regeneration
schemes developed by authors [197] favor stability of the initial caryotype. This fact is very
important for the purposes of restoration of the species and for creating nurseries and fields
of Golden root serving the pharmacological needs.

Another very important fact is the ability of *in vitro* obtained plants to synthetize salidroside
what was confirmed by the analysis of one and two years old regenerants. Salidroside content

in all the samples taken from the roots of regenerants reintroduced in nature was higher than those in plants, which developed from seeds in the mountains [198].

Roots and rhizome from one year old plant regenerants growing in the green house have lower salidroside content compared to the plants growing in the experimental field in the mountains at an altitude of 560 m. However, at the same conditions high levels of rosavin 3.2 % and 3.3 % were detected in green house plants and in mountain plants, respectively (unpublished data). Golden root extracts used in major part of the clinical research are standardized to 3.0 % of rozavins and 0.8% salidroside, which is a ratio of 3:1. This ratio was 10.75:1 in the experiments of the authors (unpublished data) for green house one year old regenerants. Similarly one and three years old regenerants growing in the mountains had higher portion of rozavins compared to salidroside (1 : 8.6 and 1 : 3.75, respectively) which was very positive fact (unpublished data)

Recently replanting of *Rhodiola rosea* regenerants in natural conditions was reported from other authors, too [199] but unlike the previous report [43] reintroduced regenerants differ morphologically. Several types of explants and nutrient media were used to reveal the morphogenic potential suitable for elaborating shemes for micropropagation [199]. The most efficient combinations were when explants from shoot nodes and apices were cultured on MS medium containing 2.0 mg/l NAA, followed by hormone free MS, then KN (1 mg/l kinetin and 0.5 mg/ l NAA), and AZ (0.2 mg/l IAA and 2 mg/l zeatin). The *in vitro* generated neo plantlets reached survival rate over 90% after transfer to septic environment in a hydroponic system for 5-7 days. After acclimatization, the regenerants were potted into soil until the first summer when they were transferred to their native habitat (at 1750 m altitude in Ceahlău Mountains, Romania). During the next summer about 73.5 % of the few dozens of reintroduced regenerants survived. This percentage dropped at 57 % during the third year. It was observed that the *in vitro* regenerants of *Rh. rosea* developing in their natural habitat differed in leaf color (light green), compared to the native individuals of this region (green- grey).

For the first time an original protocol for *in vitro* micropropagation of *Rhodiola rosea* in a RITA bioreactor system was reported [200]. Three clones were obtained from *in vitro* germinated seedlings of wild Finland golden root. Stimulation of organogenesis was studied using thidiazuron and zeatin. Two to four µM thidiazuron stimulated shoot induction but inhibited shoot growth while 1-2 µM zeatin favored shoot growth and leaf number per shoot. Multiplication rate of the clones differed significantly but the most efficient was obtained on solidified medium enriched with 2 µM zeatin. In the bioreactor 0.5 µM thidiazuron maintained rapid shoot proliferation but induced hyperhydracity at higher concentrations. However, hyperhydracity was abolished when shoots were transferred for 4 weeks on gelled medium enriched with 1-2 *µM* zeatin. Shoots formed roots for 5-6 weeks on medium without phytoregulators. Regenerants transferred to soil in the green hause surved at high rate (85–90%) and after acclimatization had normal shoot and leaf morphology.

After establishment of reliable *Rhodiolain vitro* cultures, research has continued for their implementation for practical use like production of valuable secondary metabolites in bioreactors, for biotransformation, for manipulation of the metabolic pathways and metabolic engineering. Biotransformation is a key mechanism to increase production of the biologically

Figure 3. *In vitro* regenerants of *Rhodiola rosea* Bulgarian ecotype: (a) and (b) – propagated plants on MS medium enriched zeatin; (c) – two years old regenerants growing in green house.

active compounds in callus cultures. There are few reports on golden root callus cultures with acompaning analysis of their biologically active metabolites and description of the parameters for their efficient synthesis *in vitro*. The first attempts dated a decade ago [201]. Callus was induced on leaf explants of *Rhodiola rosea* and transferred into MS liquid medium. Thus obtained suspension culture was used to to study the possibility to increase synthesis of rosavin and other cinnamyl glycosides. In the cells for about 3 days, more than 90% of the added transcinnamyl alcohol (optimal concentration of 2.5 mM) was transformed into various unidentified products. However, one of them, 3-phenyl-2-propenyl-O-(6'-O-α'-L-arabinoryranosyl)-β-D-glucopyranoside, found in the intracellular spaces, both of green and yellow strains of cell cultures, was defined as potential rozavin by very precise methods.

Biotransformation was used for increasing of biologically active substances production in callus culture in *Rhodiola rosea*. The effect of different precursors of biologically active substances on the biomass and the metabolite production was studied in *Rhodiola rosea* compact callus aggregates in liquid medium [202, 203]. Cinnamyl alcohol concentrations up to 0.1 mM in media did not bring to a significant deviation from the control; 2 to 5 mM changed slightly callus color from dark to light green. In these cultures rosin content was elevated to 1.25 % dry weight while rosavin was 0.083% dry weight. Cinnamyl alcohol induced synthesis of four new products, too. Tyrosol from 0.05 mM and 2 mM did not influence callus growth while concentrations of 3 mM up to 9 mM caused decrease in biomass production. Two mM of tyrosol were the optimal levels for salidroside production reaching 2.72 % dry weight. Addition of glucose had no positive effect on salidroside accumulation but doubled the rosin production.

Callus tissues cultivated on solid media could produce active substances characteristic for the species [204] *Rh. rosea* Addition of yeast extract in the media doubled salidroside content (from 0.8 % to 1.4) and was twice as high as in five-year-old roots of the intact plants. In the later experiments [205] *Rh. rosea* callus induced from axillary buds or from seedling hypocotyls transformed exogenous cinnamyl alcohol into rosin. However, the biotransformation process was more efficient in the hypocotyl callus where the application of 2.5 mM cinnamyl alcohol resulted in the increase of rosin content up to 1056.183 mg/100 g on solid medium and 776.330 mg/100 g in liquid medium. Callus tissue obtained from axillary buds and treated in the same way produced rosavin in a higher concentration of 92.801 mg/100 g and reached 20% of the amount produced by roots [206].

Krajewska-Patan et al. and György et al. [205, 202, 203] obtained and maintained callus from *Rh. rosea* in liquid medium adding different precursors of the biologically active substances to increase the synthesis of the substances from the main biologically active complex.

The same Bulgarian group successfully established callus cultures, too [207]. Induction of callogenesis was achieved from leaf explants, isolated from *in vitro* propagated plants, on MS media enriched with BAP in concentration from 0.5 mg/l to 2.0 mg/l; 2-iP—0.3 and 3.0 mg/l; 2,4-D—from 0.1 to 2.0 mg/l; IAA—0.2, 0.3 and 1.0 mg/l; NAA—0.5, 1.0, 1.5 mg/l; glutamine—150 mg/l and casein hydrolysate 1000 mg/l. The highest response of 62.85 % and 73.17 % formation of callus was observed on two media, both containing 1 mg/l BAP and either 1 mg/l or 0.5 mg/l 2,4-D (Figure 4 a, b, c, d, e, f, g, h). The authors observed (unpublished data) that when calli were cultured on media with the same phytoregulators as mentioned above but with higher content of sucrose (3 % instead of 2 %) the induction of of callogenesis was several folds lower and variations in callus structure and color were noted. Sucrose concentration influenced synthesis of biologically active substances. Phytochemical analysis revealed that at 2 % sucrose in the medium salidroside and rozavins were not detected in the calli (unpublished data)

Similar investigations were performed with other *Rhodiola* species. *Rh. sachalinesis* calli cultured with 5% sucrose produced high salidrosid content (0.41 % on the basis of dry wt) than normal root (0.17 %) [208]. A compact callus aggregate strain and culturing system for high yield salidroside production was established in *Rhodiola sachalinensis* [209].

Organogenic callus was obtained from leaves with efficiency of 88.33 % [210]. Among the yellow, green, and red colored calli, only green callus formed buds though with poor efficiency. Despite this, regenerated plantlets were rooted on half strength MS medium. Experiments with *Rhodiola sachalinesis* proved that cryopreservation of calli is possible followed by successful recovery of fresh and green tissues for 6 weeks. Isolation of protoplasts was also reported for this species [211].

in vitro cultures were obtained from *Rh. crenulata, Rh. yunnanensis, Rh. fastigata* [212, 213] and *Rh. quadrifida* [214] proving the role and interactions of the explant type, genotype and phytohormones for the efficiency of *in vitro* response and regeneration was also function of the genotype and the phytohormones. The authors underlined the role of 2,4-D and BA for production of biologically active substances. Similar observations about the role of the explant,

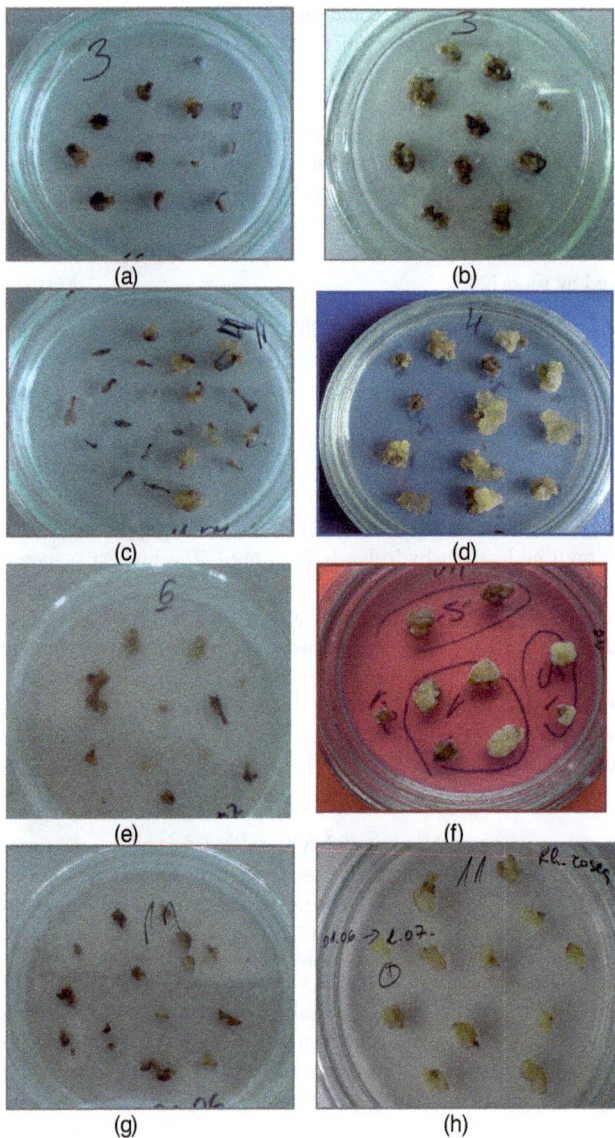

Figure 4. Various callus cultures induced on MS basal medium enriched with: (a) – BAP (1 mg/l), 2,4-D (1 mg/l) and 3% sucrose; (b) - BAP (1 mg/l), 2,4-D (1 mg/l) and 2% sucrose; (c) – BAP (1 mg/l), 2,4-D (0.5 mg/l) and 3% sucrose; (d) – BAP (1 mg/l), 2,4-D (0.5 mg/l) and 2% sucrose; (e) – BAP (1 mg/l), 2,4-D – 1 mg/l, Casein hydrolysate 1000 mg/l and 3 % sucrose; (f) – BAP (1 mg/l), 2,4-D – 1 mg/l, Casein hydrolysate 1000 mg/l and 2 % sucrose (g) – BAP (1 mg/l), NAA (0.5 mg/l), Casein hydrolysate 1000 mg/l and 3% sucrose; (h) BAP (1 mg/l), NAA (0.5 mg/l), Casein hydrolysate 1000 mg/l and 3% sucrose;

the temperature of cultivation and the pretreatment duration on salidroside synthesis in *Rhodiola kirilowii* callus were made by others [215].

Genetic transformation opens new perspectives for production of biologically active compounds. Hairy roots induced by *Agrobacterium rhizogenes* grow faster accumulating greater biological material. Genetic transformation of *Rhodiola sachalinensis* was performed with *Agrobacterium rhizogenes* [216, 217]. The authors studied conditions for high salidroside production (the major compounds from the roots of *Rhodiola sachalinensis*) when precursors (tyrosol, tyrosine, and phenylalanine) and elicitors (*Aspergillus niger, Coriolus versicolor,* and *Ganoderma lucidum*) were added into the medium. For high salidroside production, the optimal light intensity, pH value and nitrogen levels were determined, too. The optimal concentration for the elicitor was 0.05 mg/l while the optimal concentration of the precursor was 1 mmol/l. The 1000 lx scatter light, pH 4.5 - 4.8, and nitrogen (NH_4^+: NO_3^- =1:1) concentration of 80 mmol/l were the optimal condition for salidrosid production. Authors conclude that hairy roots can be used as alternative material for the production of secondary metabolites of pharmaceutical value in *Rhodiola*.

Examples, given here, though covering a small part of the enormous and tedious work on medicinal plants, and more particularly on representatives of the genera of *Gentiana, Leucojum* and *Rhodiola,* which are protected in Bulgaria, could give impression on the potential of different spheres of plant biotechnology (Table 1). The most promising ones being *in vitro* clonal propagation of endangered species to create *in vitro* and *ex situ* collections, and for obtaining of raw material and valuable compounds (Figure 5).

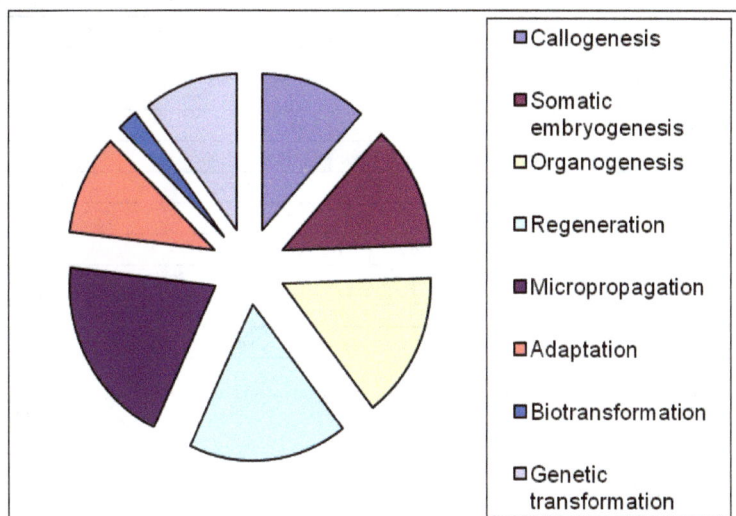

Figure 5. Relative share of the achievements in different spheres of biotechnology in the three genera: *Gentiana, Rhodiola, Leucojum.*

Plant Species	Calluso-genesis	Somatic embryo-genesis	Organo-genesis	Regene-ration	Microprop agation	Adapta-tion	Biotrans-forma-tion	Genetic transfor-mation
G. lutea	yes	yes	Yes	yes	yes	yes		yes
G. kurroo	yes	yes	Yes	yes	yes	yes		yes
G. cruciata		yes	Yes	yes	yes			yes
G. pannonica		yes						
G. punctata			Yes	yes	yes	yes		yes
G. straminea Maxim.		yes						
G. crassicaulis,		yes						
G. dinarica Beck,			Yes	yes	yes	yes		
G. corymbifera,			Yes	yes	yes			
G. pneumonanthe		yes			yes	yes		
G. purpurea,		yes	Yes	yes				yes
G. davidii var. Formosana		yes						
G. scabra	yes			yes				
G. acaulis			Yes	yes				yes
G. tibetica		yes						
G. dahuria			Yes	yes	yes	yes		yes
G. triflora					yes			
G. ligularia					yes			
G. cerina					yes	yes		
G. asclepiadea			Yes	yes				
Rh. rosea	yes		Yes	yes	yes	yes	yes	
Rh. crenulata					yes			
Rh. kirilowii	yes							
Rh. quadrifida	yes							
Rh. sachalinensis	yes				yes			yes
Rh. yunnanensis					yes			
Rh. iremelica	yes		Yes	yes				
Rh. fastigata					yes			
Rhodiola coccinea	yes			yes				
Leucojum aestivum	yes	yes	Yes	yes	yes	yes	yes	yes

Table 1. Examples of biotechnological achievements in *Gentiana*, *Rhodiola* and *Leucojum* species.

9. Conclusions

Presented data and results in this chapter aimed at enlightening the potential of plant bio-technologies in protection of valuable plant species, including the medicinal ones, which have become rare or are close to extinction as a result of the intensive industrialization, urban economy and climatic changes. One of the measures for overcoming this global problem could be the cultivation of valuable medicinal plants in experimental conditions. For this purpose along with the traditional methods for cultivation fields and nurseries, "green" biotechnologies can be used. Many scientists have realized that plant biotechnology is an important tool for multiplication and conservation of the endangered and rare populations of medicinal plants. Using environmental friendly *in vitro* technologies a great number of identical plants, can be propagated, regenerated and transferred back in nature thus restoring and expanding wild habitats. From another hand, the areas of the medicinal plants will be less subjected to vulnerable exploitation if the valuable raw material could be obtained by alternative means. In this sense by micropropagation of plants, enormous amounts of biomass can be produced continuously and/or for short period of time. In addition, production of biologically active substances in laboratory conditions contributes to less utilization of the natural resources and thus protecting the species. The fact that *in vitro* cultures, cells, tissues, organs and plantlets can produce metabolites, specific for the intact donor plant, is of tremendous importance for production of desired compounds. Development of more sophisticated instrumentation and original approaches allowing biotransformation and metabolic engineering is a revolutionary step for high technological production of valuable substances and biologically active compounds demanded from the food, nutraceutical, pharmaceutical and cosmetic industries.

Nomenclature

MS – Murashige and Skoog medium, 1962; BAP – N^6-benzylaminopurine; IAA – Indolyl-3-acetic acid; 2-iP – 6-(y,y-dimethylallyl amino) purine; 2,4-D – 2,4- dichlorophenoxyacetic acid; NAA - α- naphthyl acetic acid; TDZ – Thidiazuron; Kin – Kinetin; GA_3 – Gibberellic acid; IBA – Indole 3-butyric acid

Acknowledgements

Research was supported by National Science Fund of Bulgaria—Project for Junior Scientists DMU 03/55 (leader Dr. K. Tasheva).

Author details

Krasimira Tasheva* and Georgina Kosturkova

*Address all correspondence to: krasitasheva@yahoo.com

Regulation of Plant Growth and Development Department, Institute of Plant Physiology and Genetics, Bulgarian Academy of Sciences, Sofia, Bulgaria

References

[1] Report of the European Commission, 2008.

[2] Korver O. Functional foods: the food industry and functional foods: some European perspectives. In: Shibamoto T., Terao J., Osawa T. (eds.) Functional Foods for Disease Prevention: II. Medicinal Plants and Other Foods. (American Chemical Society, Washington, DC), 1998; p 22–25

[3] Donald P. Briskin. Medicinal Plants and Phytomedicines. Linking Plant Biochemistry and Physiology to Human Health. Plant Physiology 2000; 124 507-514

[4] Matthys K., Julsing, Wim J. Quax, Oliver Kayser. The Engineering of medicinal plants: Prospects and limitations of medicial plant biotechnology. In: Oliver Kayser and Wim J. Quax. (eds) Medicinal Plant Biotechnology from basic Research to Industrial Applications. WILEY-VCH Vergal GmbH & Co. KGaA, Weinheim., 2007; p. 1-8.

[5] World Health Organization. National policy on traditional medicine and regulation of herbal medicines, Report of a WHO global survey, Geneva, Switzerland, http://apps.who.int/medicinedocs/pdf/s7916e/s7916e.pdf (accessed May 2005).

[6] Lange D. 1998, Europe's Medicinal and Aromatic Plants: Their use, trade and conservation, IUCN, A Traffic Network Report. p. 77

[7] Lange D. The role of East and Southeast Europe in the medicinal and aromatic plants trade. Medicinal Plant Conservation 2002;8 14-18.

[8] Lange D. Medicinal and Aromatic Plants: Trade, Production, and Management of Botanical Resources Proc. XXVI IHC – Future for Medicinal and Aromatic Plants. Acta Hort. 2004;629 177-197

[9] Lange D. Chapter 11: International trade in medicinal and aromatic plants. In: R.J. Bogers, L.E. Craker and D. Lange (ed.) Medicinal and Aromatic Plants. Springer: Printed in the Netherlands, p. 155 – 170.

[10] Glaser V. Billion-dollar market blossoms as botanicals take root. Nat Biotechnol 1999;17(1) 17-18

[11] Inamul Haq. Safety of medicinal plants. Pakistan Journal of Med. Res. 2004;43(4) 203-210.

[12] World Health Organization. http://www.who.int/mediacentre/factsheets/fs-134/en/. (2010).

[13] World Health Organization. The World Medicines Situation 2011, *Traditionl medicines: Global situation, issues and challenges*, http://www.who.int/medicines/areas/policy/world medicines situation/ WMS ch18wTraditionalMed.pdf. (Geneva, Switzerland, 3rd edition, 2011).

[14] Edwards R. No remedy in sight for herbal ransack. New Science, 2004;181 10–11

[15] Vines G. 2004. Herbal harvests with a future: towards sustainable sources for medicinal plants, Plantlife International; www.plantlife.org.uk

[16] Peter H.C., Thomas H., Ernst E. Bringing medicinal plants into cultivation: Opportunities and challenges for Biotechnol 2005; 297 1-5.

[17] Kozuharova E. New Ex Situ Collection of Rare and Threatened Medicinal Plants in the Pirin Mts. (Bulgaria). Ekoloji 2009;18(72) 32-44.

[18] Petkov, V. Modern Phytotherapy. Sofia: Publ. Medicina i Fizkultura; 1982

[19] Hardalova, R., L. Evstatieva, Gusev Ch. 1994. Wild medicinal plant resources in Bulgaria and recommendation for their long-term development. In Meine C. (ed.) Bulgaria's biological diversity:conservation status and needs assesssment Sofia: Pensoft; 1994. p527-561

[20] Stoeva, T. 2000. Cultivation of Medicinal and Essential Oil Plants in Bulgaria – Traditions and Prospects - In: Sekulovic D., Maksimovic S., Kisgeci J. (eds.): proceedings of the First Conference on Medicinal and Aromatic Plants of Southeast European Countries & VI Meeting "Days of Medicinal Plants 2000", May 29-June 3, 2000, Arandelovac, Yugoslavia.

[21] Varabanova K. Medicinal and aromatic plant diversity in Bulgaria – protection, collection, study, use and conservation. Report of a working group on medicinal and aromatic plants. First Meeting, 12-14 September, 2002, Gozd Martuljek, Slovenia.

[22] Evstatieva L., Hardalova R., Stoyanova K. Medicinal plants in Bulgaria: diversity, legislation, conservation and trade. Phytologia Balcanica 2007;13(3) 415–427.

[23] Nedelcheva A. Traditional knowledge and modern trends for Asian medicinal plants in Bulgaria from an ethnobotanical view. EurAsian Journal of BioSciences 2012;6 60-69.

[24] Stoeva, T. Traditional medicine and medicinal plant use in Bulgaria. 3rd Conference on Medicinal and Aromatic Plants of Southeast European Countries: conference proceedings, September 5 – 8, 2004, Nitra, Slovak Republic

[25] Mladenova M. Bulgaria—the most exporter of medicinal plants in Europe: proceedings of the International Interdisciplinary Conference on Medicinal Plants—Solution 2000, June, 1999, Sofia, Bulgaria.

[26] Vitkova A., Evstatieva L. Spread and resources of medicinal plants in NP 'Rila. proceedings of the International InterdisciplinaryConference on Medicinal plants—Solution,June, 1999, Sofia, Bulgaria.

[27] Evstatieva L, Hardalova R. Conservation and sustainable use of medicinal plants in Bulgaria. Medicinal Plant Conservation 2004;9(10) 24-28.

[28] Evstatieva L. A review of the cultivation of endangered medicinal plants in Bulgaria. Annuire de l'Universite de Sofia "St. Kl. Ohridski" Faculte de Biologie 2006;2(97) 45-52.

[29] Kathe W. 2006. Chapter 14: Conservation of Eastern-European medicinal plants: Arnica montana in Romania. In: R.J. Bogers, L. E. Craker and D. Lange (eds.) Medicinal and Aromatic Plants, Netherlands: Springer, 2006. p.203 – 211.

[30] Food and Agriculture Organization (FAO). http://www.fao.org/biodiversity/2010-international-year-of-biodiversity/en/. (2010).

[31] Purohit S.D., Dave A., Kukda G. Micropropagation of safed mulsi (*Chlorophytum borivilianum*), a rare medicinal herb. Plant Cell Tissue Organ Culture 1994;39 93-96.

[32] Sudha C.G., Seeni S. *In vitro* propagation of *Rauwolfia micrantha*, a rare medicinal plant. Plant Cell Tissue and Organ Culture 1996;44(3) 243 – 248.

[33] Khan Mohamed Yassen, Saleh Aliabbas, Vimal Kumar, Shalini Rajkumar. Resent advances in medcinal plant biotehology. Indian Journal of Biotechnology 2009;8 9–22.

[34] Tasheva K., Kosturkova G. The role of biotechnology for conservation and biologically active substances production of *Rhodiola rosea* – endangered medicinal species. *The Scientific World Journal* 2012;2012 13pages,

[35] Butenko R.G. Cell biology of higher plants in vitro and biotechnology. Moskva: FBK–PRESS 160, 1999. (in Russian)

[36] Verpoorte R. Biotechnology and its role in pharmacognosy. 136-th Brit. Pharm. Conf., Proc. J. Pharm and Pharmacol., September 13-16, 1999, Cardiff, Wales, UK.

[37] Tripathi L., Tripathi J. N. Role of biotechnology in medeicnal plants. Tropical Journal of Pharmaceutical Research 2003;2(2) 243–253.

[38] Stanilova M., Ilcheva V., Zagorska N. Morphogenetic potential and *in vitro* micropropagation of endangered plant species *Leucojum aestivum*L. and *Lilium rhodopaeum* Delip. Plant Cell Reports 1994;13 451-453.

[39] Berkov S., Pavlov A., Ilieva M., Burrus M., Popov S., Stanilova M. CGC-MS of alkaloids in *Leucojum aestivum* plants and their *in vitro* cultures. Phytochemical Analysis 2005;16(2) 98-103.

[40] Petrova M., Zagorska N., Tasheva K., Evstatieva L. *In vitro* propagation of *Gentiana lutea* L. Genetics and Breeding 2006;35(1-2) 63-68.

[41] Atanassov A., Batchvarova R., Djilianov D. Strategic vision for plant biotechnology and genomics development. Biotechnology&Biotechnological Equipment 2007;21(1) 1-7. Thesis

[42] Jonkova I. Pharmaceutically important biologically active substances from sources wit optimization phytochemical potencial. DSc thesis. Sofia, Medical University Sofia, 2007 (in Bulgarian)

[43] Tasheva K., Kosturkova G. Bulgarian golden root *in vitro* cultures for micropropagation and reintroduction. Central European Journal of Biology, 2010;5(6) 853–863.

[44] Gorgorov R., Stanilova M., Vitkova A. *In vitro* cultivation of some endemic and rare *Alchemilla* species in Bulgaria. ACRomanian Biotechnological Letters 2011;16(6) 65 – 70.

[45] Rao Ramachandra S., Ravinshankar G.A. Plant cell cultures: Chemical factories of secondary metabolites. Biotechnology Advances 2002;20 101–153.

[46] Misawa M. Plant tissue culture: an alternative for production of useful metabolite. FAO Agricultural Services Bulletin No. 108. Roma, Italy: Food and Agriculture Organization of the United Nations. 1994.

[47] Verpoorte R., Contin A., Memelink J. Biotechnology for the production of plant secondary metabolites. Phytochem Rev. 2002;1 13–25.

[48] Julsing K.M., Wim J. Quax, Kayser O. 2006. The Engineering of Medicinal Plants: Prospects and Limitations of Medicinal Plant Biotechnology. In: Oliver Kayser, WIm J. Quax (eds) Medicinal Plant Biotechnology: From Basic Research to Industrial Applications, 2006

[49] Griga M, Kosturkova G, Kuchuk N, Ilieva-Stoilova M., 2001. Biotechnology. In: Hedley C.L. (ed) Carbohydrates in Grain Legume Seeds. Improving Nutritional Quality and Agronomic Characteristics. Wallingford, UK, CAB International, 2001. p. 145–207.

[50] Altman A. Plant biotechnology in the 21st century: the challenges ahead. Electronic Journal of Biotechnology 1999;2(2) 51–55.

[51] Sajc L., Grubisic D., Vunjak-Novakovic G. Bioreactors for plant engineering: an outlook for further research. Biochemical Engineering Journal 2000;4(2) 89–99.

[52] Zhu W., Lockwood G.B. Biotransformation of volatile constituents using plant cell cultures: a review. In: Singh S., Govil J.N., Singh V.K. (eds.) Recent Progress in Medicinal Plants, vol 2: Phytochemystry and pharmacology, 2003, p.307 – 319.

[53] Anming Wang, Fangkai Zhang, Lifeng Huang, Xiaopu Yin, Haifeng Li, Qiuyan Wang, Zhaowu Zeng, Tian Xie. New progress in biocatalysis and biotransformation of Flavonoids. Journal of Medicinal Plants Research 2010;4(10) 847-856.

[54] Vanisree Mulabagal, Hsin-Sheng Tsay. Plant Cell Cultures - An Alternative and Efficient Source for the Production of Biologically Important Secondary Metabolites. International Journal of Applied Science and Engineering 2004;2(1) 29-48

[55] Charlwood Barry V., Marcia Pletsch. Manipulation of natural product accumulation in plants through genetic engineering. Journal of Herbs, Spices & Medicinal Plants 2002;9 (2-3) 139–151.

[56] Chilton M.D., Tepfer D.A., Petit A., David C., Casse-Delbart F., Tempé J. *Agrobacterium rhizogenes* inserts T-DNA into the genome of the host plant root cells. Nature 1982;295 432-434.

[57] Archana Giri, Lakshmi Narasu M. Transgenic hairy roots: recent trends and applications. Biotechnology Advances 2000;18 1–22

[58] Sevon N., Oksman-Caldentey K.M. *Agrobacterium rhizogenes* - mediated transformation: root cultures as a source of alkaloids. Planta Medica 2002;68(10) 859-868.

[59] Terryn N., Van Montagu M., Inze D., Goossens A. Chapter 21: Functional genomic approaches to study and engineer secondary metabolism in plant cell cultures. In: Bogers R.J., Craker L.E., Lange D. (eds) Medicinal and Aromatic Plants, 2006, p. 291-300.

[60] Guillon S., Tremouillaux-Guiller J., Pati P.K., Rideau M., Gantet P. Hairy root research: recent scenario and exciting prospects. Curr. Opin. Plant Biology 2006;9 341–346.

[61] McCown B.H., McCown D. D. Workshop on micropropagation; A general approach for developing a commercial micropropagation system. *In Vitro* Cell and Dev. Biol. – Plant 1998;35(4) 276–277.

[62] Rout G.R., Samantaray S., Das P. *In vitro* manipulation and propagation of medicinal plants. Biotechnology Advances 2000;18 91-120.

[63] Henry Y., Vain P., Buyser J. Genetic analysis of *in vitro* plant tissue culture responses and regeneration capacities. Euphytica 1994;79(1-2) 45–58.

[64] Smith S.M., Street H.E. The decline of embriogenic potential as callus and suspension cultures of carrot are serially subculture. Annals of Botany 1974;38 223–241.

[65] Morgan E.R., Butler R.M., Bicknell R.A. *In vitro* propagation of *Gentiana cerina* and *Gentiana corymbifera*. New Zealand Journal of Crop and Horticultural Science 1997;25 1–8.

[66] Momcilovic I., Grubisik D., Neskovic M. Micropropagation of four *Gentiana species* (*G. lutea, G. cruciata, G. purpurea and G. acaulis*). Plant Cell, Tissue and Organ Culture 1997a;49 141-144.

[67] Migranova I.G., Leonova I.N., Salina E.A., Churaev R.N., Mardamshin A.G. Influence of the genome and of the explants tissue type to possibilities callus tissue to long-term cultivation *in vitro*. Biotechnology 2002;2 37–41(in Russian)

[68] Kosturkova G.P., Mehandjiev A.D., Dobreva I., Tsvetkova V. Regeneration systems from immature embryos of Bulgarian pea genotypes. Plant Cell, Tissue and Organ Cultures 1997;48(2) 139-142.

[69] Masaru Nakano, Miho Nagai, Sgigefumi Tanaka, Masashi Nakata, Toshinari Godo. Adventitious shoot regeneration and micropropagation of the Japanese endangered *Hylotephium sieboldii* (Sweet ex Hook) *H. Ohba* and *H. sieboldii* var. ettyuense (Tomida) H. Ohba. Plant Biotechnology 2005;22(3) 221–224.

[70] Mohammed Shafi Ullah Bhuiyan, Tehryung Kim, Jun Gyo In, Deok Chun Yang, Kwan Sam Choi. Plant regeneration from leaf explants of kalanchoe daigremontiana Hamet & Perrier. Korean J. Medcinal Crop Science 2006;14(5) 293–298.

[71] Avksentyeva O.A., Petrenko V.A., Tishchenko A.A., Zhmurko V.V. Callus initiation and morphogenesis in *in vitro* culture of isogenic on gene type and rate of development in winter wheat lines. Annual Wheat News letter 2007;54 150–152.

[72] Schulz J. Improvements in Cereal Tissue Culture by Thidiazuron: A Review. Fruit, Vegetable and Cereal Science and Biotechnology 2007;1(2) 64–79.

[73] Karuppusamy S. A review on trends in production of secondary metabolites from higher plants by *in vitro* tissue, organ and cell cultures. J. of Medicinal Plants Research 2009;3(13) 1222-1239.

[74] Murashige, T., Skoog F. A revised medium for rapid growth and bioassays with tobacco tissue cultures. Physiol Plant 1962;15 473-497.

[75] Gamborg O.L., Miller R.A., Ojima K. Nutrient requirements of suspension cultures of soybean root cells. Exp Cell Res 1968;50(1) 151–158.

[76] Gamborg O.L., Murashige T., Thorpe T.A., Vasil I.K. Plant-tissue culture media. Journal of the Tissue Culture Association 1976;12 473-478.

[77] Huang L., Murashige T. Plant tissue culture media: major constituents; their preparation and some applications. Tissue Culture Assoc 1977;3 539-548.

[78] Murashige T. Plant propagation through tissue cultures. Annual Review of Plant Physiology 1974;25 135-166.

[79] Choi Y. E., Yang D.C., Yoon E.S., Choi K. T. Plant regeneration via adventitions buds formation from cotyledon explants of *Panax ginseng*. Plant Cell reports 1998;17(9) 731–736.

[80] Zhao Wen Jun, Wang Yi, Jiang ShiCui, Xu Yuan, Sun ChunYu, Zhang MeiPing. Establishment and optimization of *in vitro* regeneration system for *Panax ginseng*. Journal of Jilin Agricultural University 2009;31(1) 41–44.

[81] Zhihua Liao, Min Chen, Feng Tan, Xiaofen Sun & Kexuan Tang. Microprogagation of endangered *Chinese aloe*. Plant Cell, Tissue and Organ Culture 2004;76(1) 83–86.

[82] Satish M. Nalawade, Abhay P. Sagare, Chen-Yue Lee, Chao-Lin Kao and Hsin-Sheng Tsay. Studies on tissue culture of Chines medicinal plant resources in Taiwan and their sustainable utilization. Bot. Bull. Acad. Sin 2003;44 79–98.

[83] Sharma U., Mohan J.S. *In vitro* clonal propagation of *Chlorophytum borivilianum* Sant. et Fernand., a rare medicinal herb from immature floral buds along with inflorescence axis. Indian J Exp Biology 2006;44(1) 77-82.

[84] Rizvi Zahid Mohd., Arun Kumar Kukreja, Suman Preet Singh Khanuja. *In vitro* culture of *Chlorophytum borivilianum* Sant. Et Fernand in liquid culture medium as a cost-effective measure. Current Science 2007;92(1) 87–90.

[85] Nurashikin Kemat, Mihdzar Abdul Kadir, Nur Ashikin Psyquay Abdullah and Farshad Ashraf. Rapid multiplication of Safed musli (*Chlorophytum borivilianum*) through shoot proliferation. African Journal of Biotechnology 2010;9(29) 4595-4600.

[86] Kumar A., Aggarwal D., Gupta P., Reddy M.S. Factors affecting *in vitro* propagation and field establishment of *Chlorophytum borivilianum*. Biologia plantarum 2010;54(4) 601 – 606.

[87] Mohammed Faisal and Mohammad Anis. Rapid mass propagation of *Tylophora indica* Merrill via leaf callus culture. Plant Cell, Tissue and Organ Culture 2003;75(2) 125-129.

[88] Mohammed Faisal, Naseem Ahmad and Mohammad Anis. An efficient micropropagation system for *Tylophora indica*: an endangered, medicinally important plant. Plant Biotechnology Reports 2007;1(3) 155-161.

[89] Taha H.S., El-Bahr M.K., Seif-El-Nasr M.M. *In vitro* studies on Egyptian *Catharanthus Roseus* (L.) G. Don.:1-calli production, direct shootlets regeneration and alkaloids determination. J. Appl. Sci. Res 2008;4(8) 1017–1022.

[90] Geetha S. Pillai, Raghu A.V., Gerald M., Satheesh G., Balachandran I., 2009. *In Vitro* Propagation of Two Tuberous Medicinal Plants: *Holostemma ada-kodien* and *Ipomoea mauritiana*. Protocols for *In Vitro* Cultures and Secondary Metabolite Analysis of Aromatic and Medicinal Plants, Methods in Molecular Biology 2009;547(1) 81-92

[91] Bin Guo, Min Gao, and Chun-Zhao Liu. *In vitro* propagation of an endangered medicinal plant *Saussurea involucrata* Kar. et Kir. Plant Cell Reports 2007;26(3) 261–265.

[92] McCarten S. A., Van Staden J. Micropropagation of the endangered *Kniphofia leucocephala* Baijnath. *In vitro* Cell Development Biology – Plant 2003;39(5) 496-499.

[93] Nadeem M., L. M. S. Palni, A. N. Purohit, H. Pandey and S. K. Nandi. Propagation and conservation of *Podophyllum hexandrum* Royle: an important medicinal herb. Biological Conservation 2000;92(1) 121–129

[94] Joshi M. and U. Dhar. *In vitro* propagation of *Saussurea obvallata* (DC.) Edgew. – an endangered ethnoreligious medicinal herb of Himalaya. Plant Cell Reports 2003;21(10) 933–939.

[95] Beena M.R., Martin K.P., Kirti P.B., Hariharam M. Rapid *in vitro* propagation of medicinally important *Ceropegia candelabrum*. Plant Cell, Tissue and Organ Culture 2003;72(3) 285–289.

[96] Khan P.S. ShaValli, Hausman J.F., Rao K.R. Clonal multiplication of *Syzigium alternifolium* (Wight.) Walp., through mature nodal segments. Silval. Genet 1999;48(1) 45-50.

[97] Lattoo S.K., Bamotra S., Sapru Dhar R., Khan S., Dhar A.K. Rapid plant regeneration and analysis of genetic fidelity of *in vitro* derived plants of *Chlorophytum arundinaceum* Baker—an endangered medicinal herb. Plant Cell Reports 2006;25(6) 499-506.

[98] Martin K.P. Rapid *in vitro* multiplication and *ex vitro* rooting of *Rotula aquatica*. Lour., a rare rhoeophytic woody medicinal plant. Plant Cell Reports 2002;21(5) 415–420.

[99] Martin K.P. Rapid *in vitro* multiplication and *ex vitro* rooting of *Rotula aquatica* Lour., a rare rhoeophytic woody medicinal plant. Plant Cell Rep 2003;21(5) 415–420

[100] Nishihara M., Nakatsuka T., Mizutani-Fukuchi M., Tanaka Y., Yamamura S. Gentians: from gene cloning to molecular breeding. In: Jaime A. Teixeira da Silva (ed) Part 2 Cut flowers and flower colourFloriculture, Ornamental and Plant Biotechnology, Advances and Topical Issues, first edition, volume V, Global Science Books, Ltd., 2008. p.57-67

[101] Jensen, S. R., Schripsema J. Chemotaxonomy and pharmacology of Gentianaceae. In Struwe, L., Albert V. (eds) Gentianaceae Systematics and Natural History. Cambridge University Press:2002. p.573-631.

[102] Raina R., Behera M.C., Chand R., Sharma Y. Reproductive biology of *Gentiana kurroo* Royle. Current Science 2003;85(5) 667–670.

[103] Tao He, Lina Yang and Zhigang Zhao. Embryogenesis of *Gentiana straminea* and assessment of genetic stability of regenerated plants using inter simple sequence repeat (ISSR) marker. African Journal of Biotechnology 2011;10(39) 7604-7610.

[104] Butiuc-Keul A., Şuteu A., Deliu C.. *In vitro* organogenesis of *Gentiana punctata*. Not. Bot. Hort. Agrobot XXXIII/2005, 2005;33(1) 38–41

[105] Wesolowska, M., Skrzypczak, L., Dudzinska, R.: Rodzaj *Gentiana* L. w kulturze *in vitro*. Acta Pol. Pharm 1985;42(1) 79-83

[106] Yamada Y., Shoyama Y., Nishioka I., Kohda H., Namera A., Okamoto T. Clonal micropropagation of *Gentiana scabra* BUNGE var, buergeri Maxim and examination of the homogeneity concerning the gentiopicroside content. Chem Pharm Bulletin 1991;39(1) 204–206

[107] Sharma N., Chandel K.P.S., Paul A. *In vitro* propagation of *Gentiana kurroo*: an indigenous threatened plant of medicinal importance. Plant Cell Tissue Organ Culture 1993;34(3) 307–309.

[108] Vinterhalter B., Vinterhalter D. *In vitro* propagation of spotted gentian *Gentiana punctata* L. Arch Biol Sci 1998;50(3) 177–182.

[109] Zhang Z., Leung D.W.M. A comparision of *in vitro* and *in vivo* flowering in *Gentian*. Plant Cell, Tissue and Organ Culture 2000;63(3) 223-226.

[110]] Zhang Z. and Leung D.W.M. Factors influencing the growth of micropropagated shoots and in vitro flowering of gentian. Plant Growth Regul 2002;36(3) 245–251.

[111] Liu S.L., Qi H.Y., Qi H., Zhang M., Wang Z.T. Species of ligularia in the northwestern China and their medicinal uses. China J Chin Mater Med 2006;31(10) 780–797.

[112] Zeleznik A., Baricevic D., Vodnik D. Micropropagation and acclimatization of yellow gentian *(Gentiana lutea* L.). Zbornik Biotehniske fakultete Univerze v Ljubljani 2002;79(1) 253–259.

[113] Cao J.P., Liu X., Hao J.G., Zhang X.Q. Tissue culture and plantlet regeneration of *Gentiana macrophylla*. Bot Boreali Occident Sin 2005;25(6) 1101–1106.

[114] Fiuk A., Rybczynski J.J. Morphogenic capability of *Gentiana kurroo* Royle seedling and leaf explants. Acta Physiol Plantarum 2008a;30(2) 157–166.

[115] Vinterhalter B., Krstić Milošević D., Janković T., Milojević J., Vinterhalter D. *In vitro* propagation of *Gentiana dinarica* Beck. Central European Journal of Biology 2012;7(4) 690–697.

[116] Bach A., Pawłowska B. Somatic embryogenesis in *Gentiana pneumonanthe* L. Acta Biologica Cracoviensia - Botanica 2003;45(2) 79–86

[117] Wen Wei, Yang Ji. Study on the tissue culture and propagation system of *Gentiana scabra* Bunge. Medicinal Plant 2010;1(4) 13–15.

[118] Sadiye Hayta, Ismail Hakki Akgun, Markus Ganzera, Erdal Bedir, Aynur Gurel. Shoot proliferation and HPLC – determination of iridoid in clones of *Gentiana cruciata* L. Plant Cell, Tissue and Organ Culture 2011;107(1) 175–180.

[119] Mencovic N., Savikin-Feduluvic K., Momcilovic I., Grubisic D. Qantitative Determination of Secoiridoid and gamma-Pyrone Compounds in *Gentiana lutea* Cultured *in vitro*. Planta Medica 2000;66(1) 96-98.

[120] Dević M., Momcilovic I., Kristic D., Maksimovic V., Konjevic R. *In vitro* multiplication of willow gentian (*Gentiana asclepiadea* L.) and the production of gentiopicrine and mangiferin. Phyton 2006;46(1) 45-54.

[121] Kaur R., Neelam Panwar, Brawna Saxena, Raina R., Bharadwaj S.V. 2009. Genetic stability in long-term micropropagation plants of *Gentiana kurroo* – an endangered medicinal plant. Journal of New Seeds 2009;10(4) 236–244.

[122] Holobiuc I., Catana R. Recurrent somatic embryogenesis in long term cultures of *Gentiana lutea* as a source for synthetic seed production for medium term preservation. Arch. Biol. Sci., Belgrade, 2012;64(2) 809–917.

[123] Fiuk A., Rybczyn´ski J.J. Genotype and plant growth regulator-dependent response of somatic embryogenesis from *Gentiana* spp. leaf explants. *In Vitro* Cellular & Developmental Biology-Plant 2008b;44(2) 90–99.

[124] Mikula A., Tykarska T., Kuras M., Rybczynski J. Somatic embryogenesis of *Gentiana cruciata* L.: Histological and ultrastructural changes in seedling hypocotyls explant. *In vitro* cell & Dev Biology – Plant 2005;41(5) 686–694

[125] Fiuk A., Rybczyn´ski J.J. Factors influencing efficiency of somatic embryogenesis of *Gentiana kurroo* (Royle) cell suspension. Plant Biotechnol Rep. 2008c;2(1) 33–39

[126] Fiuk A., Rybczyn´ski J.J. The effect of several factors on somatic embryogenesis and plant regeneration in protoplast cultures *of Gentiana kurroo* (Royle). Plant Cell Tiss Organ Culture 2007;91(3) 263–271

[127] Mikula A, Rybczynski J.J. Somatic embryogenesis of *Gentiana* genus I: the effect of the preculture treatment and primary explant origin on somatic embryogenesis of *Gentiana cruciata* (L.), *G. pannonica* (Scop.), and *G. tibetica* (King). Acta Physiol Plantarum 2001;23(1) 15–25.

[128] Fu-Shin Chueh, Chung-Chuan Chen, Hsin-Sheng Tsay. Studies on Factors Affecting the Establishment of *Gentiana davidii* var. *formosana* (Hayata) T. N. Ho Cell Suspension Cultures. Journal of Food and Drug Analysis 2000;8(4) 297-303.

[129] Cai YunFei, Liu YanLing, Liu ZhenHua, Zhang Feng, Xiang FengNing, Xia Guang-Min. High-frequencyembryogenesis and regeneration of plants with high content of gentiopicroside from the Chinese medicinal plant *Gentiana straminea* Maxim. *In vitro* Cell & Dev Biology – Plant 2009;45(6) 730–739.

[130] Rybczynski J.J., Mikula A., Fiuk A. Endangered species – model plants for experimental botany and biotechnology. Bulletin of Botanical Gardens 2004;13 59–63.

[131] Rybczynski J.J., Borkowska B., Fiuk A., Gawronska H., Sliwinska E., Mikuła A. Effect of sucrose concentration on photosynthetic activity of *in vitro* culture *Gentiana kurroo* (Royle) germlings. Acta Phys Plantarum 2007;29(5) 445–453.

[132] Mikula A., Olasa M., Sliwinska E., Rybczynski J.J. Cryopreservation by encapsulation of *Gentiana* spp. cell suspension maintains regrowth, embryogenic competence and DNA content. Cryo Letters 2008;29(5) 409-418.

[133] Mikula A., Tomiczak K., Wojcik A., Rybczynski J.J. Encapsulation – dehydration metod elevates embryogenic abilities of *Gentiana kuroo* cell suspension and carrying on

genetic stability of its regenerants after cryopreservation. Acta Horticulturae 2011;908 143–154.

[134] Momcilovic I., Grubisic D., Kojic M., Neskovic M., *Agrobacterium rhizogenes*-mediated transformation and plant regeneration of four *Gentiana* species. Plant Cell, Tissue and Organ Culture 1997b;50(1) 1-6.

[135] Budimir S., Janosevic D., Momcilovic I., Grubisic D. Morphology and anatomy of *Gentiana lutea* hairy roots. Archives of Biological Sciences 1998;50(2) 99-104.

[136] Vinterhalter B.S., Momcilovic I.D., Vinterhalter D.V. Kultura korenova *Gentiana punctata* L. transformisanih pomoću *Agrobacterium tumefaciens* C58Cl(pArA4b). Archives of Biological Sciences 2000;52(2) 83-87.

[137] Shao Bo Sun, Lai Sheng Meng. Genetic transformation of *Gentiana dahurica* Fisch by *Agrobacterium tumefaciens* using zygotic embryo – derived callus. Acta Physiologiae Plantarum 2010;32(4) 629–634.

[138] Parolo G., Abeli T., Rossi G., Dowgiallo G., Matthies D. Biological flora of Central Europe: *Leucoujum aestivum* L., Perspectives in Plant ecology, Evolution and Systematics 2011;13(4) 319-330.

[139] Zagorska N., Stanilova M., Ilcheva V., Gadeva P. 1997. Micropropagation of *Leucojum aestivum* L. (*Summer snowflake*). In: Bajaj Y.P.S. (ed) Biotechnology in Agriculture and Forestry, vol. 40, VI, High-Tech and Micropropagation, Springer: 1997. p.178-192.

[140] Bogdanova Y., Stoeva T., Yanev S., Pandova B., Molle E., Burrus M., Stanilova M. Influence of plant origin on propagation capacity and alkaloid biosynthesis during long-term *in vitro* cultivation of *Leucojum aestivum* L. *In vitro* Cell and dev Biology – Plant 2009;45(4) 458–465.

[141] Karaog'lu C. *In vitro* propagation of summer snowflake. MSc thesis, 2004, (http://72.14.221.104/search?q=cache:X_xsbdsosl4J:papirus.ankara.edu.tr/tez/FenBilimleri/Yuksek_Lisans_Tezleri/2004/FY2004_184/Ozet.pdf+Leucoju   m+in+vitro&hl=hu&gl=hu&ct=clnk&cd=5)

[142] Kohut E., Ördögh M., Jámbor-Benczúr E. & Máthé Á. Results with the establishment of *in vitro* culture of *Leucojum aestivum*. International Journal of Horticultural Science 2007;13(2) 67–71.

[143] Georgieva L., S. Berkov, V. Kondakova, Jaume Bastidab, Francesc Viladomat, A. Atanassov, Carles Codinab. Alkaloid Variability in *Leucojum aestivum* from Wild Populations. Z. Naturforsch 2007;62(c) 627–635.

[144] Georgieva L., Atanassov A., Davidkova L., Kondakova V. Long-term *in vitro* storage and multipli cati on of *Leucojum aestivum* L. Biotechnol.&Biotechnol. Eq. 2010;24(3) 1950-1954.

[145] Pavlov A., Berkov S., Courot E., Gocheva T., Tuneva D., Pandova B., Georgiev M., Georgiev V., Yanev S., Burruse M., Ilieva M. Galanthamine production by *Leucojum aestivum in vitro* systems. Process Biochemistry 2007;42(4) 734–739.

[146] Ptak A., Tahchy A. El., Wy_zgolik G., Henry M., Laurain-Mattar D. Effects of ethylene on somatic embryogenesis and galanthamine content in *Leucojum aestivum* L. cultures. Plant Cell Tiss Organ Culture 2010;102(1) 61–67.

[147] Berkov S., Pavlov A., Ilieva M., Burrus M., Popov S., Stanilova M. CGC-MS of alkaloids in *Leucojum aestivum* plants and their *in vitro* cultures. Phytochem Anal. 2005;16(2) 98–103.

[148] Berkov S., Pavlov A., Georgiev V., Bastida J., Burrus M., Ilieva M., Codina C. Alkaloid synthesis and accumulation in *Leucojum aestivum in vitro* cultures. Nat Prod Commun 2009;4(3) 359–364.

[149] Georgiev V., Ivanov I., Berkov S., Ilieva M., Georgiev M., Gocheva T., Pavlov A. Galanthamine production by *Leucojum aestivum*L. shoot culture in a modified bubble column bioreactor with internal sections. Eng. Life Sci. 2012 doi:10.1002/elsc.201100177.

[150] Ivanov I., Georgiev V., Berkov S., Pavlov A. Alkaloid patterns in *Leucojum aestivum* shoot culture cultivated at temporary immersion conditions. J. Plant Physiology 2012;169(2) 206–211.

[151] Schumann A., Berkov S., Claus D., Gerth A., Bastida J., Codina C. Production of Galanthamine by *Leucojum aestivum* Shoots Grown in Different Bioreactor Systems. Applied Biochemistry and Biotechnology 2012;167(7) 1907–1920.

[152] Diop M.F., Hehn A., Ptak A., Chretien F., Doerper S., Gontier E., Bourgaud F., Henry M., Chapleur Y., Laurain-Mattar D. Hairy root and tissue cultures of *Leucojum aestivum* L.—relationships to galanthamine content. Phytochemistry Reviews 2007;6(1) 137 – 141.

[153] Kelly G.S. *Rhodiola rosea*: a possible plant adaptogen. Altern Med Rev 2001;6(3) 293-302.

[154] Kajmakanova I. Ecologically and phytochemical investigation of Rhodiola rosea L. (family Crassulaceae) in Bulgaria (in natural ant cultural conditions). in Scientific publications from Student scientific conference "Conservation of the biological diversity and management of the protected areas, Sofia,Bulgaria, 2005.

[155] Zhang S, Wang J, Zhang H. Chemical constituents of Tibetan medicinal herb *Rhodiola kirilowii* (Reg.). Gansu Chung Kuo Chung Yao Tsa Chih 1991;16(8) 483-512.

[156] Wang S., Wang F.P. Studies on the chemical components of *Rhodiola crenulata*. Yao Hsueh Hsueh Pao 1992;27(2) 117-120.

[157] Wang S., You X.T., Wang F.P. HPLC determination of salidroside in the roots of *Rhodiola* genus plants. Yao Hsueh Hsueh Pao 1992;27(11) 849-52.

[158] Shi L., Ma Y., Cai Z. Quantitative determination of salidroside and specnuezhenide in the fruits of *Ligustrum lucidum* by high performance liquid chromatography. Biomedical Chromatography 1998;12(1) 27-30.

[159] Brown R.P., Gorbarg P.L., Ramazanov Z. *Rhodiola rosea* - a phytomedicinal overview. Herbal Gram 2002;56 40-52.

[160] Platikanov S, Evstatieva L. Introduction of Wild Golden Root (*Rhodiola rosea* L.) as a Potential Economic Crop in Bulgaria. Economic Botany 2008;62(4) 621-627.

[161] "Bulgarian Law Gazette (State newspaper)," vol. 77, 09.08.2002.

[162] Abidov M., Grachev S., Seifulla R. D., Ziegenfuss T. N. Extract of *Rhodiola rosea* radix reduces the level of C-reactive protein and creatinine kinase in the blood. Bulletin of Experimental Biology and Medecine 2004;138(7) 73–75.

[163] Walker B. Thomas, Stephen A. Altobelli, Arvind Caprihan, Robert A. Robergs. Failure of *Rhodiola rosea* to skeletal muscle phosphate kinetics in trained men. Metabolism Clinical and Experimental 2007;56(8) 1111–1117.

[164] Ma Li, Cai Donglian, Li Huaixing, Tong bende, Song Lihua, Wang Ying. Anti-fatigue effects of salidroside in mice. Journal of medical colleges of PLA 2008;23(2) 88–93.

[165] Chen Q. G., Y.S. Zeng, Z. Q. Qu, J. Y. Tang, Y. J. Qin, P. Chung, R. Wong, U. Hägg. The effect of *Rhodiola rosea* extract on 5-HT level, cell proliferation and quantity of neurons at cerebral hippocampus of depressive rats. Phytomedicine 2009;16(9) 830–838.

[166] Wójcik R., Siwicki A.K., Skopińska-Różewska E., Bakuła T., Furmanowa M. The *in vitro* effect of *Rhodiola quadrifida* and *Rhodiola kirilowii* extracts on pigs blood lymphocyte response to mitogen Concanavalin A. Centr Eur J Immunol 2009;34(3) 166-170.

[167] Skopińska-Różewska E., Stankiewicz W., Zdanowski R., Siwicki A.K., Furmanowa M., Buchwald W., Wasiutyński A. The in vivo effect of Rhodiola quadrifida extracts on the antibody production, on the blood leukocytes subpopulations and on the bacterial infection in mice. Centr Eur J Immunol 2012;37(2) 140-144.

[168] Wiedenfeld H., Dumaa M., Malinowski M., Furmanowa M., Narantuya S. Phytochemical and analytical studies of extracts from *Rhodiola rosea* and *Rhodiola quadrifida*. Pharmazie 2007;62 308–311.

[169] Guoying Zuo, Zhengquan Li, Lirong Chen, Xiaojie Xu. Activity of compounds from Chinese herbal medicine *Rhodiola kirilowii* (Regel) Maxim against HCV NS3 serine protease. Antiviral Research 2007;76(1) 86–92.

[170] Siwicki A.K., Skopińska-Różewska E., Wasiutyński A., Wójcik R., Zdanowski R., Sommer E., Buchwald W., Furmanowa M., Bakuła T., Stankiewicz W. The effect of Rhodiola kirilowii extracts on pigs' blood leukocytes metabolic (RBA) and prolifera-

tive (LPS) activity, and on the bacterial infection and blood leukocytes number in mice. Centr Eur J Immunol 2012;37(2) 145-150.

[171] Mizue Ohsugi, Wenzhe Fan, Koji Hase, Quanbo Xiong, Yasuhiro Tezuka, Katsuko Komatsu, Tsuneo Namba, Tomohiro Saitoh, Kenji Tazawa, Shigetoshi Kadota. Active-oxygen scavenging activity of traditional nourishing-tonic herbal medicines and active constituents of *Rhodiola sacra*. Journal of Ethnopharmacology 1999;67(1) 111 – 119.

[172] Mool-Jung Inhee, Hee Kim, Wenzhe Fan, Yasuhiro Tezuka, Shigetoshi Kadota, Hisao Nishijo, Min Whan Jung. Neuroprotective Effects of Constituents of the Oriental Crude Drugs, *Rhodiola sacra*, *R. sachalinensis* and Tokaku-joki-to, against Beta-amyloid Toxicity, Oxidative Stress and Apoptosis. Biol. Pharm. Bull. 2002;25(8) 1101—1104.

[173] Yidong Leia, Peng Nana,b, Tashi Tseringc, Zhankui Baia, Chunjie Tiana, and Yang Zhonga. Chemical Composition of the Essential Oils of Two *Rhodiola* Species from Tibet. Z. Naturforsch 2003;58c 161–164.

[174] Yidong Lei, Hong Gao, Tashi Tsering, Suhua Shi and Yang Zhong. Determination of genetic variation in *Rhodiola crenulata* from the Hengduan Mountains Region, China using inter-simple sequence repeats. Genetics and Molecular Biology 2006;29(2) 339-344.

[175] Jianfeng Xu, Su Zhiguo, Feng Pusum. Suspension culture of compact callus aggregate of *Rhodiola sachalinensis* for improved salidroside production. Enzyme and Microbial Technology 1998;23(1–2) 20–27.

[176] Rajesh Arora, Raman Chawla, Ravinder Sagar, Jagdish Prasad, Surendar Singh, Raj Kumar, Ashok Sharma, Shikha Singh and Rakesh Kumar Sharma. Evaluation of radioprotective activities of *Rhodiola imbricata* Edgew – A high altitude plant. Molecular and Cellular Biochemistry 2005;273(1-2) 209–223.

[177] Ishmuratova M.M. Rhodiola Iremelica Boriss. in Ural: ecological, biological, biochemical characteristics, tactics, strategic production and protection. DSc thesis, VAK, 03.00.05, 2004.

[178] Kaftanat V. N., Bodrug M.V., Floryia V.N. Enhanced multiplication of Rhodioloa rosea in Moldova. in proceedings of the 2ndNational Conference onMedicinal Botany, Kiev, Ukraine, 1988.

[179] Kirichenko E.B., Rudenko S.S., Baglaj B.M., Masikevich U.G. Leaf culture from *invitro* propagated *Rhodiola rosea*," Bulletin GBS, RAN, 1994;169 50–54 (in Russian).

[180] Bazuk O.F., Baraneckii G.G., Fedyaj L.V. Micropropagation of rose root. in proceedings of the Conference of Investigations of Ontogenesis Natural and Cultural Flora in Botanical Garden Eurasia, Kiev, Ukraine, 1994.

[181] Ishmuratova M.M. Clonal propagation of *Rhodiola rosea* L. and *R. iremelica* Boriss. *in vitro*. Rastitelnii resursi (Plant resources) 1998;34(1) 12-23. (in Russian)

[182] Ishmuratova M. M. Effect of *Rhodiola rosea* plant extracts on the *in vitro* development of *Rhodiola rosea* L. and *Rhodiola tremelica* Boriss Explants. Biotekhnologiya 2002;6 52–56 (in Russian).

[183] Yin W.B., Li W., Du G.S., Huang Q.N. Studies on tissue culture of Tibetan *Rhodiola rosea*. Acta Bot. Boreal. Occident. Sin. 2004;24 1506–1510.

[184] Evstatieva L. N., Revina T.A. Investigation of Polyphenols in Rhodiola rosea. Groupe polyphenols. Journees Internationales d'Etudes 1984;12 127–128.

[185] Evstatieva L., Hardalova R., Stoyanova K. Medicinal plants in Bulgaria: diversity, legislation, conservation and trade. Phytologia Balcanica 2007;13(3) 415–427.

[186] Platikanov S, Evstatieva L. Introduction of Wild Golden Root (*Rhodiola rosea* L.) as a Potential Economic Crop in Bulgaria. Economic Botany 2008;62(4) 621–627.

[187] Revina T.A., Krasnov E.A., Sviridova T.P., Stepanuk G.A. Biologically characteristics and chemical composition of *Rhodiola rosea* L., reintroducing in Tomske. Plant resources 1976;12(3) 355-360.

[188] Kapchina-Toteva V., Sokolov L. *In vitro* micropropagation of *Rhodiola rosea* L., Annuaire de L'Universite de Sofia "St. Kliment Ohridski" 1997;88(4) 222-226.

[189] Galambosi Bertalan, 2006. Demand and availability of *Rhodiola rosea* L. raw material. In: Bogers R.J., Craker L.E., Lange D. (eds) Chapter 16, Medicinal and Aromatic Plants, Netherlands: Springer; 2006. p.223-236.

[190] Dimitrov B., Tasheva K., Zagorska N., Evstatieva L. *In vitro* cultivation of *Rhodiola rosea* L. Genetics and Breeding 2003;32(1-2) 3–6.

[191] Tasheva K., Zagorska N., Dimitrov B., Evstatieva L. *In vitro* cultivation of *Rhodiola rosea* L. 2003. International Scientific Conference, proceedings of scientific papers, 75 years of the Forest Research Institute, Octobre 1-5, 2003(a).

[192] Tasheva K., M. Petrova, N. Zagorska, L. Evstatieva. *In vitro* propagation of *Rhodiola rosea*. Tenth Jubilee Scientific Session, Faculty of Biology; Sofia University, November 20 – 21, Sofia, Bulgaria, 2003(b).

[193] Tasheva K., Petrova M., Zagorska N., Evstatieva L. *In vitro* germination of three medicinal plants. An International Meeting on Seeds and the Environment. Seed Ecology 2004, April 29 – May 4, Rhodes, Greece, 2004.

[194] Tasheva K., Petrova M., Zagorska N., Georgieva E. Micropropagation *In Vitro* of *Rhodiola rosea* L. COST 843 final conference, June 28 – July 3, Stara Lesna, Slovakia, 2005.

[195] Tasheva K., Kosturkova G. Bulgarian Golden root *in vitro* cultures, micropropagation and reintroduction. Central European Journal of Biology 2010a;5(6) 853-863.

[196] Tasheva K., Kosturkova G. *Rhodiola rosea in vitro* cultures peculiarities. Scientific publications of University of Agronomical sciences and veterinary medicinal – Biotechnology 2010b 103–112.

[197] Tasheva K. Kosturkova G. Rhodiola rosea L. in vitro plants morphophysiological and cytological characteristics. Romanian Biotechnological Letters 2011;16(6) 79–85.

[198] Bozhilova M., Evstatieva L., Tasheva K. Salidroside content in *in vitro* propagated *Rhodiola rosea* L. 5th conference on medicinal and aromatic plants of Southeast European countries (5th CMAPSEEC), proceedings of scientific paper, September 2 -5, Brno Czech republic, 2008.

[199] Gogu G., Hartan M., Maftei D-E., Nicuta D. Some considerations regarding the *In Vitro* culture of *Rhodiola rosea* L. Romanian Biotechnological Letters 2011;16(1) 5902–5908.

[200] Debnath S.C. Zeatin and TDZ-induced shoot proliferation and use of bioreactor in clonal propagation of medicinal herb, roseroot (*Rhodiola rosea* L). Journal of Plant Biochemistry and Biotechnology 2009;18(2) 245-248.

[201] Furmanowa M., Oledzka H., Michalska M., Sokolnicka I., Radomska D. *Rhodiola rosea* L. (Roseroot): *in vitro* regeneration and the biological acivity of roots.In: Bajij Y. (ed) Biotechnology in Agriculture and Forestry, vol. 33 of Medicinal and Aromatic Plants VIII, Berlin, Germany, Springer; 1995. p.412–426,

[202] György Z., Tolonen A., Pakonen M., Neubauer P., Hohtola A. Enhancing the production of cinnamyl glycosides in compact callus aggregate cultures of *Rhodiola rosea* by biotransformation of cinnamyl alcohol. Plant Science 2004;166(1) 229–236.

[203] György Z., *Glycoside production by in vitro* Rhodiola rosea *cultures*, Ph.D. thesis, Acta Universitatis Ouluensis C Tehnica 244, Oulu, Finland, 2006.

[204] Krajewska-Patan A., Mscisz A., Kedzia B., Lutomski J. The influence of elicitation on the tissue cultures of roseroot (*Rhodiola rosea*). Herba Polonica 2002;48(2) 77–81.

[205] Krajewska-Patan A., Dreger M., Lowicka A. Górska-Paukszta M., Mścisz A., Mielcarek S., Baraniak M., Buchwald W., Furmanowa M., Mrozikiewicz P.M. Chemical investigation of biottransformed *Rhodiola rosea* callus tissue. Herba Polonica 2007;53(4) 77-87.

[206] Krajewska-Patan A., Dreger M., Lowicka A., Górska-Paukszta M., Przemyslaw L., Mścisz A., Buchwald W., Furmanowa M., Mrozikiewicz P.M. Preliminary pharmacological investigation of biotransformed roseroot (*Rhodiola rosea*) callus tissue. Herba Polonica 2008;54(3) 50-58.

[207] Tasheva K., Kosturkova G. Establishment of callus cultures of Rhodiola rosea Bulgarian ecotype. Acta Horticulturae 2012;955 129-135.

[208] Soo Jung Kim, Kwang Soo Kim, Sung Jin Hwang, Sang Uk Chon, Young Ho Kim, Jun Cheul Ahn, Baik Hwang. Identification of salidroside from Rhodiola sachalinen-

sis A. Bor. and its production through cell suspension culture. Korean J. Medicinal Crop Sci 2004;12(3) 203–208.

[209] Wu S, Zu Y & Wu M. High yield production of salidroside in the suspension culture of *Rhodiola sachalinensis*. Journal of Biotechnology 2003;106(1) 33-43.

[210] Jianfeng L., Xiufeng Y., Yun-Qing C., Xiao-Mei Z. Cryopreservation of calli by vitrification and plant regeneration of *Rhodiola sachalinesis*. Journal of Beijing Forestry University, 2007 (Chinese).

[211] Jian-Feng L., Yun-Qing C., Zhi-Wen C. Protoplast isolation and plant regeneration from leaves of *Rhodiola sachalinesis*. Chinese Traditional and Herbal Drugs 2009;7 2010–2014.(Chinese).

[212] Liu Hai-jun, Guo Bin, Yan Qiong, Liu Yu-jun, Liu Chun-zhao. Tissue culture of four *Rhodiola species*. Acta Botanica Boreali-Occidentalia Sinica 2006; 207-210, Doi: cnki:ISSN:1000-4025.0.2006-10-009

[213] Wang Yun-mei. Tissue culture and rapid propagation of Yunnan Wild *Rhodiola crenulata*. Journal of Anhui Agricultural Sciences 2009;17 57–61. Doi: CNKI:SUN:AHNY. 0.2009-17-025

[214] Sheng Chang-zhong, Hu Tie-qiang, BI Haoq Yuan Ying-jin, Jiang Yan. Effects of plant growth substances on induction and culture of callus from *Rhodiola quadrifida*, China Journal of Chinese Materia Medica 2005;30(16) 1237–4016 (in Chinese); DOI: cnki:ISSN:1001-5302.0.2005-16-003

[215] Li Wei, Du Gui-seng, Huang Qin-ni. Salidroside contents and related enzymatic activities in *Rhodiola kirilowii* callus. Acta Botanica Boreali-occidentalia Sinica, 2005; 2005-08, doi: cnki:ISSN:1000-4025.0.2005-08-025

[216] Xiaofu Zhou, Yuxia Wu, Xingzhi Wang, Bao Liu, and Hongwei Xu. Salidroside Production by Hairy Roots of *Rhodiola sachalinensis* Obtained after Transformation with *Agrobacterium rhizogenes* Biol. Pharm. Bull. 2007;30(3) 439—442.

[217] Xiao-fu Zhou, Xiao-wei Wei, Zhuo Zhao, Jing-di Sun, Jie Lv, Yui Cai, Hong-wei Xu. The influence of external factors on biomass and salidroside content in hairy roots of *Rhodiola sachalinensis* induced by Agrobacterium rhizogenes. 3rd International Conference on Biomedical Engineering and Informatics (BMEI), 2010.

Chapter 11

The Use of Interactions in Dual Cultures *in vitro* to Evaluate the Pathogenicity of Fungi and Susceptibility of Host Plant Genotypes

Katarzyna Nawrot - Chorabik

Additional information is available at the end of the chapter

1. Introduction

The subject of biological and biochemical bases for immune and defense reactions of an organism to e.g. pathogens is a very broadly defined question. Although the substance of immunology has common ground, it is otherwise perceived in the case of humans and animals, while different aspects are highlighted in the case of plant organisms. While in the case of human the emphasis is put on research aimed to develop a variety of therapies to heal autoimmune disorders, in plants the studies focus on the effect of various biotic, abiotic and anthropogenic stress factors on plant organisms. Biotic stress factors include: fungal infectious diseases, insect pests or excessive occurrence of herbivorous mammals while abiotic stress factors include: atmospheric conditions (extreme weather events, relative humidity deficiencies), soil properties e.g. fertility, physiographic conditions or different stress conditions: oxidative, i.e. load of oxygen, sodium chloride, water deficit or stress caused by the effect of heavy metals (lead - Pb, mercury - Hg, iron - Fe etc.), while anthropogenic stress factors include: air, water and soil pollution, forest fires or improper forest management. The impact of anthropogenic factors, which has escalated in recent decades, causes changes in the natural environment and ecosystems of certain regions of the globe. Forest trees, as important elements of the ecosystem, are vulnerable to climate and environmental changes. This suggests that it is important to thoroughly understand and explain the problem of dieback of trees as a result of the impact of stress factors on the forest environment, especially that the determination of the cause-effect relationship is more complex than previously believed.

The phenomenon of dieback of economically important forest trees due to pathogenic fungi as biotic environmental components is not fully understood yet. This can be explained by the fact that the disease process of trees depends among others on the properties of host plants as well as on characteristics of fungi – the potential causal agents of diseases. These both groups of organisms are affected by environmental factors, that may increase or decrease the susceptibility to diseases. In most cases the assessment of the host plant susceptibility and the properties of pathogenic fungi are done by evaluating the frequency of disease symptoms' outbreak in relation to the presence of a certain fungal species and based on the pathogenicity tests. The implementation of such tests is often very difficult, because in the case of forest ecosystems we are dealing with perennial plants.

A new method of research, that shows the relationship (interactions) between the two organisms involving the stimulation or inhibition of their growth was developed in the eighties. This is dual cultures research (involving two organisms), in which one of the organisms is callus (*in vitro* cultured plant tissue that covers plant wounds in embryonic or non-embryonic stadium), while the second organism is the studied fungal species. [4] from the Slovak Academy of Sciences, and [8] from the University of New Brunswick in Canada were the first ones who studied the pathogenicity in dual *in vitro* cultures at the embryonic level. Such research was conducted later in Finland [31, 21], Germany [28, 24], Norway [13, 12], Slovakia [7, 33] and in Great Britain [3].

Based on previous research conducted in Poland [Nawrot–Chorabik unpublished, 19, 20] it may be stated that such experiments are promising, because the development of reliable tests in dual cultures involving fungi and embryonic stadium of their host plant tissues may provide the basis to the evaluation of the pathogenicity and severity of hazard caused by these fungi. It may also allow the selection of plant genotypes that are more resistant to certain pathogens, which is particularly important in the cases of fungi known for their ability of epiphytic occurrence. The selection at the embryonic level of tree genotypes resistant to infection would emerge a new direction in the resistance breeding [14].

The aim of this chapter is to present in a condensed way current issues and results of research on interactions between fungi and the host plant callus in dual cultures. This would bring the attention to the usefulness of this method to study pathogenicity and to the opportunities of its application in the forest practice.

2. Materials and methods

The *in vitro* cultured tree callus is the plant material necessary to establish a dual culture and fungus is the second organism which the dual culture consists of. The fungal material growing in dual culture with callus needs to be isolated and identified. The callus initiation in the case of gymnosperm trees is based on a biotechnological method – somatic embryogenesis, involving the initiation of callus on disinfected primary explant (e.g. zygotic embryo isolated from a seed). There are also dual cultures, in which two different fungi grow in one culture.

2.1. Stages of dual cultures

Dual cultures consist of the following steps:

- Obtaining a considerable amount of non-embryogenic or embryogenic callus of the studied plant species, e.g. by using a somatic embryogenesis method.

The *in vitro* initiation of non-embryogenic (unorganized mass of dividing cells, in nature – cells that cover wounds) or embryogenic (unorganized mass of dividing cells, able to form somatic embryos and in later stages of organogenesis – to regenerate plants in the process of micropropagation) callus in sufficiently large amounts is an essential element to establish a dual culture (Fig. 1). To achieve this objective, an optimal method for obtaining callus (e.g. somatic embryogenesis) should be selected, disinfection of explants should be developed and media together with growth regulators, concentration of sugars and other components necessary for the proper growth of callus should be selected individually for each species [18].

- Selecting the best genotypes (lines) of callus, which should be proliferated until obtaining at least 15 clones of each genotype weighing 300 - 500 mg each [15, 16, 17].

- Isolation and identification of fungi.

Fungi should be properly selected, most preferably with different ecological statuses, corresponding to the studied host plant species. Moreover, they need to be isolated and identified. Traditional methods of fungal identification based on mycelial morphology may be complemented by using molecular techniques based on DNA sequences that allow rapid selection and identification of these organisms.

- Adaptation of fungi selected for dual culture to grow on an "enriched" medium, dedicated to the proliferation of plant callus.

Mycelial fragments of 0.5 x 0.5 cm are transferred from malt extract agar (MEA) onto a medium on which the callus has already been cultured, i.e. the medium with high concentration of macro- and micronutrients and other components (such as casein hydrolyzate, meso-inositol, growth regulators). Fungal identification based on their phenotypic features should be repeated after about 1- 3 weeks.

- Establishment of dual cultures (fungus – plant callus).

Dual cultures are established in sterile Petri dishes with solidified callus-proliferation medium. Fungal inoculum of 0.5 x 0.5 cm should be placed in the center of the sterile Petri dish while the host-plant callus with a diameter of 1.0 - 1.5 cm and weighing 500 mg should be put 5.0 mm from the edge of the dish [19]. Very slow-growing fungi, e.g. *Lophodermium sediosum* should be placed closer to the callus. Replicates of cultures (their number depends on the frequency of embryogenesis and proliferative capacity of callus genotypes) are stored in the dark in incubators or phytotrons at 25ºC +/- 1ºC and humidity of 40 – 50%. Moreover, anomalies in callus cells exposed to stress should be microscopically analyzed and immune proteins should be identified. In order to identify the molecular basis for fungal pathogenicity on embryonic level in dual *in vitro* cultures, genetic analyses using molecular markers are applied. Previously acquired knowledge supported by practice in dual *in vitro* cultures allowed the Author of this paper to draw attention to the aspects of the possible research directions. The analyses may therefore be directed towards: understanding the type of interactions between the tested organisms and assessment of the pathogen behavior under the influence of host plant callus and the callus behavior under the influence of the pathogen, as well as towards the selection of pathogen-resistant plants. Therefore, dual culture experiments may be conducted on one - selected callus genotype or on multiple genotypes, depending on whether we want to select genotypes resistant to the pathogen, or rather to thoroughly investigate the genotype of the pathogen.

• Conducting measurements together with macro- and microscopic observations summarized with statistical analysis.

Species of fungus:	Friedmann's test results:		Average ranks:			
	F	p	G4	G5	G6	Control
H. abietinum	11.846	0.0079*	2.95	1.77	3.14	2.14
H. parviporum	22.773	0.0001*	3.14	2.14	3.41	1.32
H. annosum	17.487	0.0006*	2.27	3.09	3.23	1.41

Table 1. Results of Friedman's ANOVA for dual cultures of *Heterobasidion* fungi with the callus of *Abies alba*. * differences statistically significant at α = 0,05

Species of fungus:	Direction of the mycelium range	Mean	Compared directions of mycelium range	Difference	p*
Heterobasidion abietinum	control	20.28	towards callus - oppositely	- 0.434	
	towards callus	20.76	towards callus - control	0.480	
	oppositely to callus	21.19	oppositely - control	0.914	
Heterobasidion parviporum	control	19.52	towards callus - oppositely	0.070	
	towards callus	24.01	towards callus - control	4.490	+
	oppositely to callus	23.94	oppositely - control	4.420	+
Heterobasidion annosum	control	20.96	towards callus - oppositely	0.029	
	towards callus	22.82	towards callus - control	1.862	+
	oppositely to callus	22.79	oppositely - control	1.939	+

Table 2. The statistically significant differences of the fungal growth of *Heterobasidion* species in dual cultures with *Abies alba* callus (towards the callus and oppositely to it) when compared to the control mycelium (t – Student's test for dependent variables).*statistically significant differences (+) in the growth of the objects compared, with confidence level 95%

Measurements of mycelial growth range are carried out in the directions towards and opposite to the callus. Macro- and microscopic observations of fungi and callus are carried out in terms of phenotypic changes in tissues with particular reference to anomalies in the callus cells subjected to stress. Measurements and observations are performed every 24 hours (in the case of fast-growing fungi) or every 48 hours (in the case of slow-growing fungi). Based on the results on the growth of tissues, a Friedman's repeated-measures analysis of variance should be performed to verify whether there is a statistically significant difference in the growth of the analyzed fungi towards the examined genotypes (lines) of callus (Tab. 1). To determine inhibition or stimulation of fungal growth it is recommended to perform a t-Student test for dependent variables based on measurements of average length of the mycelial radius (Tab. 2).

• Biochemical analyses:

- identification of PR-type immune proteins (*pathogenesis-related proteins*) in plant callus.

After establishing a dual culture, hyphae grows at different rates towards the living callus tissue. Once the mycelium approaches the distance of about 1-2 mm to the callus or stops its growth at some distance from the callus, the callus tissue should be sampled for the analysis of proteins. The callus may be stored in a freezer at - 81°C, in sterile, 0.5 ml Eppendorf tubes. The next step is the quantitative determination of fresh and dry weight of plants (callus), protein extraction and their quantitative determination using e.g. [1] using bovine albumin as a standard. This method may be modified according to the specific needs. Further biochemical analyses concern separation of proteins depending on their mass (length of polypeptide chain) using vertical electrophoresis in the presence of SDS (SDS-PAGE electrophoresis). The samples need to be properly prepared (buffer optimization, selection of centrifugation time and voltage – usually 10-30 mA/1mm of gel thickness). For proper electrophoretic separation of proteins according to their weight, the samples need to be reduced in appropriate temperature and buffer before applying to gel. Mercaptoethanol or DTT (dithiothreitol) need to be used for this purpose. Both compounds reduce the disulfide bridges (S-S) between proteins, therefore denaturizing the secondary structure of proteins. This procedure is intended to exclude that the heavier but more "packaged" protein migrates faster in the gel than the protein with lower weight but less "packaged". To obtain good electrophoretic separation, the sample volume must not exceed 10% of the path volume and the amount of proteins applied to the wells of the gel must fit within the range of 10-50 µg. Furthermore, the excessive amount of salt in the sample needs to be reduced, as its ionized form may cause smudging and heterogeneity of bands. Staining of bands on the gel requires selection of the most common staining methods such as using Coomassie Brillant Blue. The bands are visualized using a gel documentation system with a digital camera. The image of the obtained bands is used to determine the length of proteins with respect to the marker and to establish which of them may qualify as immune proteins produced by the callus tissue.

- introducing elicitors that induce plant defense responses.

Because of the need to maintain sterile conditions, defense reactions may be induced only with sterile elicitors added to culture media. In such cases specific toxins produced by a pathogen, or post-culture filtrates of the studied fungi are often applied. Mycelia are cu tured in liquid aerated media which are subsequently filtered. In this case, we examine the response of tissues or individual cells to the entire spectrum of metabolites excreted by fungi into media. These studies are often based on callus cultures as well as on cell suspensions or protoplasts.

• Inoculation of *in vitro* acquired somatic seedlings with the studied fungus.

When carrying out further steps of the somatic embryogenesis process, after a year of research we obtain somatic plant seedlings on embryogenic callus, which then may be subjected to classical pathogenicity tests.

Figure 1. Dual cultures *in vitro* (callus - fungus): a - *Pinus sylvestris - Lophodermium seditiosum*, b - *Abies alba - Heterobasidion parviporum*, c - *Abies alba - Heterobasidion abietinum*, d, e - *Pinus sylvestris - Phacidium lacerum*, f - *Abies alba - Geosmithia* sp.; Bars a, b, e = 7.0 mm

3. The possibilities of using dual cultures

Studies on dual cultures in the world started in the second half of the eighties. At that time the *in vitro* studies were conducted on endophytic fungi and host plant callus. These experiments included both forest trees and herbaceous plants [8, 27]. Analyses of dual cultures in subsequent years involved selected fungal species with pathogenic [3, 29, 13, 2], endophytic [24] and ectomycorrhizal properties [28, 22, 21]. Saprotrophic fungi were most often subjected to studies for comparative purposes (negative control) [3]. In Poland those studies were initiated in 2006 at the Department of Forest Pathology, University of Agriculture in Kraków (Nawrot–Chorabik unpublished). Defense reactions of embryogenic tissues towards various isolates of *Heterobasidion* fungi were studied on the selected genotypes of *in vitro* cultured callus of silver fir. In subsequent years the analyses were supplemented with tests on fungi with different ecological status [19, 20]. The research is continued and cover biochemical, cytological and other aspects.

3.1. Pathogenic organisms

Different organisms were taken into consideration as pathogens in dual cultures: bacteria, fungi and fungi-like organisms (*Chromista*), and were tested with different plants: agriculturally cultivated, fruit trees and parasitic flower plants [2]. Given the chronology of research in dual cultures it may be observed that they gradually become more and more important, which is reflected in the application of more complicated analyses presented by the authors of publications. One of the first research in this field was carried out by [4, 5, 6], who conducted *in vitro* studies of spruce, beech, birch and poplar. They attempted to determine the role of basidiomycete fungi (*Basidiomycota*) in the phenomenon of dying back of some forest tree species. The observed stimulation of fungal growth by the callus was significant for the necessity of studying dual *in vitro* cultures. This phenomenon was the earliest reported in 1981 by [5], who decided to introduce dual cultures into pathogenicity research. These authors showed that wood decay-causing fungi, capable of infecting living tree tissues were stimulated by the callus, while the development of fungi that are able to colonize only dead tissues was inhibited. Based on these results it occurred that the degree of growth stimulation is positively correlated with fungal virulence. [3], taking into account the changes that occur in the forest environment under the influence of different pathogenic and saprophytic fungal species in dual *in vitro* cultures with beech, considered the implications of wider use of such methods for clarifying forest pathology issues. Experiments have shown that the specificity of fungal species towards the particular plant tissue or organ may be better demonstrated using dual systems than with any other methods. It was additionally stated that the dual culture method may be appropriate in the case of forest fungi, because it provides a model of cell division which is similar to the cambium of trees [3]. In the experiment of [13] or [29] the growth of pathogenic fungi was observed in the presence of callus obtained from a few genotypes of Norway spruce (*Picea abies*) and Mediterranean cypress (*Cupressus sempervirens*). [13] demonstrated a clear effect of the spruce callus on the growth of the mycelium. This result was consistent with conclusions drawn by [4]. In the experiments of [13] two genotypes of plant callus differed significantly in their susceptibility to pathogenic fungi. The growth of *Heterobasidion annosum* cultured with the spruce callus of one genotype was significantly inhibited after 60 hours, while after the same time of culture with another genotype – it was not inhibited. Changes involving the reduction of matter of living callus while increasing the share of dead callus were observed in tissue cultures of sweet chestnut (*Castanea sativa*) affected by the virulent strain of *Cryphonectria parasitica*. In this case, the research in dual cultures fully confirmed the *in vivo* observations [25]. Many studies show that virulent pathogens are stimulated by the callus in dual cultures, while saprotrophic fungi or fungi that re-colonize wood are inhibited by the callus [5, 3, 13, 33]. This phenomenon was previously noted by [5], who decided to introduce dual cultures into the study of pathogenicity, including basidiomycete fungi causing wood decay of living trees. Among fungi occurring on fruit trees, considerable attention was given to the application of dual cultures in the selection of apple trees resistant to apple scab caused by *Venturia inaequalis*. However considering bacteria and fruit trees combination, the majority of studies were focused on *Agrobacterium tumefaciens* - the cause of root galls and *Erwinia amylovora* - the cause of fire blight. On the other hand, among the supergroup *Chromista* the genus *Phytophthora*, which includes the causal agents of phytophthoroses of different trees, shrubs and ornamental plants [23], was mainly taken into consideration. An example of the earliest practical use of tissue cultures in plant protection was the cultivation of virus-free poplar hybrids [23]. The usefulness of dual cultures to evaluate the pathogenicity of fungi was also confirmed with regard to parasitic flowering plants. Hemlock dwarf mistletoe (*Arceuthobium tsugense*) occurring on *Tsuga* was colonized *in vivo* by two fungi: *Cylindrocarpon cylindroides* and *Colletotrichum gloeosporioides*. The dual culture analysis with fungi and the mistletoe tissue showed high pathogenicity of both fungal species, which colonized the plant tissue inter- and intracellularly causing cell walls' degradation [2].

3.2. Endophytic fungi

Some of the earliest papers on endophytic fungi in dual cultures were focused on cultures of bigleaf maple (*Acer macrophyllum*) [27] and herbaceous plants: red deadnettle (*Lamium purpureum*) and wood sage (*Teucrium scorodonia*) [24]. The effects of interaction between the endophyte *Cryptodiaporthe hystrix* and the callus of its host – bigleaf maple were studied in dual cultures as mutual interactions. All *C. hystrix* isolates inhibited the growth of callus (the opposite situation was in the case of fungus, whose growth was always strongly stimulated by the callus presence). The Authors suggest that *C. hystrix* and its host (*Acer macrophyllum*) exist in a near-equilibrium state for a certain time, i.e. in natural conditions maple is able to prevent the extensive growth of *C. hystrix* in its tissue. But eventually *C. hystrix* not only did inhibit the growth of callus, but finally overgrew and killed it [27]. This may indicate that *C. hystrix* represents the group of endophytes that may become pathogenic when the plant is under stress conditions. The dual culture analyses confirm the results of the studies conducted *in vivo*, emphasizing that when the plant is under stress conditions (industrial immissions, parasitic insects, adverse weather conditions) and when aging of the plant organs, endophytes may become pathogenic [9, 10, 11]. Studies on dual *in vitro* cultures may contribute to obtaining more detailed results, that provide the basis for determining which species of endophytic fungi are able to cause the tree disease under stress [27]. Among herbaceous plants, the studies in dual cultures with endophytes were conducted on the host plant and non-host plants (red deadnettle - *Lamium purpureum* and wood sage - *Teucrium scorodonia*) [24]. Endophytic fungi: *Coniothyrium palmarum, Geniculosporium* sp. and *Phomopsis* sp. under the influence of the herbaceous host plant callus were characterized by increased growth – as opposed to the behavior of the same fungi in dual cultures with the callus of non-host plants [24].

3.3. Ectomycorrhizal fungi

The first research on ectomycorrhizal fungi in dual cultures was initiated in Germany [28] at the Institute of Botany in Tübingen, where Norway spruce (*Picea abies*) callus and mycorrhizal fungi: *Amanita muscaria, Lactarius deterrimus, Hebeloma crustuliniforme, Suillus variegatus* were studied. Pathogenic fungus, that does not stimulate mycorrhization, i.e. *Heterobasidion annosum* was used as control. The results were unambiguous: only two of the studied mycorrhizal fungi (*S. variegatus* and *L. deterrimus*) caused distinct response of spruce cells, which - wrapped with Hartig net - were characterized by better nutrient exchange, especially in the interaction zone of dual cultures. Additionally, irregularities in cell structures of the callus were observed, including the cell cytoplasm, which withdrew in favor of the fungal hyphae. Similar studies were conducted at the Institute of Botany in Finland in cooperation with the Forest Research Institute in Slovakia [22]. The research was conducted on three genotypes of Scots pine (*Pinus sylvestris*) callus and ectomycorrhizal fungi: *Laccaria bicolor, L. proxima, Pisolithus tinctorius, Paxillus involutus* and *Suillus variegatus*. Research in dual cultures was conducted on the effect of these fungi on initiation and proliferation of embryogenic callus [22]. Depending on the fungal species, positive (better callus growth) or negative (browning, necrosis of the callus) embryogenic tissue reaction was observed. Observations of callus cells' behavior under the influence of stress factor (fungi) showed that the hyphae of *Laccaria bicolor* "penetrated" embryogenic cells of Scots pine and by entwining them, it caused positive growth reaction [22]. A few years later [21] retook the research in dual cultures of fungus *Pisolithus tinctorius* and Scots pine. This time it was demonstrated that this fungus improves the *in vitro* germination of somatic embryos of pine. To accomplish this task, somatic embryos of Scots pine, induced *in vitro* on embryogenic callus, were subjected to mycorrhization. The Authors, through multi-faceted research methodology, using a variety of media containing different sugar concentrations and other necessary breeding procedures, obtained the desired effect in dual cultures. Using this method they improved one of the most difficult stages of somatic embryogenesis of Scots pine, i.e. germination of somatic embryos into seedlings.

4. Protein as a symptom of fungal virulence

Proteins produced by the host plant cells are important determining factors as to whether plant tissues defend themselves against the pathogen attack, or do not exhibit defensive characteristics. Dual cultures (fungus – plant callus tissues) may be helpful in protein analysis, because *in vitro* conditions are favorable both to research of proteins as well as oth compounds produced by plants under stress conditions (phenols, reactive oxygen forms, carbohydrates). Moreover, *in vitro* studies on embryonic level may be applied to *in vivo* conditions, in which obtaining relevant results seems difficult to achieve, especially since the resistance protein synthesis is observed not only during the pathogen attack, but also during the exposure of plant cells to abiotic stress factors such as drought or frost. Protein analyses were conducted in dual cultures by [27]. It was found that water-soluble metabolites, produced by the bigleaf maple callus could have been responsible for stimulating the mycelial growth. Similar results were presented by [24], who argued that the metabolites, excreted by three endophytes studied by them in dual cultures both with the callus of herbaceous host plant and with the callus of non-host plant, caused necroses and death of callus cells. These metabolites were nonspecific, as they inhibited the growth of not only one of the hosts - *Lamium purpureum,* but also *Lepidium sativum* and *Cantharis fusca.* The current molecular analyses on DNA and protein cooperation are among the most promising research areas. The processes such as transcription, replication and reparation of DNA engage proteins that function as transcription factors (activators, repressors). Because many transcription factors are responsible for regulation of specific groups of genes (e.g. related to cell cycle), the activity of these proteins has a significant impact on the basic cellular processes, such as callus proliferation, differentiation, organogenesis and others, e.g. response to stress factors (including plant pathogens). The fact that the plants show a high degree of specificity in the detection of the pathogen leads to varying degrees of resistance, which involves the entire mechanism of biochemical reactions. Genes trigger the corresponding proteins that are designed to eliminate the pathogen. In recent years studies have been aimed at the so-called pathogenesis-related proteins, which have been described for various plant-pathogen interactions. In such situation we speak of PRs (*pathogenesis-related proteins*), whose functions made them the potential antimicrobial proteins. PRs were determined and identified biochemically for 14 different plant families. The following enzymatic proteins were identified biochemically: α-1,3 – glucanase (PR-2); chitinase (PR-3, PR-4, PR-8, PR-11); proteinase (PR-6); peroxidase (PR-9) [32]. These proteins, encoded by the corresponding genes, occurred in tissues infected by pathogens [34].

5. Phenolic compounds involved in defense reactions

Phenolic compounds are the most numerous group of secondary metabolites with many different properties. Especially phytoalexins (Greek: *phyton* - plant, *alexein* - defend) are of great importance in the immune responses of plants to abiotic and biotic stresses. [30] proposed to refer to the products of higher plants, that are absent or present in trace amounts in healthy tissues, but accumulated in large quantities after the pathogen attack, as phytoalexins. Currently, phytoalexins are defined as low-molecular derivatives of phenols, accumulated in plants under the influence of various stress factors, such as heavy metals, UV light, drought, frost, pathogens, elicitors and fungicides. They are nonspecific chemical compounds, products of the host plant cells and are intended to stop the pathogen growth after the host plant comes into contact with the parasite, and cause defense reactions only in living cells. More-

over, plants resistant and susceptible to the pathogen should differ in the phytoalexin formation rate and defense reaction should be limited only to the affected tissues and their
nearest neighborhood. In order to produce alexins, a signal is needed from reactive forms of
oxygen. Elicitors, triggering phytoalexins' production, include: arachidonic acid produced
by *Phytophthora infestans*, β-1,3- β-1,6-hepta-glucoside produced by *Phytophthora megasperma*
and protein cryptogein produced by *Phytophthora cryptogea* [26]. The other phenolic compounds exhibiting direct antibiotic effect include chlorogenic acid, which is a strong fungi-
cide, bacteriocide and virucide. Plant tissues which contain higher concentrations of this
acid are more resistant to pathogens.

6. Callus tissue necrosis as the effect of hypersensitive response to stress factors

The hypersensitive response is the effect of stress applied to e.g. callus tissue growing in dual culture with the pathogenic fungus. It involves killing the plant's own cells infected by
the pathogen, therefore the infection is limited and isolated from the healthy tissue, which
prevents the replication and spreading the infected cells in the plant organism. The dying
back of cells, which results from the hypersensitivity, is accompanied by necrosis of tissues
infected by the pathogen (Fig. 1). However, this type of necrotizing involves different program than the necrosis caused by the change in pH, temperature or hypoxia. Both processes
- necrosis (localized cell death within a living organism) and apoptosis (programmed cell
death) differ in morphological, biochemical and physiological changes. Necrosis consists in
a loss of membrane integrity, cell swelling and disintegration of organelles. This leads to in-
flammation (rapid increase of heat emission by the advantage of catabolic processes over
anabolic ones). Necrosis is not associated with the plant development, it does not involve
proteases or nucleases, does not require signal transduction or calcium ions and protein
phosphorylation. During apoptosis, due to the adverse impact of external conditions, the
cell disintegrates into small apoptotic bodies, that retain cell organelles. These bodies are absorbed by other neighboring cells, which is why apoptosis is not accompanied by inflamma-
tion [26]. The plant tissue infected by the pathogen triggers a series of active defense-related
processes. This is why the mechanism and effect of pathogenesis is complicated. During the
defense reaction plant cells may simultaneously launch several signal transduction pathways, which finally results in visible necroses, reflected in the subsequent physiological condition of plants subjected to stress factors.

7. Conclusions

Many experiments were carried out using pathogenic, endophytic, ectomycorrhizal or saprotrophic fungi and the embryonic stadium of *in vitro* cultured host plant tissue. Results of
these experiments indicated diverse applicability of dual cultures in forest practice. According to the Author of this paper, developing reliable *in vitro* assays may provide a basis for

the evaluation of the pathogenicity and the degree of threat posed by fungi. This may also enable the selection on embryonic level the plant genotype that is more resistant to the pathogen, which is particularly important for fungi known for their ability of epiphytic occurrence. Moreover, the studies of dual *in vitro* cultures may provide more detailed results which would be the basis for determining which species of endophytic fungi are able to cause the disease in trees under stress conditions. Interactions observed in dual cultures may be used to assess the biotrophic properties of various fungal species or lower taxa (strains, varieties, physiological races). These studies may provide the basis for more proper determining of etiology of many diseases and to determine the correlation between the type of protein produced by the callus and the degree of fungal virulence. Conditions, which plant tissues (callus) are subject to during conducting dual cultures induce anatomical changes in callus cells, which may be used in identifying the mechanism of stress factor effect on cyto-physiological changes in plant cells. In the case of positive results of analyses, the dual cultures may give rise to new techniques used in pathogenicity studies in forest pathology.

Acknowledgements

The Author would like to thank Professor Tadeusz Kowalski for providing fungi used in the study of dual cultures, which is why they could have been documented in this paper and to thank for any substantive assistance on the phytopathological aspects of the study. This work was supported by the research project from the Polish Ministry of Science and Higher Education (N N309 705740) for 2011–2014.

Author details

Katarzyna Nawrot - Chorabik*

Address all correspondence to: rlnawrot@cyf-kr.edu.pl

Department of Forest Pathology, Faculty of Forestry, University of Agriculture in Kraków, Poland

References

[1] Bradford, M. M. (1976). A rapid and sensitive for the quantization of microgram quantities of protein utilizing the principle of protein-dye binding. *Analytical Biochemistry*, 72, 248-254.

[2] Deeks, S. J., Shamoun, S. F., & Punja, Z. K. (2002). Histopathology of callus and germinating seeds of Arceuthobium tsugense subsp. tsugense infected by Cylindrocar-

pon cylindroides and Colletotrichum gloeosporioides. *International Journal of Plant Sciences*, 163, 765-773.

[3] Hendry, S. J., Boddy, L., & Lonsdale, D. (1993). Interactions between callus cultures of European beech, indigenous ascomycetes and derived fungal extracts. *New Phytologist*, 123, 421-428.

[4] Hřib, J., & Rypáček, V. (1978). The growth response of wood- destroying fungi to the presence of spruce callus. *Mycology*, 32, 55-60.

[5] Hrib, J., & Rypacek, V. (1981). A simple callus test to determine the aggressiveness of wood destroying fungi. *European Journal of Forest Pathology*, 11, 270-274.

[6] Hrib, J., & Rypacek, V. (1983). In vitro testing for the resistance of conifers to the fungus Phaeolus schweinitzii on callus cultures. *European Journal of Forest Pathology*, 13, 86-91.

[7] Hřib, J., Vookova, B., Salajová, T., Bolvanský, M., & Flak, P. (1995). Testing of embryogenic and non-embryogenic calli of European black pine (Pinus nigra Arn.) for defence reactions to the fungus Phaeolus schweinitzii. *Biologia, Bratislava*, 50(4), 403-410.

[8] Johnson, J. A., & Whitney, N. (1988). A preliminary study of conifer tissue reaction to the presence of endophytic fungi using issue culture, light and electron microscopy techniques. *Canadian Journal of Plant Pathology*, 10, 366.

[9] Kowalski, T. (1996). Grzyby endofityczne. III. Możliwości ograniczania populacji Mikiola fagi (Hartig) przez Apiognomonia errabunda (Rob. ex Desm.) Höhn. [Endophytic fungi. III Possibilties of reducing Mikiola fagi (Hartig) population by Apiognomonia errabunda (Rob. ex Desm.) Höhn.]. Proceedings of the Symposium "Plant diseases and the environment". Poznań June 27th-28th; Polish Phytopathological Society. Poznań. , 83-89.

[10] Kowalski, T. (1999). Grzyby endofityczne a choroby zgorzelowe drzew w warunkach oddziaływania imisji przemysłowych. [Endophytic fungi and diseases of stems and branches of trees growing under the influence of industrial emissions]. *Scientific Papers of the Academy of Agriculture in Cracow*, 348, 83-99.

[11] Kowalski, T., & Sadłowski, W. (1993). Grzyby endofityczne II. Znaczenie dla roślin i możliwości ich wykorzystania. [Endophytic Fungi II. Their Importance for Plants and Possibilites of Use]. *Sylwan*, 10, 9-15.

[12] Kvaalen, H., Christiansen, E., Johnsen, Ř., & Solheim, H. (2001). Is there a negative genetic correlation between initiation of embryogenic tissue and fungus susceptibility in Norway spruce? *Canadian Journal of Forest Research*, 31, 824-831.

[13] Kvaalen, H., & Solheim, H. (2000). Co-inoculation of Ceratocystis polonica and Heterobasidion annosum with callus of two Norway spruce clones with different in vivo susceptibility. *Plant Cell, Tissue and Organ Culture*, 60, 221-228.

[14] Minter, D. W., & Millar, C. S. (1980). Ecology and biology of three Lophodermium species on secondary needles of Pinus sylvestris. *European Journal of Forest Pathology,* 10, 169-181.

[15] Nawrot-Chorabik, K. (2007). Induction and development of Grand Fir (Abies grandis Lindl.) callus in tissue cultures. *Electronic Journal of Polish Agricultural Universities,* 10(4), 1-11.

[16] Nawrot-Chorabik, K. (2008). Embryogenic callus induction and differentiation in silver fir (Abies alba Mill.) tissue culture. *Dendrobiology,* 59, 31-40.

[17] Nawrot-Chorabik, K. (2009). Somaclonal variation in embryogenic cultures of silver fir (Abies alba Mill.). *Plant Biosystems,* 143, 377-385.

[18] Nawrot-Chorabik, K. (2011). Somatic embryogenesis in forest plants. *Ken-ichi Sato (ed.) Embryogenesis, Rijeka: InTech,* 20, 423-446.

[19] Nawrot-Chorabik, K., & Jankowiak, R. (2010). Interakcje pomiędzy kalusem trzech genotypów Abies alba a grzybami o różnym statusie ekologicznym. Interactions among three genotypes of Abies alba callus and fungi with different ecological status. *Forest Research Work,* 71(4), 381-389.

[20] Nawrot-Chorabik, K., Jankowiak, R., & Grad, B. (2011). Growth of two blue-stain fungi associated with Tetropium beetles in the presence of callus cultures of Picea abies. *Dendrobiology,* 66, 41-47.

[21] Niemi, K., & Häggman, H. (2002). Pisolithus tinctorius promotes germination and forms mycorrhizal structures in Scots pine somatic embryos in vitro. *Mycorrhiza,* 12, 263-267.

[22] Niemi, K., Krajnakova, J., & Häggman, H. (1998). Interaction between embryogenic cultures of Scots pine and ectomycorrhizal fungi. *Mycorrhiza,* 8, 101-107.

[23] Ostry, M. E., & Skilling, D. D. (1992). Applications of tissue culture for studying tree defense mechanisms. *Defense mechanisms of woody plants against fungi (ed. Blanchette R. A., Biggs A. R.), New York, Springer Verlag,* 405-423.

[24] Peters, S., Draeger, S., Aust, H. J., & Schulz, B. (1998). Interactions in dual cultures of endophytic fungi with host and nonhost plant calli. *Mycologia,* 90(3), 360-367.

[25] Piagnani, C., Faoro, F., Sant, S., & Vercesi, A. (1997). Growth and ultrastructural modifications to chestnut calli induced by culture filtrates of virulent and hypovirulent Cryphonectria parasitica strains. *European Journal of Forest Pathology,* 27, 23-32.

[26] Płażek, A. (2011). Patofizjologia roślin. Patophysiology of plants. *Publishing House of the University of Agriculture in Cracow,* 1-139.

[27] Sieber, T. N., Sieber-Canavesi, F., & Dorworth, C. E. (1990). Simultaneous stimulation of endophytic Cryptodiaporthe hystrix and inhibition of Acer macrophyllum callus in dual culture. *Mycologia,* 82, 569-575.

[28] Sirrenberg, A., Salzer, P., & Hager, A. (1995). Induction of mycorrhiza-like structures and defence reaction in dual cultures of spruce callus and ectomycorrhizal fungi. *New Phytologist*, 130, 149-156.

[29] Spanos, K. A., & Woodward, S. (1997). Responses of Cupressus and Chamaecyparis callus tissues to inoculation with Seiridium cardinale. *European Journal of Forest Pathology*, 27, 13-21.

[30] Stoessl, A. (1981). Structure and biogenetic relation: fungal nonhost-specific. *Toxins in plant disease. R.D. Durbin (editor) Academic Press, Now York- London- Toronto- Sydney-San Francisco*, 109-219.

[31] Terho, M., Pappinen, A., & von Weissenberg, K. (2000). Growth reactions of a Gremmeniella abietina isolate and Scots pine embryogenic tissue cultures differ in a host-parasite in vitro system. *Forest Pathology*, 30, 285-295.

[32] Van Loon, L. C., & Van Strien, E. A. (1999). The families of pathogenesis-related proteins, their activities, and comparative of PR-1 type proteins. *Physiological and Molecular Plant Pathology*, 55, 85-97.

[33] Vookova, B., Hŕib, J., Kormutak, A., & Adamec, V. (2006). Defence reactions of developing somatic embryos of Algerian fir (Abies numidica). *Forest Pathology*, 36, 215-224.

[34] Walters, D., Newton, A., & Lyon, G. (2007). Induced Resistance for Plant Defence. Mechanism of defence to pathogens: biochemistry and physiology. *Blackwell Publishing book. Oxford*, 6, 109-131.

Chapter 12

The Extracellular Indolic Compounds of *Lentinus edodes*

Olga M. Tsivileva, Ekaterina A. Loshchinina and
Valentina E. Nikitina

Additional information is available at the end of the chapte

1. Introduction

In macrobasidial fungi, the properties of compounds of a phytohormonal nature, which are well known in higher plants and are intensively studied in soil associative microorganisms, are only described in an unsystematic manner and, apparently, to an insufficient degree. Auxins are the most studied group of phytohormonal substances. As an object of research, along with other mycological objects of industrial cultivation, the higher fungus–xylotrophic basidiomycete *Lentinus edodes* (Berk.) Sing (*Lentinula edodes* (Berk.) Pegler or shiitake), which is of high practical importance and the physiological and biochemical characteristics of which are obviously insufficiently studied, is of particular interest.

For long enough, there have been speculations that phytohormones, including representatives of the auxin group, are involved in the processes of cell growth and cytodifferentiation not only in plants, but also in fungi. Nevertheless, this issue still remains practically unstudied.

Of particular interest are the effects and mechanisms of the action of biologically active substances at low doses. At small and ultrasmall concentrations ($10^{-20} - 10^{-13}$ M), there is a manifestation of the activity of many natural chemomediators - toxins and antidotes, substances warning of danger, pheromones, cryoprotectants, and other compounds, including phytohormones [1]. There is a description of the paradoxical nature of the effect of low concentrations of toxic substances and drugs, which is particularly expressed in the bimodal or polymodal dependence dose–effect. It is noted in [2] that the consequences of the effects of small doses of xenobiotics may be no less serious than the consequences of high single doses: under their influence, essential links may change and some adaptation systems may fail, because the body is only able to adapt to effects, which are in the usual range of action.

There are two main ways of biosynthesis of phytohormone of indole-3-acetic acid (IAA), namely tryptophan-dependent (Trp-dependent), in the case of which amino acid tryptophan serves as a precursor to IAA, and tryptophan-independent (Trp-independent), in the case of which IAA is produced from indole, anthranilic acid, and indole-3-glycerophosphate [3]. Trp-dependent synthesis of IAA by microorganisms can take place in one of the following four ways (see the scheme below): by indole-3-pyruvic acid and indole-3-acetaldehyde (the most common way), by tryptamine and indole-3-acetaldehyde, by indole-3-acetamide, and by indole-3-acetonitrile. According to some reports, indolylacetaldoxime may also be converted into IAA through indole acetaldehyde [4, 5].

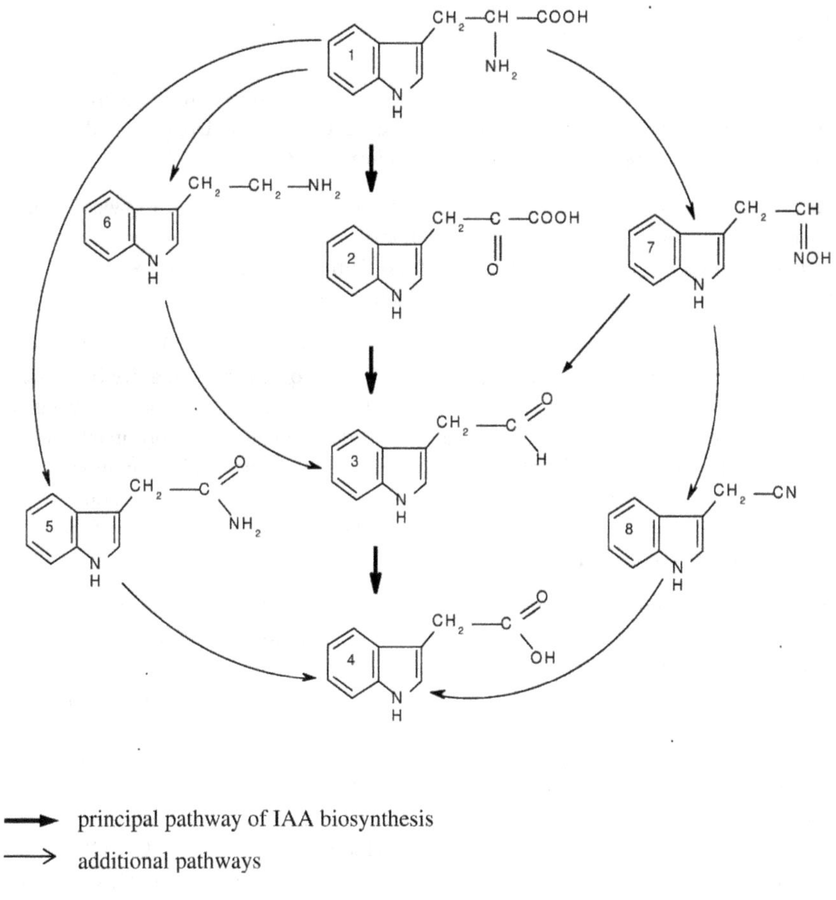

→ principal pathway of IAA biosynthesis

⟶ additional pathways

Scheme 1. Tryptophan-dependent biosynthesis of indolylacetic acid in microorganisms. 1 – tryptophan, 2 – indolylpyruvic acid, 3 – indolylacetaldehyde, 4 – indolylacetic acid, 5 – indolylacetamide, 6 – tryptamine, 7 – indolylacetaldoxime, 8 – indolylacetonitrile

282 Environmental Biotechnology

Many phytopathogenic fungi and bacteria have multiple ways of IAA biosynthesis. The ways of IAA biosynthesis in macrobasidiomycetes have been at best considered only at the level of qualitative effect of tryptophan supplementation to a growth medium.

The purpose of the work is to study the composition of a group of indole metabolites, which accompany the production of IAA by the basidiomycete *Lentinus edodes*, and to establish whether this way of biosynthesis of IAA is Trp-dependent or there is a switching to the Trp-independent way during growing xylotroph in the presence of exogenous synthetic analogs of compounds - precursors to IAA.

2. Experimental

A culture of *Lentinus edodes* (strain F-249) obtained from the collection of macrobasidiomycetes of the Department of Mycology and Algology, Moscow State University (Russia) was used. This fungal culture was maintained on wort agar at 4°C.

As an inoculum, a 14-day culture of *L. edodes* grown on beer-wort agar (4°Bx) was used. The temperature of growing was 26°C. The agar blocks with mycelium were cut out in sterile conditions using a metal punch with a diameter of 5 mm, and served for inoculation of liquid nutritive media at a rate of two blocks for 20 ml of medium.

The submerged culture of the fungus was grown in synthetic medium (9 g/l of glucose and 1.5 g/l of L-asparagine), as well as in beer-wort (1.2°Bx). To determine the dry biomass, the mycelium was filtered, weighed on an analytical balance, and dried to a constant weight. To study the effect of compounds of indolic nature, we added them as solutions in ethanol-H_2O (1 : 1, *v/v*) mixtures to the medium subjected to autoclave glucose and asparagine immediately before planting in sterile conditions. The concentrations of indolic compounds in the culture medium were 0.1, 1, 10, and 100 mg/l. The effect of IAA on the growth of the culture was studied in the range of $10^{-8} - 10^{-1}$ g/l.

The indolic compounds were determined in the culture fluid by high performance liquid chromatography (HPLC) using pure commercial preparations of IAA, Trp, tryptamine (TAM), indolylacetamide (IAAm), indolylpyruvic acid (IPyA), indolylacetaldehyde (IAAld), indole, indole-3-acetonitrile, anthranilic acid, and 5-hydroxy-indole-3-acetic acid (5-hydroxy-IAA) as standards. For the identification and quantification of indolic compounds, samples of the culture fluid were collected under sterile conditions during its growth; then, they were filtered by membrane filters type 0.22 μM GVPP (Millipore, Ireland) and analyzed. Junction reverse-phase HPLC was performed on a medium with chemically bound hydrophobic residues of C18 (5 μM). The column (150 × 4.6 mm) Luna 5μ C18 (2) (Phenomenex, United States) equipped with a precolumn (type of "security guard") of the same brand. As an eluent, a mixture of methanol and water was used (36 : 64 or 50 : 50, *v/v*). A UV absorbance detector operating in the wavelength range of 250-300 nm was used. The sample volume was 20 μl, and the pressure was 12 MPa.

Mass (m/V) expression of concentration of indolic compounds in liquid media was used. The selected method of expressing concentrations is used in most published articles relevant to the subject presented in this work and, therefore, allows for comparisons in the most convenient form [3-5].

3. Results and discussion

3.1. Intermediate products of biosynthesis of IAA in *L. edodes*

Our assumptions about the existence of Trp-dependent synthesis of IAA in the fungus were based on the following observations:

1. A phenomenon of biosynthesis of extracellular Trp by the studied strain of *L. edodes* was discovered. In a synthetic medium, initially not containing this amino acid, the concentrations of Trp ranged from 14 mg/l on the 7th day to 24 mg/l on the 21st day. A reduction in the concentration was observed on the 7th and 14th days. For all studied ages of the culture, the introduction of Trp additives (10 and 100 mg/l) into the medium resulted in a significant increase in the content of this substance in the culture fluid compared to the initial one. The maximum quantity of Trp (about 330 mg/l) was observed on the 14th day in the medium supplemented with 100 mg/l of this acid.

2. It was revealed that the submerged culture of *L. edodes* F-249, growing on a glucose and asparagine medium, was capable of forming extracellular IAA. In a control experiment, the highest concentration of auxin (about 7.5 mg/l) was observed on the 21st day. At an exogenous introduction of Trp into the culture medium, the content of IAA increased and a maximum (9.4 mg/l) was reached on the 14th day in the medium supplemented with 100 mg/l of amino acid, i.e., with the greatest concentration of Trp.

Thus, at the moment when the method for phytohormone formation by the fungus culture started to become clear, we discovered the biosynthetic ability of the fungus with regard to IAA and its precursor Trp.

The main known methods of IAA biosynthesis are associated with Trp. A method, which is independent of tryptophan (Trp-independent), occurs in plants and among the bacteria detected in azospirilla and cyanobacteria. To date, the prevailing view is that the contribution of the Trp-independent way to IAA biosynthesis is not significant; the mechanism of auxin biosynthesis has not been studied. Nevertheless, researchers' opinions are divided. Thus, although previous works prove the existence of an indolylacetamide way in *Azospirillum brasilense* [6]; give biochemical and genetic grounds for the usage of the way via IPyA by azospirilla [7, 8]; and, at the same time, make an assumption that 90% of the IAA in *Azospirillum* is biosynthesized in the tryptophan-independent way [6].

The data we obtained earlier [9] showed that in the *L. edodes* F-249 culture medium, there were intermediate formations as a result of three ways of Trp-dependent biosynthesis of IAA - through TAM, IAAm, and IPyA. This is consistent with the known data, according to which the ability to synthesize IAA simultaneously in several different reactions is found in various microorganisms, including some fungi. Interestingly, indolylacetaldehyde, which is an intermediate in the IAA synthesis from both IPyA and TAM, is only found in media supplemented with IPyA, where IAAld is accumulated in large quantities, but the level of IAA is at the same time very low. The way via IPyA is not completely realized. None of the samples studied had indole-3-acetonitrile, which is another intermediate of IAA synthesis from Trp. This is consistent with the data of literature, according to which cases of IAA synthesis by this intermediate have not yet been identified in fungi [5].

3.2. Prerequisites of the Trp-independent way of IAA synthesis in Shiitake

Our assumptions about the possibility of the existence of an IAA biosynthesis way distinct from the Trp-dependent one in the studied fungal culture were initially based on the following: It is considered in [10, 11] that the bacterial synthesis of IAA is a way to detoxify tryptophan. According to some authors [12], in bacteria, such as *A. brasilense*, there is no way of Trp degradation, which is toxic to them, but it can be transformed into IAA. Therefore, for bacterial producers, Trp is the most effective and "rational" precursor to IAA [13].

For the higher fungi under our study, in contrast to azospirilla, Trp is not toxic at least up to relatively high concentrations, which a fungal culture creates during submerged cultivation, i.e., up to 330 µg/ml (see above).

The Trp-independent way is connected to the synthesis of IAA from indole, anthranilic acid, and indole-3-glycerophosphate. One of the hardest things in proving the Trp-independence of IAA biosynthesis is that indole is a substance, which can serve as both a precursor to IAA in Trp-independent biosynthesis and a precursor to tryptophan. Then, Trp synthesized from indole can also serve as a precursor for IAA [14].

As a result of the study of IAA biosynthesis by bacteria, many authors have come to a conclusion that allows them to make judgments concerning the preferred substances-precursors of IAA in the context of its biosynthesis. The arguments are as follows:

1. Bacteria cannot produce IAA by the Trp-independent way when cultured with indole as a precursor to this phytohormone, because IAA synthesis is not stimulated while growing bacteria in media containing indole [13]. This suggests that indole is not a preferred precursor to IAA compared to Trp.

2. The use of anthranilic acid or indole for the synthesis of IAA by the non-tryptophan way using microorganisms is unlikely, because the presence of Trp in the culture fluid is demonstrated in all experiments.

On the basis of the experimental data obtained in the present paper, one can provide arguments for and against the coexistence of two alternative ways of IAA biosynthesis in *L. edodes*, Trp-dependent and Trp-independent. The media with the addition of indole were characterized by fairly high values of IAA (up to 9 mg/l) on the 14th - 21st days of cultivation, and the amounts of formed IAA were not dependent on the initial concentration of indole. The medium with 100 mg/l of indole was an exception, where the level of IAA decreased on the 21st day and the concentration of IPyA at this point increased dramatically. The data are presented in the Table.

Experimental conditions		Final concentration in the culture fluid, mg/l				
Indolic additive, mg/l	Cultivation time, days	Tryptamine	Indolyl-acetamide	Indolyl-acetic acid	Indolyl-pyruvic acid	Tryptophan
Control (medium without indole)	3	3.7	3.3	3.7	n/d	19.5
	7	3.8	3.2	n/d	n/d	13.8
	10	4.2	3.1	n/d	n/d	18.6
	14	3.4	2.3	3.7	0.9	16.9
	21	3.9	3.7	7.4	7.7	23.9
0.1	3	3.1	3.4	4.0	1.0	17.5
	7	1.3	n/d	n/d	0.3	9.5
	10	1.9	n/d	n/d	0.7	10.8
	14	2.5	2.6	6.9	5.3	16.1
	21	3.3	3.4	9.0	11.6	20.5
1	3	2.5	2.7	5.5	n/d	14.8
	7	1,3	n/d	n/d	0.4	8.3
	10	1.5	n/d	n/d	0.9	9.8
	14	2.3	2.3	5.6	2.8	16.2
	21	2.9	2.7	7.4	5.9	16.7
10	3	2.9	3.0	3.4	0.5	17.1
	7	1.9	n/d	n/d	0.3	9.9
	10	1.8	n/d	n/d	0.9	7.9
	14	2.4	2.2	6.2	4.7	15.6
	21	3.7	2.7	8.1	7.7	17.2
100	3	3.1	3.5	n/d	0.2	13.9
	7	1.2	n/d	n/d	0.4	13.9
	10	1.2	n/d	n/d	0.4	10.1
	14	2.4	2.2	6.1	4.2	15.5
	21	3.3	2.2	2.2	16.7	19.1

Note: n/d - not detectable

Table 1. Effect of indole on the formation of extracellular indole compounds by the mushroom *Lentinus edodes* F-249

Thus, the indole stimulated the synthesis of IAA. In our experiments on a synthetic medium, initially not containing this substance, the concentrations of extracellular Trp ranged from 13.8 mg/l on the 7th day to 9.23 mg/l on the 21st day. This means that there were no formally noted cases of IAA synthesis in the absence of tryptophan and the Trp-dependent way of IAA biosynthesis took place in the studied fungal culture. Nevertheless, one can assume that there is a transfer to the Trp-independent way or a connection between this alternative ways (which is likely), realizing in the presence of exogenous indole within the concentration range of $1 \bullet 10^{-3} - 1 \bullet 10^{-4}$ g/l (Table).

It is necessary to note the following about the additions of indole: in this case, a background level of Trp also exists, but, firstly, it is virtually unchanged compared to the control experiment and the IAA synthesis is by a factor of 1.5–1.9 greater (Fig. 1a).

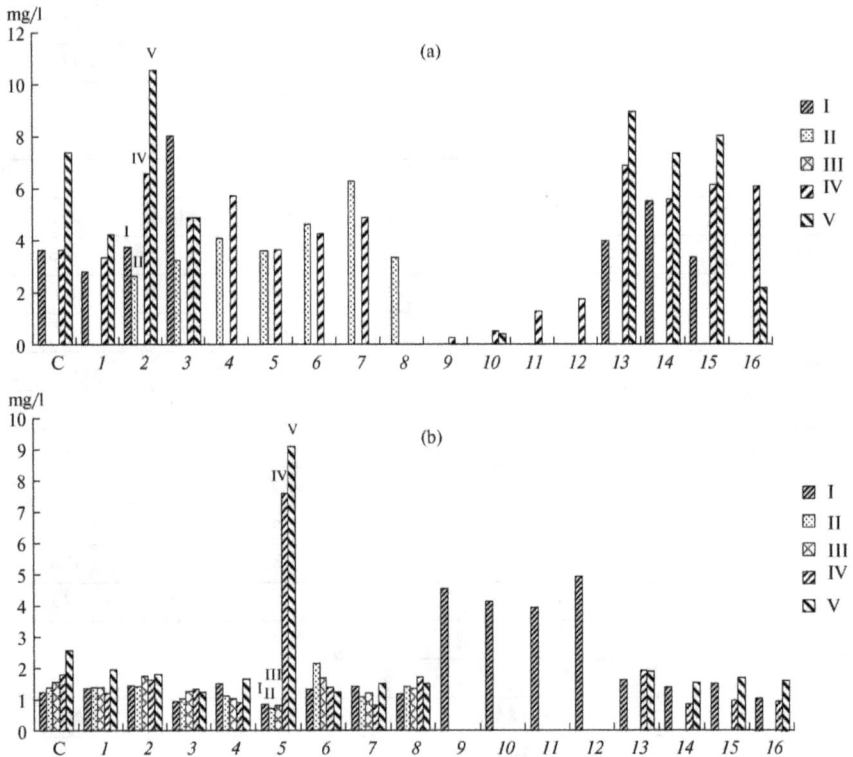

Figure 1. Synthesis (mg/l) of (a) extracellular indolylacetic acid, and (b) 5-hydroxy-indolylacetic acid by submerged cultures of *Lentinus edodes* F-249 on media with an addition of indolic compounds at different durations of cultivation (days): (I) 3, (II) 7, (III) 10, (IV) 14, and (V) 21. C - control; tryptamine of (1) 0.1, (2) 1, (3) 10, and (4) 100 mg/l; indolylacetamide of (5) 0.1, (6) 1, (7) 10, and (8) 100 mg/l; indolylpyruvic acid of (9) 0.1, (10) 1, (11) 10, and (12) 1000 mg/l; indole of (13) 0.1, (14) 1, (15) 10, and (16) 100 mg/l.

This happens in spite of the reduced fungal biomass under the influence of indole, which clearly had no positive effect on the growth rates of *L. edodes*. It caused a significant decrease in the biomass compared to the control (up to 34% in the medium with 100 mg/l of indole) and strongly inhibited the growth of the culture (Fig. 2).

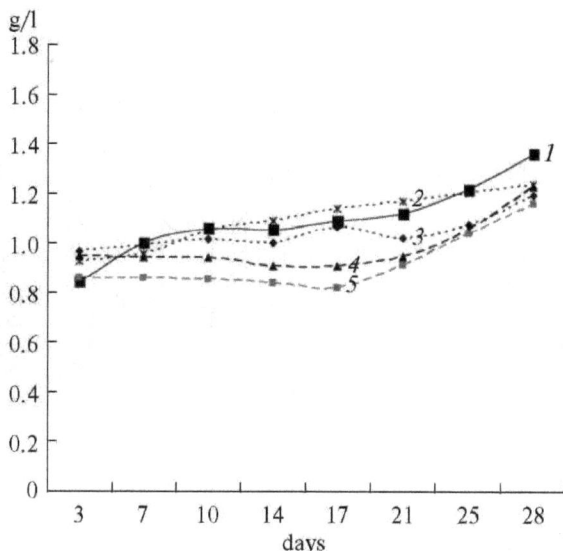

Figure 2. Mycelial biomass accumulation (g/l) by submerged cultures of *Lentinus edodes* F-249 of different ages (days) on media with an indole addition (mg/L): (1) 0, (2) 0.1, (3)1, (4) 10, and (5) 100.

Secondly, the concentration of Trp in the medium completely remains the same when the level of exogenous indole is increased by 1000 times (from 0.1 to 100 mg/l), and, therefore, indole is not presumably a significant precursor to Trp in this case (with further transformation of the latter into IAA). The elevated level of indole in the culture medium was increasing the biosynthesis of IAA, but not tryptophan. Thus, we are talking about the connection of the Trp-independent way.

Not only in the case of exogenous indole there are signs of this way, but also while inducing IAA biosynthesis with its exogenous microadditives ($1 \cdot 10^{-5} - 1 \cdot 10^{-8}$ g/l). For example, at an initial IAA concentration of $1 \cdot 10^{-7}$ g/l on the tenth day of growth, the level of phytohormone was about $4 \cdot 10^{-4}$ g/l; *i.e.*, it had increased by a factor of 4000 (Fig. 3a).

The appearance of anthranilic acid (up to 1.5 mg/l) as a sign of the Trp-independent path was only marked by us under these experimental conditions (Fig. 3b). At the same time, Trp itself was not detected in the culture fluid in any of the eight tested concentrations of IAA additive.

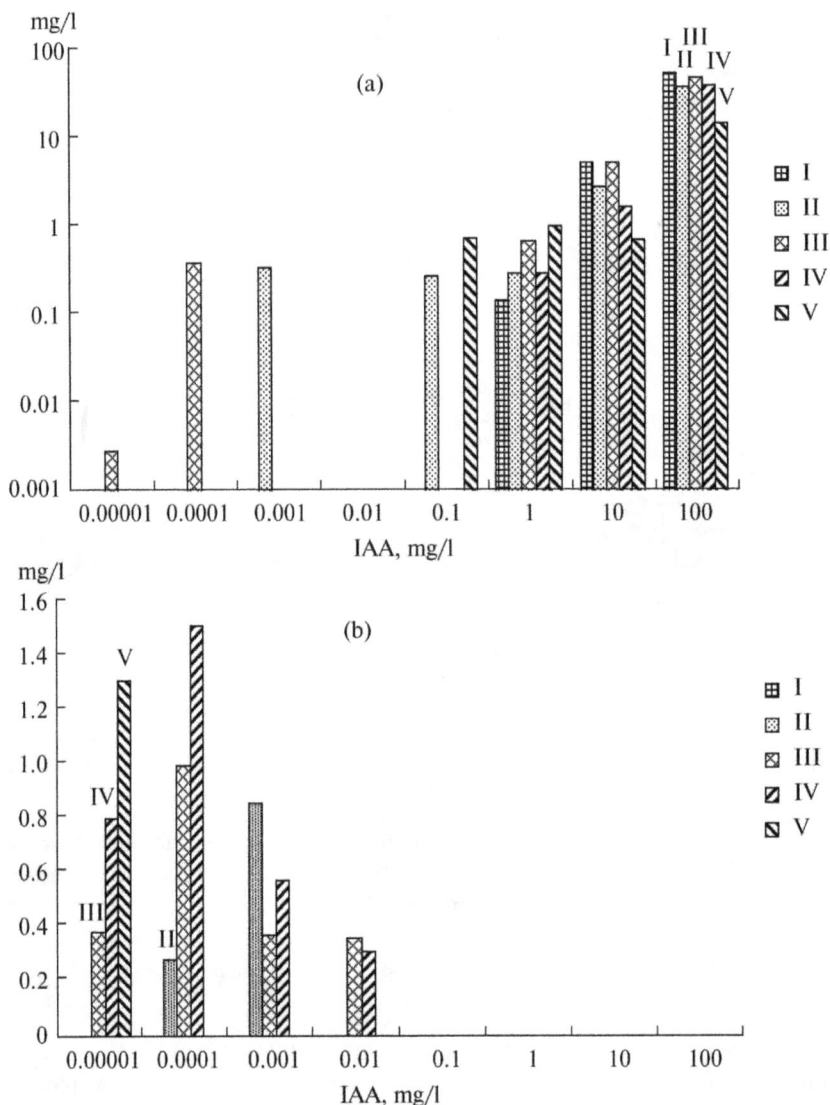

Figure 3. Effect of exogenous indolylacetic acid additives (mg/l) on the content (mg/l) of (a) indolylacetic acid, and (b) anthranilic acid in the cultural fluid of *Lentinus edodes* F-249 of different ages (days): (I) 3, (II) 7, (III) 10, (IV) 14, and (V) 21.

Besides anthranilic acid, IPyA is also synthesized in the presence of low concentrations of IAA ($1\bullet10^{-4} - 1\bullet10^{-8}$ g/l), but the appearance of IAAld - a product of the conversion of IPyA during Trp-dependent IAA synthesis - was not observed. This situation with IAAld changes at the higher concentrations ($1 \bullet 10^{-4} - 1 \bullet 10^{-1}$ g/l) of exogenous IAA, when on the tenth day of growth, for example, there accumulated from 3.6 to 8.7 mg/l of IAAld (Fig. 4a).

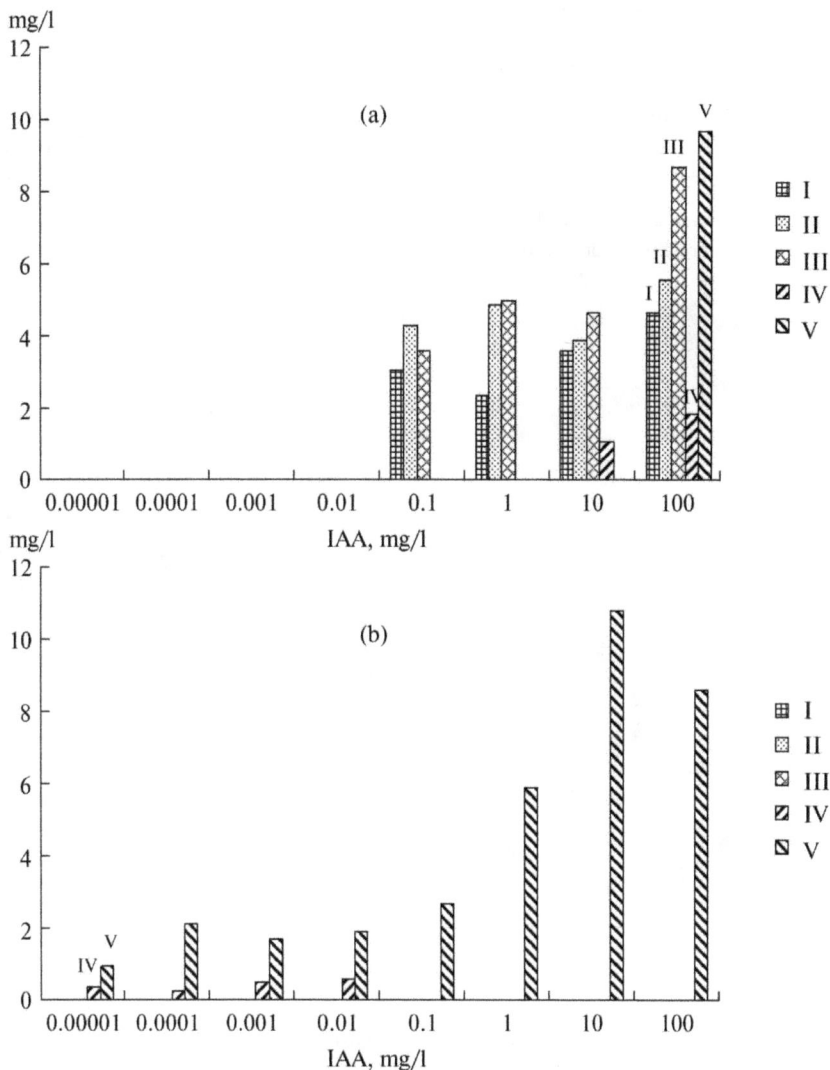

Figure 4. Effect of exogenous indolylacetic acid additives (mg/l) on the content (mg/l) of (a) indolylacetaldehyde, and (b) indolylpyruvic acid in the cultural fluid of *Lentinus edodes* F-249 of different ages (days): (I) 3, (II) 7, (III) 10, (IV) 14, and (V) 21.

Consequently, the reasons for the lack of Trp-independency of IAA biosynthesis in bacteria generally described in the literature do not take place with respect to *L. edodes*. The obtained results suggest that anthranilic acid or indole is a quite effective precursor to IAA as compared to Trp. It also allows for the detection of the effect of low concentrations of exogenous indolic compounds in the studied fungal culture.

3.3. Effect of small doses of IAA in the submerged culture of Shiitake

The biological activity of IAA at low concentrations has not been studied, the molecular mechanism of the phytohormonal action of IAA is not finally determined, and there is no explanation to the two-phase effects of heteroauxin.

In our work, where a culture of the shiitake mushroom was used as a biological object of study, it was interesting to detect the "effect of small doses" of biologically active substances of indolic nature.

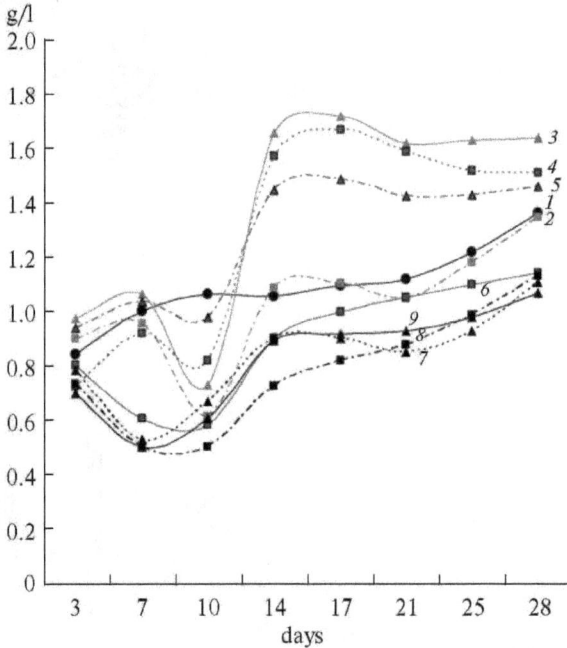

Figure 5. Mycelial biomass accumulation (g/l) by submerged cultures of *Lentinus edodes* F-249 of different ages (days) on media with IAA additives (g/l): (1) 0, (2) 10^{-8}, (3) 10^{-7}, (4) 10^{-6}, (5) 10^{-5}, (6) 10^{-4}, (7) 10^{-3}, (8) 10^{-2}, and (9) 10^{-1}

Under the influence of exogenous IAA, we observed growth stimulation in the submerged mycelium within a certain concentration range of auxin. It can be noted that, virtually for the entire range of values on the x axis, the set of curves describing the growth of the fungus (dependence of the accumulation of dry biomass on the duration of cultivation) (Fig. 5) is divided into two groups of curves with the boundary curve corresponding to an IAA concentration of 10^{-4} g/l.

Above the latter, there is the area of growth activated by auxin (10^{-8} – 10^{-5} g/l of IAA). Below is a region with dependences with opposite properties, where the level of IAA additives is 10^{-1} – 10^{-3} g/l. During the cultivation periods of 3–7 and 12–28 days (*i.e.*, at all studied ages of the culture, except for the interval of 8–11 days), the optimal concentration of IAA was 10^{-7} g/l.

During studying the effect of the additives of exogenous IAA on its content in the culture liquid of *L. edodes*, the concentration of auxin (10^{-4} g/l) was a turning point in the sense that the level of IAA in the medium was higher than originally (Fig. 3a). The concentration of IAA additives of 10^{-7} g/l ($0.57 \bullet 10^{-9}$ M) was notable for the fact that it induced a 4100-fold increase in the level of phytohormone.

When adding IAA, very significant stimulation of IAAld biosynthesis during the cultivation of the shiitake mushroom was observed (up to 9.7 mg/l) on the 21st day (Fig. 4a). However, this only occurred up to 10^{-4} g/l of IAA, below which the effect was absent.

Indole-3-pyruvic acid accumulated in the medium after 21 days of cultivation of the fungus at all concentrations of auxin (Fig. 4b), but it was only below an exogenous IAA concentration of 10^{-4} g/l that we managed to find significant quantities of IPyA as soon as on the 14th day (0.56 mg/l).

Synthesis of extracellular anthranilic acid by submerged cultures of *L.edodes* F-249 of various ages on media with IAA was not observed up to a concentration of 10^{-4} g/l (Fig. 3b), below which the concentration of anthranilate was almost 0.4 mg/l as early as the tenth day of cultivation. The maximum output of anthranilate (1.5 mg/l) was detected for the experimental variant with an inducing concentration of IAA of 10^{-7} g/l ($0.57 \bullet 10^{-9}$ M).

In these omnidirectional effects of exogenous IAA, the abovementioned effect of small doses of phytohormone had a further illustration. Indeed, just as in [1], a type of a point of sign change of the biological effect of IAA was the concentration of $5.7 \bullet 10^{-7}$ M ($1.0 \bullet 10^{-4}$ g/l). In addition, when IAA was used at a concentration of 10^{-9} M ($1.75 \bullet 10^{-7}$ g/l), which, according to the just aforementioned work, promotes a more effective interaction between auxin and the membrane or the receptor, we observed absolute maxima in the respective series of experiments (the formation of IAA and anthranilate).

3.4. Effect of IAA and its predecessors on the pigmentation of submerged mycelia of *L. edodes* F-249

The value of phytohormones for the morphogenesis of higher xylotrophic fungi has practically not been studied. Some evidence of the involvement of IAA phytohormone in the process of morphogenetic differentiation of mushrooms of the *Lentinus* genus obtained through simple experiments has existed for a long time, although they are only described in a few papers [15, 16]. These authors suggest that phytohormones play an important role in the differentiation of the fungal culture and that the process of morphogenesis is closely related to the dynamics of the level of endogenous growth regulators, including IAA.

Investigation of the role of phytohormonal substances in the metabolism of *L. edodes* is of particular relevance at the stage prior to fruiting, which is characterized in shiitake by the development of a specialized vegetative formation - brown mycelial film (BMF), because the biochemical conditions for the genesis of such morphogenetic structures are still poorly understood.

In studying the effects of IAA and its precursors, as tryptophan, tryptamine, indolylacetamide, indolylpyruvic acid, and indole, on the growth of submerged mycelia of *L. edodes* F-249, in most cases, there was no change in the morphology of the culture. The only exception was the medium with 0.1 mg/l of IAAm. In this medium, there was a decrease in the time required for the appearance of a brown mycelial film. In the medium with this additive, the film appeared as soon as the 17th day, whereas in the control version with other concentrations of IAAm and in the media with the remaining indole compounds, its formation was not observed.

In a study of the intensity of the growth processes of *L. edodes* in the presence of indolylacetamide, precisely the level of 0.1 mg/l of IAAm in the original culture medium corresponded to the largest positive effect in regard to the accumulation of mycelial biomass (Fig. 6). At the mentioned optimal concentration of IAAm, in turn, the increase in biomass was the greatest precisely on the 17th day (24%).

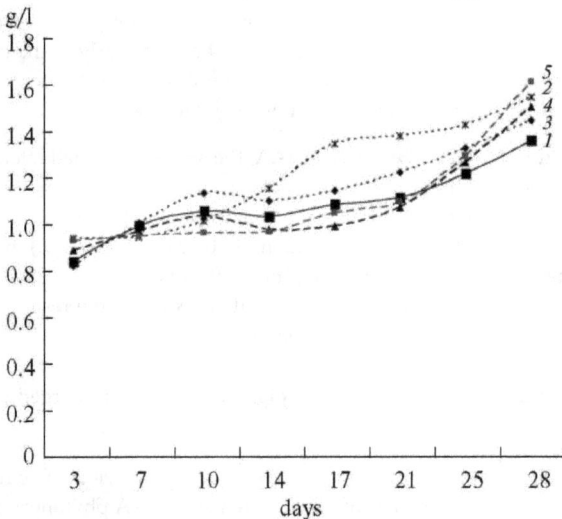

Figure 6. Mycelial biomass accumulation (g/l) by submerged cultures of *Lentinus edodes* F-249 of different ages (days) on media with indolylacetamide additives (mg/l): (1) 0, (2) 0.1, (3) 1, (4) 10, and (5) 100.

In the studied samples of culture fluid, 5-hydroxy-IAA, an oxidized form of IAA, was present in the control and in the media with TAM and IAAm at all ages of the mycelium, and its concentration in this case ranged from 0.78 to 2.63 mg/l (Fig. 1b). The only exception was the medium with 0.1 mg/l of IAAm, where on the 14th–21st days the level of 5-hydroxy-IAA increased dramatically. As was mentioned above, this version of the experiment differed from the rest by an early appearance of a brown mycelial film. On the 14th day, there was observed a significant pigmentation of the mycelium, and the film had completely formed by 17th day. We can assume that 5-hydroxy-IAA is involved in the formation of the BMF.

The existence of extracellular hydroxylating enzymes in mushrooms may be of ecological and environmental as well as biotechnological relevance. Biocatalytic oxygen transfer by isolated enzymes or whole microbial (fungal) cells is an elegant and efficient way to achieve selective hydroxylation. Selective hydroxylation of aromatic compounds is among the most challenging chemical reactions in synthetic chemistry and has gained steadily increasing attention during recent years, particularly because of the use of hydroxylated aromatics as precursors for pharmaceuticals [17].

The accumulation of a large biomass of mycelium is one of the factors necessarily required to make a transition to the morphogenetic stage prior to fruiting (BMF in the shiitake fungus), but not a sufficient condition for the development of a BMF. In the formation of mycelium pigmentation, oxidized indole derivatives should be also involved, the participation of which in the conversion of 5,6-dihydroxyindole-indole → 5,6-quinone → melanochrome → melanin in the case of catalysis by fungal tyrosinases leading to the formation of melanin pigments of fungi has been known for a long time [18].

Thus, during the submerged cultivation of the xylotrophic basidiomycete L. edodes, a group of metabolites of indolic nature was revealed, the composition and quantitative ratio of which indicated the coexistence of two alternative routes of IAA biosynthesis in L. edodes. These routes are Trp-dependent (mainly via tryptamine) and Trp-independent, the latter being not only implemented in the presence of exogenous indole within the concentration range of $1 \cdot 10^{-3} - 1 \cdot 10^{-4}$ g/l, but also at inducing the biosynthesis of IAA by its exogenous microadditives. The involvement of indole-3-acetamide and 5-hydroxy-IAA in the morphogenetic processes of the shiitake culture has been revealed. There was established the interrelationship of the level of IAAm and 5-hydroxy-IAA with the formation of a brown mycelial film in the submerged culture and with the processes of growth and development of fungal mycelia during submerged cultivation.

Despite several biocatalytic processes which have successfully put on the market, much research remains to be done before enzymes can be used routinely throughout the chemical industry [19, 20]. Enzymes as catalysts in chemical syntheses make chemical reactions possible under mild, environmentally friendly conditions in aqueous reaction mixtures and mostly, enzymes show specificity and selectivity that cannot be achieved by traditional chemical methods [21]. Enzymes implemented in the indolic compounds' biosynthesis not only in plants and bacteria, but also in mushrooms, are capable of catalyzing regio-and stereoselective transformations leading to products that are useful as fine chemicals or pharmaceuticals.

4. Conclusions

An enormous body of knowledge demonstrates that auxin indolyl-3-acetic acid can be synthesized from tryptophan by plants and bacteria. Only few experiments are done with fungi, therewith the published works on phytohormonic potentialities of xylotrophic mushrooms are virtually absent.

The characterization of the group of extracellular indolic compounds of basidiomycete *Lentinus edodes* in relation to the submerged cultivation conditions has been presented. The *L. edodes* F-249 culture has been stated to synthesize the indolic-nature compounds when being grown in submerged culture. The group of extracellular indolic compounds of shiitake includes, in different proportions, the following components: L-tryptophan, β-indolyl-3-acetic acid, β-indolyl-3-acetaldehyde, β-indolyl-3-acetamide, indolyl-3-pyruvic acid, tryptamine, 5-hydroxy-β-indolyl-3-acetic acid.

The time-course production of indolic derivatives in culture liquid of shiitake has been studied comparatively to reveal correlations with the growth rates. It has been established that shiitake is capable of producing indolylacetic acid rather actively in the culture-age dependent manner. Maximal indolylacetic acid content we marked in experiments was 10.6 mg/l provided that the initial tryptamine level in the medium was 1 mg/l. The highest levels of tryptamine (22.8 mg/l), indolylpyruvic acid (13.9 mg/l), indolylacetaldehyde (27.9 mg/l), 5-hydroxy-indolylacetic acid (9.2 mg/l) have been observed at different additives of the indole-group substances, and varied with the fungal culture age. The indolylacetamide level changed only slightly as compared to reference probe. Starting from the synthetic tryptophan-free medium, the concentrations of extracellular tryptophan became from 13.8 mg/l (7th day) to 23.9 mg/l (21st day).

On the basis of literature data on physiologically active concentration values for auxins in plants, we use the concentration range of 10^{-1}- 10^{-8} g/l of additives to explore the effects of indolylacetic acid and its precursors upon the shiitake submerged culture. The positive influence of exogenic auxin within its concentration range of $2 \bullet 10^{-7}$ to $2 \bullet 10^{-4}$ g/l upon the *L. edodes* biomass accumulation has been found, a minimal growth-inhibiting phytohormone concentration being about $5 \bullet 10^{-4}$ g/l on mineral medium. When inducing the indolylacetic acid biosynthesis by its exogenous micro-additives ($1 \bullet 10^{-5}$ to $1 \bullet 10^{-8}$ g/l), the increase in phytohormone level in the medium (up to 4000-fold) accompanied by the anthranilic acid appearance (up to 1.5 mg/l) have been revealed under these experimental conditions exclusively.

The studies of biosynthetic routes for indolylacetic acid realized by the mushroom culture under question have been attempted in order to conclude whether the above route is tryptophan-dependent, or tryptophan-independent pathway becomes also involved when the mushroom grows in the presence of exogenous synthetic analogs of the auxin precursors. The experimental evidences in favor of co-existence of two alternative pathways for indolylacetic acid production by *L. edodes* have been obtained. Those routes are: tryptophan-dependent (mainly via tryptamine) and tryptophan-independent, the latter being realized in the presence of exogenous indole within the concentration range $1 \bullet 10^{-3} - 1 \bullet 10^{-4}$ g/l or the auxin micro-additives.

The induction of generative developmental stage by indolic derivative has been revealed for shiitake. It has been stated that among the compounds - indolylacetic acid precursors, solely indolylacetamide at a concentration of about 10^{-4} g/l in *L. edodes* culture liquid exerts the explicitly marked stimulating effect on the occurrence of morphological structure - brown mycelial film.

More thorough investigations and specific search for the extracellular indolic compounds among the huge number of basidiomycetous fungi colonizing litter or lignicelluloses will surely result in the discovery of further fungal enzymes, as well as may help to understand better the biological formation of the mushroom growth promoting substances and polycyclic aromatic compounds in terrestrial ecosystems.

Acknowledgements

This work was supported in part by the "OOO Sibirskoye zdorov'e" Russian Foundation, grant of the series "Useful properties of plants and minerals for the medicinal cosmetics and natural prophylactic preparations production - 2012".

Author details

Olga M. Tsivileva*, Ekaterina A. Loshchinina and Valentina E. Nikitina

*Address all correspondence to: tsivileva@ibppm.sgu.ru

Laboratory of Microbiology, Institute of Biochemistry and Physiology of Plants and Microorganisms, RAS, Saratov, Russia

References

[1] Rogacheva, S.M., Rol' vodnoi komponenty i polisakharidov kletochnoi poverkhnosti v protsessakh kommunikatsii zhivykh sistem: analiz molekulyarnykh modelei (The Role of Aqueous Components and Cell Surface Polysaccharides in the Communication Processes of Living Systems: Analysis of Molecular Models), Voronezh: Izd. Voronezh. GOs. Univ., 2008. http://vak.ed.gov.ru/ru/dissertation/index.php?id54=1099

[2] Burlakova, E.B., Konradov, A.A., and Mal'tseva, E.L. Extremely small actions of chemical compounds and physical factors on biological systems. Biophysics, 2004, vol. 49, no. 3, pp. 522-534.

[3] Baca, B.E. and Elmerich, C. Associative and Endophytic Nitrogen-Fixing Bacteria, and Cyanobacterial Associations, in Series: Nitrogen Fixation: Origins, Applications, and Research Progress, V. 5, Elmerich, C. and Newton, W.E., Eds., Dordrecht, The Netherlands: Springer, 2007, chapter 6, pp. 113-143.

[4] Eckardt, N.A. New Insights into Auxin Biosynthesis. Plant Cell, 2001, vol. 13, no. 1, pp. 1-3.

[5] Reineke, G., Heinze, B., Schirawski, J., Buettner, H., Kahmann, R., and Basse, C.W. Indole-3-acetic acid (IAA) biosynthesis in the smut fungus Ustilago maydis and its relevance for increased IAA levels in infected tissue and host tumour formation. Mol. Plant Pathol., 2008, vol. 9, no. 3, pp. 339-355.

[6] Prinsen, E., Costacurta, A., Michiels, K., Vanderleyden, J., and Van Onckelen, H. Azospirillum brasilense indole-3-acetic acid biosynthesis: evidence for a non-tryptophan dependent pathway. Mol. Plant–Microbe Interact., 1993, vol. 6, no. 5, pp. 609-615.

[7] Bar, T. and Okon, Y., in Azospirillum VI and Related Microorganisms: Genetics, Physiology, Ecology, Fendrik, I., Gallo, M. D., Vanderleyden, J., and Zamaroczy, M., Eds., Berlin: Springer, 1995, pp. 347-359.

[8] Costacurta, A., Keijers, V., and Vanderleyden, J. Molecular cloning and sequence analysis of an Azospirillum brasilense indole-3-pyruvate decarboxylase gene. Mol. Gen. Genet., 1994, vol. 243, no. 4, pp. 463-472.

[9] Loshchinina, E.A., Tsivileva, O.M., Nikitina, V.E., and Makarov, O.E., Vavilovskie Chteniya-2008: Materialy mezhd. nauchno-prakt. konf (Vavilov Reading – 2008: Proceedings of the Int. Scientific and Practical. Conf.), Saratov: ITs Nauka, 2008, pp. 30-32.

[10] Lebuhn, M., Heulin, T., and Hartmann, A. Production of auxin and other indolic and phenolic compounds by Paenobacillus polymyxa strains isolated from different proximity to plant roots. FEMS Microbiol. Ecol., 1997, vol. 22, no. 4, pp. 325-334.

[11] Bar, T. and Okon, Y. Induction of indole-3-acetic acid synthesis and possible toxicity of tryptophan in Azospirillum brasilense Sp 7. Symbiosis, 1992, vol. 13, nos. 1–3, pp. 191-198.

[12] Lebuhn, M. and Hartmann, A. Production of auxin and ltryptophan related indolic and phenolic compounds by Azospirillum brasilense and Azospirillum lipoferum. in Improving Plant Productivity with Rhizosphere Bacteria, Ryder, M.H., Stephens, P.M., and Bowen, G.D., Eds., Australia: CSIRO, 1994, pp. 145-147.

[13] Zakharova, E.A., Shcherbakov, A.A., Brudnik, V.V., Skripko, N.G., Bulkhin, N.Sh., and Ignatov, V.V. Biosynthesis of indole-3-acetic acid in Azospirillum brasilense. Insights from quantum chemistry. Eur. J. Biochem., 1999, vol. 259, no. 3, pp. 572-576.

[14] Dewick, P.M. The biosynthesis of shikimate metabolites. Nat. Prod. Rep., 1998, vol. 15, no. 1, pp. 17-58.

[15] Rypacek, V. and Sladky, Z. The character of endogenous growth regulators in the course of development in the fungus Lentinus tigrinus. Mycopathol. Mycol. Applic., 1972, vol. 46, no. 1, pp. 65-72.

[16] Rypacek, V. and Sladky, Z. Relation between the level of endogenous growth regulators and the differentiation of the fungus Lentinus tigrinus studied in a synthetic medium. Biologia Plantarum (Praha), 1973, vol. 15, no. 1, pp. 20-26.

[17] Ullrich, R. and Hofrichter, M. Enzymatic hydroxylation of aromatic compounds. Cell. Mol. Life Sci., 2007, vol. 64, no. 3., pp. 271-293.

[18] Choi, S.W. and Sapers, G.M. Purpling Reaction of Sinapic Acid Model Systems Containing L-DOPA and Mushroom Tyrosinase. J. Agric. Food Chem., 1994, vol. 42, no. 5, pp. 1183-1189.

[19] Schoemaker, H.E., Mink, D., and Wubbolts, M.G. Dispelling the myths - biocatalysis in industrial synthesis. Science, 2003, vol. 299, pp. 1694-1697.

[20] Poliakoff, M, Fitzpatrick, J.M., Farren, T.R., and Anastas, P.T. Green chemistry: science and politics of change. Science, 2002, vol. 297, pp. 807-810.

[21] Hofrichter, M., and Ullrich, R. Heme-thiolate haloperoxidases: versatile biocatalysts with biotechnological and environmental significance. Appl. Microbiol. Biotechnol., 2006, vol. 71, no. 3, pp. 276-288.

Permissions

The contributors of this book come from diverse backgrounds, making this book a truly international effort. This book will bring forth new frontiers with its revolutionizing research information and detailed analysis of the nascent developments around the world.

We would like to thank Prof. Marian Petre, for lending his expertise to make the book truly unique. He has played a crucial role in the development of this book. Without his invaluable contribution this book wouldn't have been possible. He has made vital efforts to compile up to date information on the varied aspects of this subject to make this book a valuable addition to the collection of many professionals and students.

This book was conceptualized with the vision of imparting up-to-date information and advanced data in this field. To ensure the same, a matchless editorial board was set up. Every individual on the board went through rigorous rounds of assessment to prove their worth. After which they invested a large part of their time researching and compiling the most relevant data for our readers. Conferences and sessions were held from time to time between the editorial board and the contributing authors to present the data in the most comprehensible form. The editorial team has worked tirelessly to provide valuable and valid information to help people across the globe.

Every chapter published in this book has been scrutinized by our experts. Their significance has been extensively debated. The topics covered herein carry significant findings which will fuel the growth of the discipline. They may even be implemented as practical applications or may be referred to as a beginning point for another development. Chapters in this book were first published by InTech; hereby published with permission under the Creative Commons Attribution License or equivalent.

The editorial board has been involved in producing this book since its inception. They have spent rigorous hours researching and exploring the diverse topics which have resulted in the successful publishing of this book. They have passed on their knowledge of decades through this book. To expedite this challenging task, the publisher supported the team at every step. A small team of assistant editors was also appointed to further simplify the editing procedure and attain best results for the readers.

Our editorial team has been hand-picked from every corner of the world. Their multi-ethnicity adds dynamic inputs to the discussions which result in innovative outcomes. These outcomes are then further discussed with the researchers and contributors who give their valuable feedback and opinion regarding the same. The feedback is then collaborated with the researches and they are edited in a comprehensive manner to aid the understanding of the subject.

Apart from the editorial board, the designing team has also invested a significant amount of their time in understanding the subject and creating the most relevant covers. They scrutinized every image to scout for the most suitable representation of the subject and create an appropriate cover for the book.

The publishing team has been involved in this book since its early stages. They were actively engaged in every process, be it collecting the data, connecting with the contributors or procuring relevant information. The team has been an ardent support to the editorial, designing and production team. Their endless efforts to recruit the best for this project, has resulted in the accomplishment of this book. They are a veteran in the field of academics and their pool of knowledge is as vast as their experience in printing. Their expertise and guidance has proved useful at every step. Their uncompromising quality standards have made this book an exceptional effort. Their encouragement from time to time has been an inspiration for everyone.

The publisher and the editorial board hope that this book will prove to be a valuable piece of knowledge for researchers, students, practitioners and scholars across the globe.

List of Contributors

Marian Petre
Department of Natural Sciences, Faculty of Sciences, University of Pitesti, Romania

Violeta Petre
Department of Fruit Growing, Faculty of Horticulture, University of Agronomic Sciences and Veterinary Medicine-Bucharest, Romania

Amy Philbrook, Apostolos Alissandratos and Christopher J. Easton
CSIRO Biofuels Research Cluster, Research School of Chemistry, Australian National University, Canberra ACT, Australia

Prihardi Kahar
Program in Environment and Ecology, School of Science and Engineering, Meisei University, Hino-shi, Tokyo, Japan

Nidal Madad, Latifa Chebil, Hugues Canteri, Céline Charbonnel and Mohamed Ghoul
Laboratoire d'Ingénierie des Biomolécules, ENSAIA-INPL, Vandoeuvre-lès-Nancy, France

Sonja Nybom
Department of biosciences, Biochemistry, Åbo Akademi University, Finland

Shouji Takahashi, Katsumasa Abe and Yoshio Kera
Department of Environmental Systems Engineering, Nagaoka University of Technology, Kamitomioka, Nagaoka, Niigata, Japan

Bruno Alexandre Quistorp Santos
Faculty of Engineering, Department of Chemical Engineering, Cape Peninsula University of Technology, Cape Town, South Africa

Seteno Karabo Obed Ntwampe and James Hamuel Doughari
Faculty of Applied Science, Department of Agriculture and Food Science, Biotechnology programme, Cape Peninsula University of Technology, Cape Town, South Africa

Olusola Solomon Amodu and Tunde Victor Ojumu
Faculty of Engineering, Department of Chemical Engineering, Cape Peninsula University of Technology, Cape Town, South Africa

Hitoshi Miyasaka, Hiroshi Okuhata and Satoshi Tanaka
The Kansai Electric Power Co., Environmental Research Center, Seikacho, Japan

Takuo Onizuka and Hideo Akiyama
Toray Research Center, Inc., Kamakura, Japan

Krasimira Tasheva and Georgina Kosturkova
Regulation of Plant Growth and Development Department, Institute of Plant Physiology and Genetics, Bulgarian Academy of Sciences, Sofia, Bulgaria

Katarzyna Nawrot - Chorabik
Department of Forest Pathology, Faculty of Forestry, University of Agriculture in Kraków, Poland

Olga M. Tsivileva, Ekaterina A. Loshchinina and Valentina E. Nikitina
Laboratory of Microbiology, Institute of Biochemistry and Physiology of Plants and Microorganisms, RAS, Saratov, Russia

www.ingramcontent.com/pod-product-compliance
Lightning Source LLC
Chambersburg PA
CBHW070736190326
41458CB00004B/1186